PARTICLE STRENGTHENING OF METALS AND ALLOYS

PARTICLE STRENGTHENING OF METALS AND ALLOYS

Eckhard Nembach
Institut für Metallforschung
University of Münster

A WILEY-INTERSCIENCE PUBLICATION
JOHN WILEY & SONS, INC.
New York • Chichester • Brisbane • Toronto • Singapore • Weinheim

This text is printed on acid-free paper.

Library of Congress Cataloging in Publication Data:

Nembach, E. (Eckhard)

Particle strengthening of metals and alloys / Eckhard Nembach.

p. cm.

Includes bibliographical references and index.

ISBN 0-471-12072-3 (cloth : alk. paper)

1. Metals—Effect of radiation on. 2. Strengthening mechanisms in solids. 3. Particle beams. I. Title.

TA460.N35 1996 96-14932

620.1′6—dc20 CIP

Printed in the United States of America

10 9 8 7 6 5 4 3 2 1

Die lohnendsten Forschungen sind diejenigen,
welche, indem sie den Denker erfreu'n, zugleich der Menschheit nützen.

The most rewarding researches are those
which give joy to the thinker, and at the same time benefit mankind.

Christian Doppler (1803–1853)

CONTENTS

SYMBOLS AND ABBREVIATIONS

LATIN

a	lattice constant
a_p	unconstrained lattice constant of the particle/nucleus (Section 3.2.1.1)
a_s	unconstrained lattice constant of the solid solution/matrix (Section 3.2.1.1)
A	parameter defined in Eq. (3.27a)
A	area swept out by a dislocation which bows out under the stress τ_{ext} (Section 4.2.1.4.3)
A	area [Eqs. (5.1) and (5.2)]
A	parameter defined in Eq. (5.45)
A	A-atom: atom preferentially occupying A-sites of the $L1_2$-crystal structure (Section 5.5.1)
A_c	area swept out by a dislocation that has reached its critical configuration, τ_{ext} = CRSS (Section 4.2.1.4.3)
A_n	normalization constant in Eqs. (4.4) and (4.6)
A_{AB}	area between the x-axis and the dislocation arc over the length AB (Figure 4.13)
A_F	area (Figure 4.13, Section 4.2.1.3.1)
A_1, A_2	parameters appearing in Eq. (5.65)
b	length of the Burgers vector
$\mathbf{b}$	Burgers vector
B	drag coefficient [Eq. (2.14)]
B	B-atom: atom preferentially occupying B-sites of the $L1_2$-crystal structure (Section 5.5.1)
B	numerical constant [Section 6.1; Eqs. (6.2), (6.4), etc.]
B_e, B_m, B_{ph}	electron/magnon/phonon contribution to the drag coefficient (Section 4.2.2.2)
B_s	contribution of solute atoms to the drag coefficient (Section 4.2.2.2)
c	solute concentration = atomic fraction
c	half the cube length [Section 5.2.1; Eqs. (5.13), (5.15), and (5.16)]
c_p	solute concentration in the particle/nucleus (Section 3.2.1.2)

c_s	solute concentration in the solid solution/matrix (Section 3.2.1.2)
C	constant involving the shear modulus (Section 4.2.1.1)
C_1, C_2	numerical constants in Eqs. (4.62), (4.63), and (4.67)
C_{ik}	elastic stiffness
CMAX	strongest dislocation configuration in a planar random array of obstacles (Section 4.2.1.4.1)
CRSS	critical resolved shear stress
d_p	diameter of the particle measured in the glide plane (Section 6.1.1)
d_p, d_s	width of the stacking fault ribbon in the particle/solid solution matrix (Section 5.3)
ds	$\|d\mathbf{s}\|$ (Section 2.1)
$d\mathbf{s}$	dislocation segment of infinitesimal length, $d\mathbf{s}$ is parallel to $\mathbf{s}$ (Section 2.1)
dF_{obst}	force exerted by the obstacles/particles on the dislocation segment ds; dF_{obst} acts in the glide plane in the glide direction (Section 2.5)
d_1, d_2	mean length of the dislocation D1/D2 lying inside of a long-range-ordered particle (Figure 5.23)
D	effective length [Eqs. (4.53) and (6.3)]
D1, D2	leading/trailing dislocation of a pair gliding in a material strengthened by coherent long-range-ordered particles (Section 5.5.1)
e_m	energy density stored in a 180° domain wall [Eq. (8.4)]
e_{st}	energy density stored in the strain field of a dislocation (Section 5.2.1)
E_A	energy per unit area of a 180° domain wall (Section 8.1)
E_l	energy per unit length of a perfect straight dislocation (Section 2.2)
E_m	interaction energy between a nonmagnetic particle and a 180° domain wall [Eq. (8.5a)]
E_μ	particle–dislocation interaction energy due to a modulus mismatch of the particle [Eq. (5.10)]
f	volume fraction of the particles (Section 3.1)
f_j'	fraction of obstacles of class j in a random planar array of point obstacles of various strengths (Section 4.2.2.1)
f_0	$F_0/(2S)$ [Eq. (4.58a)]
F_{arc}	force exerted by a stress on a dislocation arc [Eq. (2.9c)]
F_{obst}	force exerted by the obstacles/particles on a dislocation
$F(y - h_y, -h_z)$	force exerted by an obstacle/particle centered at $(0, h_y, h_z)$ on a straight infinitely long dislocation that is parallel to the x-axis and that is positioned at y [Eq. (4.2)], the glide plane is characterized by $z = 0$

F_γ	F due to a mismatch of the stacking fault energy (Section 5.3) or to the long-range order of the particle (Section 5.5.3)
F_Γ	F due to chemical strengthening (Section 5.1)
F_ε	F due to a lattice mismatch of the particle (Section 5.4.1); the dislocation is undissociated
$F_{\varepsilon\,\mathrm{dis}}$	F_ε for a dissociated edge dislocation [Eq. (5.48)]
F_μ	F due to a modulus mismatch of the particle (Section 5.2.1)
F_0	maximum of $\lvert F(y - h_y, -h_z)\rvert$ [Eq. (4.7a)]/maximum of the absolute value of the interaction force between a particle and a 180° domain wall (Section 8.1)
$F^*(y^*)$	force profile [Eqs. (4.4)]
$g(\rho/r)$	general distribution function of particle radii (Section 3.1)
g_p	Gibbs free energy per unit volume of the particle/nucleus (Section 3.2.1.1)
g_s	Gibbs free energy per unit volume of the solid solution/matrix (Section 3.2.1.1)
$g_{\mathrm{WLS}}(\rho/r)$	Wagner–Lifshitz–Slyozov distribution function of particle radii [Eq. (3.7)]
$g^*(\lambda/l)$	distribution function of interparticle spacings (=center-to-center spacings of the plane–particle intersections) in parallel planes (Section 3.3)
$g_0(\rho/r)$	distribution function of particle radii; all particles have the same radius r [Eq. (3.6)]
h_x, h_y, h_z	coordinates of the center of a spherical particle (Section 4.1)
h_0, h_1	parameters defined in Sections 5.3 and 5.4.2
H	magnetic field
H_c	coercive field, coercivity
k	exponent in Eqs. (4.57), (5.58), and (7.5)
k_{opt}	optimum value of the exponent k in Eqs. (4.57), (5.58), and (7.5) (Section 7.2.1)
k_B	Boltzmann constant
k_s, k_{si}, k_{st}, k_p	exponents in Eqs. (7.11)
$\mathbf{k}, \mathbf{k}'$	wave vector of the incident/scattered electrons or neutrons
$K_E(\theta_d)$	elastic prefactor of the dislocation line energy [Eq. (2.2)]
K_{Eg}	geometric mean of $K_E(\theta_d = 0°)$ and $K_E(\theta_d = 90°)$ [Eq. (6.12b)]
$K_S(\theta_d)$	elastic prefactor of the dislocation line tension [Eq. (2.6)]
l	mean nearest-neighbor spacing (=center-to-center spacing of the plane–particle intersections) of particles in parallel planes (Section 3.3)
l	length defined in Eq. (5.35d)
l_0	length defined in Eq. (5.28a)
L	particle spacing along the dislocation

L_c particle spacing along the dislocation when it breaks free from a particle — that is, in the critical moment (Section 4.2.1.2)
L_{cF} Friedel length [Eqs. (4.27)]
L_{cL} Labusch length [Eq. (4.67)]
L_{max} mean particle spacing along parallel straight lines (Section 3.1)
L_{min} center-to-center nearest-neighbor particle spacing in parallel planes, if the particles form square arrays (Section 3.1)
L_p free interparticle spacing (Figure 6.1)
L_z length of the 180° domain wall (Figure 8.2)
L_1, L_2 mean spacing of long-range-ordered particles along the dislocation D1/D2 (Figure 5.23)
m_{eff} effective mass per unit length of the dislocation, inertial mass [Eq. (2.13)]
m_p magnetic moment of a superparamagnetic particle (Section 3.3.5)
M magnetization (Section 3.3.5 and Chapter 8)
M_s saturation magnetization (Chapter 8)
M_{ps} saturation magnetization of a superparamagnetic particle (Section 3.3.5)
n exponent in Eqs. (4.4d) and (4.6)
n_s number of particles intersecting unit area of parallel planes (Section 3.1)
n_{sj} number (per unit area) of obstacles of class j in a random planar array of obstacles of various strengths (Section 4.2.2.1)
n_v number of particles per unit volume (Section 3.1)
n_A true number of particles per unit area of the thin foil [Eq. (3.21)]
n_{AM} number of *visible* TEM images of particles per unit area of the thin foil [Eq. (3.23b)]
$\mathbf{n}$ normal of the glide plane, $|\mathbf{n}| = 1$ (Section 2.1)
n_0 number of obstacles in a simulated planar random array (Section 4.2.2.1)
ODS oxide dispersion strengthened (Section 3.2.2)
O_i obstacle i in the glide plane (Section 4.2.1.4.1)
O_i obstacle of type i (Chapter 7)
p subscript standing for particle/nucleus
P1, P2 leading/trailing partial dislocation of a dissociated dislocation (Sections 5.3 and 5.4.1)
q mean area of the intersection of parallel planes and spherical particles (Section 3.1)
q $= h_z/\rho$ (Section 5.4)
q width of a 180° domain wall [Eq. (8.4)]

r — mean particle radius in a specimen (Section 3.1)
r_r — mean radius of the intersections of parallel planes with spherical particles (Section 3.1)
$2r_d$ — mean linear traverse length of parallel straight lines over spherical particles (Section 3.1)
R_c — radius of the dislocation core
R_i — inner cutoff radius of a dislocation (Section 2.2)
R_l — radius of the local curvature of the dislocation [Eqs. (2.16)]
R_{loop} — radius of a dislocation loop (Section 2.2)
R_o — outer cutoff radius of a dislocation (Section 2.2)
s — subscript standing for solid solution/matrix
$\mathbf{s}$ — vector of unit length parallel to the local direction of the dislocation line (Section 2.1)
S — dislocation line tension (Section 2.2)
$S_{eff\ 1}$ — effective dislocation line tension [Eq. (4.50a)]
SANS — small-angle neutron scattering (Section 3.3.3)
SEM — scanning electron microscope/microscopy (Section 3.3.2)
t — time
T_D — temperature of deformation
T_H — temperature of heat treatment
TEM — transmission electron microscope/microscopy (Section 3.3.1)
v_n — velocity of the dislocation normal to $\mathbf{s}$ [Eqs. (2.17)]
w_x, w_y — range in the x/y-direction of the particle–dislocation interaction force (Section 4.1.1)
y^* — $(y - h_y)/w_y$ [Eq. (4.4c)]
z — thickness of the thin foil in the TEM (Section 3.3.1)
Z — atomic number

GREEK

$\alpha_\varepsilon, \alpha_{\gamma'}$ — parameters defined in Eq. (5.52a) and Eq. (5.61c), respectively
α_1, α_2 — coefficients appearing in Eq. (5.18) and Eq. (5.20), respectively
β_1, β_2 — coefficients appearing in Eqs. (5.18), (5.20), and (5.49)
γ — shear strain (Section 4.2)
γ — normalized drag coefficient [Eq. (4.58k)]
γ — specific antiphase boundary energy of long-range-ordered particles (Section 5.5.1)
γ — matrix phase of nickel-base superalloys (Section 5.5.1)
γ_p, γ_s — specific stacking fault energy of the particles/solid solution matrix (Section 5.3)
$\dot{\gamma}$ — shear strain rate (Section 4.2)
γ' — γ'-phase = precipitated phase in nickel-base superalloys
Γ — specific interface energy between particle and matrix (Sections 3.2.1.1 and 5.1)
δ — unconstrained lattice mismatch [Eq. (3.16)]

$\delta(x)$ Dirac's δ-function
δ length defined in Eq. (5.36d)
ΔG Gibbs free energy difference [Eq. (3.17)]
$\Delta\mu$ $=\mu_p - \mu_s$ [Eq. (5.14)]
ε constrained lattice mismatch [Eq. (3.32)]
η normalized y-coordinate [Eq. (4.58c)]
η_0 normalized range of the particle–dislocation interaction force [Eq. (4.58g)]
θ normalized time [Eq. (4.58h)]
θ_d angle between the direction of a dislocation and its Burgers vector (Section 2.1)
θ_{d0} θ_d of a dislocation before it bows out
θ_k $2\theta_k$ = angle between incident and scattered electrons/neutrons
λ individual nearest-neighbor spacing (=center-to-center spacing of the plane–particle intersections) of particles in parallel planes (Section 3.3)
μ shear modulus
μ_F, μ_K, μ_R shear moduli [Eq. (5.9a), Eqs. (5.9c) and (6.5a), and Eq. (5.9b), respectively]
μ_p, μ_s shear modulus of the particles, shear modulus of the solid solution matrix (Section 5.2)
μ_0 permeability of vacuum $= 4\pi \times 10^{-7}$ Vs/(Am)
ν Poisson's ratio
ν_K anisotropic Poisson's ratio [Eq. (6.5b)]
ξ coefficient defined in Eq. (3.29)
ξ normalized x-coordinate [Eq. (4.58b)]
ξ coefficient defined in Eq. (5.66)
ρ radius of an individual particle (Section 3.1)
ρ_{bp} average neutron scattering length density of the particles (Section 3.3.3)
ρ_{bs} average neutron scattering length density of the solid solution/matrix (Section 3.3.3)
ρ_c critical radius of a nucleus [Eq. (3.18a)]
ρ_d density of mobile dislocations (Section 4.2)
ρ_m density of the material [Eq. (2.15)]
ρ' radius of the particle–glide-plane intersection [Section 5.1, Eq. (5.27)]
$\boldsymbol{\sigma}$ stress tensor (Section 2.3)
$\boldsymbol{\sigma}_{\text{obst}}$ tensor of the stresses exerted by the obstacles/particles on the dislocation (Section 2.5)
τ resolved shear stress
$\tau^{\otimes}$ τ, normalized [Eq. (4.58j)]
$\tau_{\text{back}}, \tau'_{\text{back}}$ resolved shear stress due to the dislocation line tension, back stress [Eqs. (2.11a) and (2.12), respectively]

τ_{drag} resolved shear stress needed to overcome the viscous friction [Eq. (2.14)]
τ_{ext} externally applied resolved shear stress (Section 2.5)
τ_{ext}^* τ_{ext}, reduced [Eq. (4.35)]
$\tau_{\text{ext}}^{\otimes}$ τ_{ext}, normalized [Eq. (4.58j)]
τ_i contribution of obstacles of class i to the total CRSS (Section 7.1)
τ_{inert} resolved shear stress due to the finite inertia of the dislocation [Eq. (2.13)]
τ_{obst} resolved shear stress exerted by the obstacles/particles on the dislocation (Section 2.5)
τ_p critical resolved shear stress contribution of the particles, particle hardening (Chapter 4)
τ_p^* τ_p, reduced [Eq. (4.18)]
$\tau_p^{\otimes}$ τ_p, normalized [Eq. (4.58j)]
τ_{pj} contribution of obstacles of class j to the total CRSS (Sections 4.2.2.1 and 7.2.2)
τ_{pj}^* τ_{pj}, reduced [Eq. (4.18)]
τ_{pu}, τ_{pl} upper/lower CRSS (Section 4.2.2.2)
$\tau_{pu}^{\otimes}, \tau_{pl}^{\otimes}$ τ_{pu}, τ_{pl}, normalized [Eq. (4.58j)]
τ_{pF} τ_p, calculated on the basis of Eqs. (4.26); the subscript F stands for "Friedel"
τ_{pL} τ_p, calculated on the basis of Eqs. (4.62)–(4.64). $\tau_{pL} = \tau_{pu}$ (Section 4.2.2.2); the subscript L stands for "Labusch"
τ_s critical resolved shear stress contribution of the solute atoms, solid solution hardening (Chapter 4)
τ_{self} self-stress introduced by Brown (Section 4.2.1.4.3)
τ_t total measured critical resolved shear stress, comprising τ_p and τ_s (Chapter 4)
ϕ bowing angle of dislocations (Figure 4.9)
ϕ_c critical value of ϕ, describing the strength of obstacles [Eq. (4.8b)]
ω normalized dislocation velocity [Eq. (4.58i)]
ω_d r_d/r [Eq. (3.5c)]
ω_n mean of (ρ^n/r^n) [Eq. (3.5b)]
ω_q $q/(\pi r^2)$ [Eq. (3.5d)]
ω_r r_r/r [Eq. (3.5a)]

PREFACE

The systematic investigation of particle strengthening of metals and alloys started with an observation that Wilm [1.10] made in the first decade of this century: The hardness of aluminum–magnesium–copper alloys that had been quenched from elevated temperatures increased during their exposure to ambient temperature. At that time it was impossible to understand the underlying physical processes: Dislocations were not known, and the experimental techniques needed for the characterization of the fine particles were not available.

Particles act as obstacles to the glide of dislocations and thus reduce their mobility. Two avenues are open to dislocations to overcome these obstacles, either by shearing them (Figure 1.4b) or by bypassing them (Chapter 6). The latter mechanism has been suggested by Orowan [1.21] and was subsequently named after him. Only coherent particles can be shared. Models that relate the critical resolved shear stress of materials strengthened by shearable particles to the latters' properties involve two steps: (i) the determination and characterization of the relevant particle–dislocation interaction mechanism and (ii) calculating the number of particles which simultaneously interact with the dislocation. Friedel's [1.22] analytical solution to this statistical problem and Foreman and Makin's [1.23] computer simulations of the glide of a dislocation through a planar random array of obstacles have become the bases of all later developments. In Chapter 4 a unified view of strengthening by shearable particles is presented, and in Chapter 5 these general models are applied to different strengthening mechanisms. In Chapter 6 the Orowan process is analyzed, in Chapter 7 the superposition of various strengthening effects and the resulting synergisms are treated. In Chapter 8 the models presented in Chapter 4 are adapted to describe particle hardening of ferro- and ferrimagnetic materials. Basic physical processes are emphasized, and technically important particle–strengthened materials are discussed exemplarily.

Some derivations that are mathematically somewhat more involved, but that are not essential for the understanding of the following, have been set in smaller print. Thus the reader can recognize this material and perhaps skip it.

In conclusion I would like to extend my sincere thanks to all those who helped in the preparation of this book: to Dr. D. Rönnpagel and to my wife Keiko, M.Sc., for reading the whole manuscript and for many helpful comments and criticisms; to colleagues and to former and present coworkers who provided micrographs and line drawings, to Dipl.-Phys. R. Amelung, Dr. K.

Hilfrich, and Mr. M. Mevenkamp for preparing most of the line drawings; and — above all — to Mrs. A. Fröse for deciphering my manuscript and for patiently typing and retyping it.

Münster, Germany ECKHARD NEMBACH

CHAPTER 1

Introduction

Structural materials are supposed to combine high strength with at least some ductility. If they were brittle, there would be the danger of catastrophic fracture in case the applied stress accidentally exceeded the design limits. Because both strength and ductility are governed by the mobility of dislocations, the bases for structural materials are ductile metals (e.g., aluminum or nickel) that are strengthened by the introduction of obstacles to the glide of dislocations. Any inhomogeneity that interacts with dislocations can be used as a hardener. The four best-known ones are

- Grain boundaries
- Other dislocations produced by cold-working
- Single solved atoms
- Particles of secondary phases

In most cases, various types of obstacles to the glide of dislocations are simultaneously present.

The strong effect of particle hardening is demonstrated here for the commercial nickel-base superalloy NIMONIC PE16 [1.6] (Section 5.5). Its main constituents are nickel, iron, chromium, aluminum, and titanium. Above 1150 K, it is a homogeneous solid solution. If this state is preserved by quenching, the critical resolved shear stress (CRSS) is 69 MPa at room temperature. Heat-treating this alloy at 1029 K leads to the precipitation of fine particles of the γ'-phase, which is rich in nickel, aluminum, and titanium. After annealing for 330 h at 1029 K, the particles' volume fraction and radii are 0.09 and 25 nm, respectively. In this state the room-temperature CRSS is 170 MPa; that is, the precipitates more than double the CRSS.

1.1 DISCOVERY OF PARTICLE STRENGTHENING

During the first decade of this century, Wilm [1.10] observed that the hardness of aluminum-base aluminum–magnesium alloys quenched from elevated temperatures increased during their exposure to room temperature. This effect is

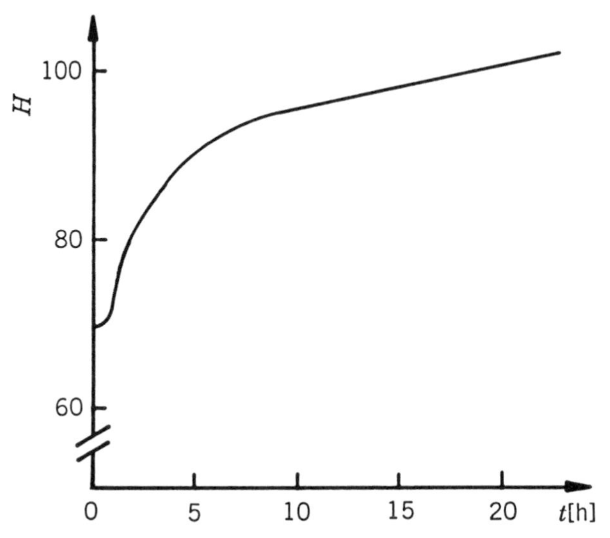

Fig. 1.1. Age-hardening curve of an aluminum-base aluminum–magnesium alloy, after Wilm [1.10]. Hardness H versus aging time t at ambient temperature.

called *age hardening*. Figure 1.1 shows some of Wilm's results. This author also investigated the effect of copper on the hardness of aluminum base alloys. An understanding of the relevant processes was not possible at that time for the following reasons:

1. It was not before 1934 that Orowan [1.11], Polanyi [1.12], and Taylor [1.13] independently developed the concept of dislocations.
2. The size of the precipitates that strengthened Wilm's aluminum alloys was probably below 10 nm. Thus they were far too small to be resolved by optical microscopy, the only means for studying microstructures available to him. Nowadays such particles are investigated by small-angle scattering of X-rays or neutrons (SANS) [1.14, 1.15], conventional or high-resolution transmission electron microscopy (TEM) [1.16], or even field ion microscopy [1.17]. Guinier [1.18] and Preston [1.19] were the first to present evidence that in the early stages of the decomposition of aluminum-rich aluminum–copper alloys, disc-shaped copper clusters are formed; they are one to several atomic layers thick. These clusters have been named Guinier–Preston zones, after their discoverers. High-resolution transmission electron micrographs of such zones are shown in Figure 1.2 [1.20]. In Figure 1.2b the {002}-lattice planes are resolved, their spacing is 0.2 nm.

1.2 COHERENT AND INCOHERENT PARTICLES

Coherent particles have a crystal structure that is the same as, or very similar to, that of the matrix; the lattice planes are continuous across the phase boundary, and their spacings differ only slightly in the two phases. Consequently the slip geometry, including the Burgers vectors of dislocations, is nearly the same in both phases, and matrix dislocations can also glide in the particles. A coherent particle is sketched in Figure 1.3a. Figure 1.4a gives an example of coherent precipitates: particles of the γ'-phase in the nickel-base superalloy NIMONIC PE16. Figure 1.4b shows such γ'-particles in a deformed specimen; some of them have been sheared by dislocations. Dislocation processes in NIMONIC PE16 are discussed in detail in Section 5.5.

In many cases, coherent precipitates are only metastable. During prolonged aging they are replaced by incoherent particles of the thermodynamic equilibrium phase (Section 3.2.1.1). This holds, for example, for the Guinier–Preston zones mentioned above.

There is no continuity of the lattice planes across incoherent phase boundaries (Figure 1.3b). Matrix dislocations cannot enter incoherent particles, but must circumvent them. A mechanism for this has been suggested by Orowan [1.21]; subsequently it has been named after him (Chapter 6). Hardening by incoherent oxide particles is often referred to as *dispersion hardening*.

Evidently, particles may be shearable or nonshearable, and an alternative pair of expressions is penetrable and impenetrable.

1.3 MODELING OF STRENGTHENING BY SHEARABLE PARTICLES

In relating the CRSS to the properties of the coherent shearable particles, one proceeds in two steps:

1. First the interaction force between one particle and one dislocation is calculated. To do so, one has to know the relevant interaction mechanism (e.g., the stress field surrounding the particle, if its lattice constant differs significantly from that of the matrix (Section 5.4.1)).
2. In a second step the CRSS is derived from the total force exerted by all particles interacting simultaneously with the dislocation. The statistical problem is complicated by the fact that the dislocation bows out between the particles as a stress is applied. This is demonstrated in Figure 1.5. If the dislocation were infinitely stiff, it would only touch the two particles numbered 1 and 2. Because it is actually flexible, it interacts with the six particles 1–6. The stronger is the interaction force between the particles and the dislocation, the higher the stress that has to be applied to overcome them. On the other hand, a high stress leads to strong bow-outs of the dislocation and thus raises the number of particles

Fig. 1.2. High-resolution transmission electron micrographs of Guinier–Preston zones in an aluminum-base aluminum–copper alloy [1.20]. (Courtesy of T. Sato, Tokyo.)

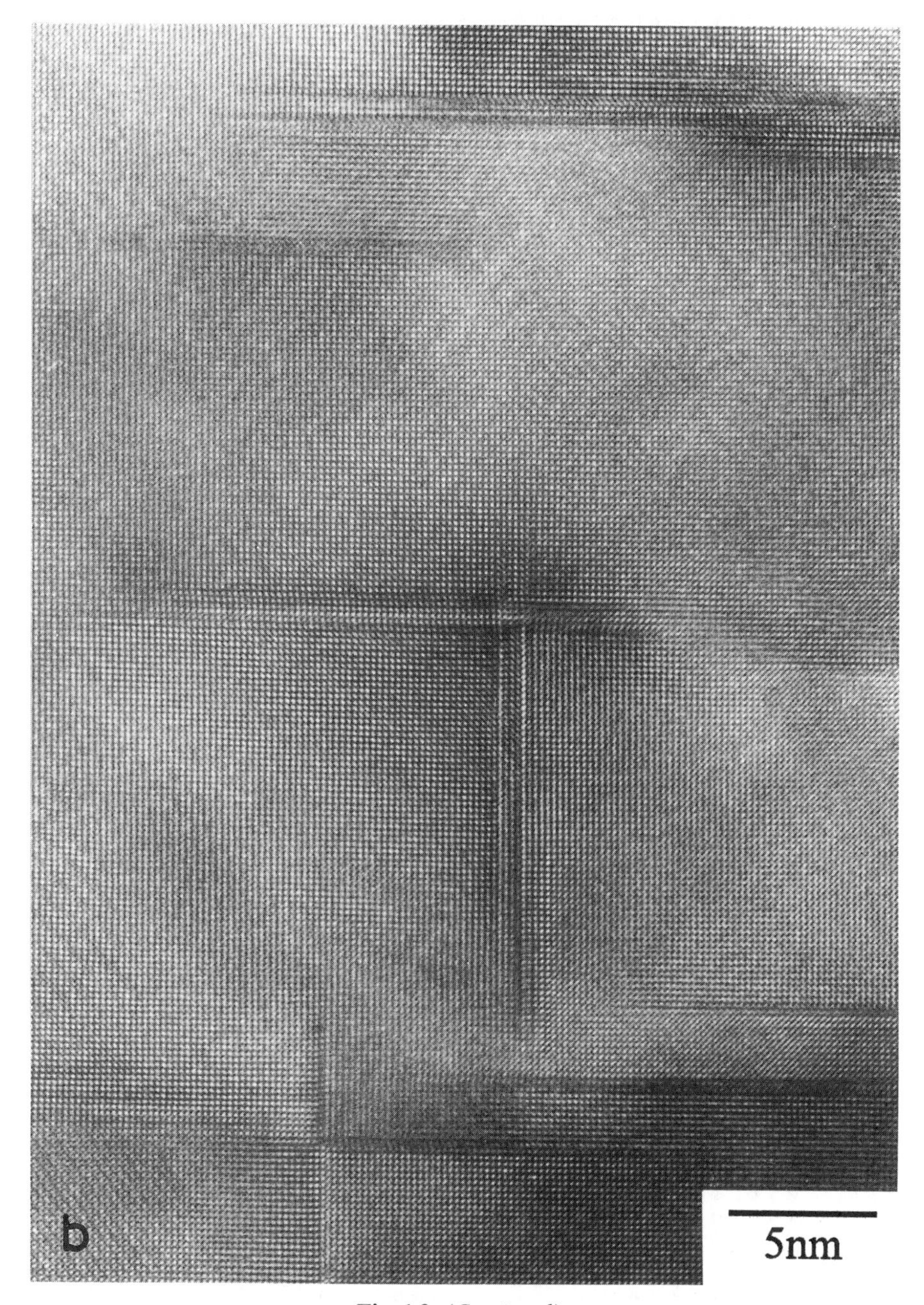

Fig. 1.2. (*Continued*).

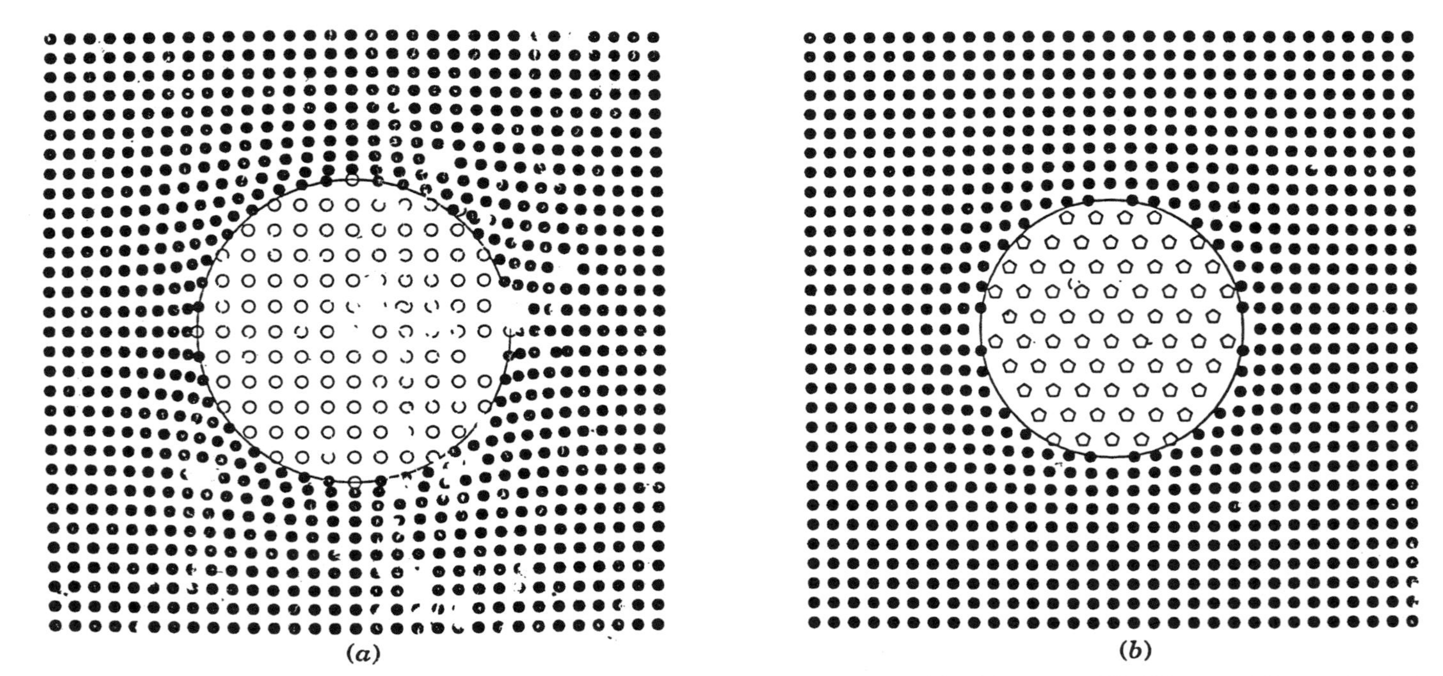

Fig. 1.3. Schematic representations of a (*a*) coherent particle and (*b*) incoherent particle. (Courtesy of D. Rönnpagel, Braunschweig.)

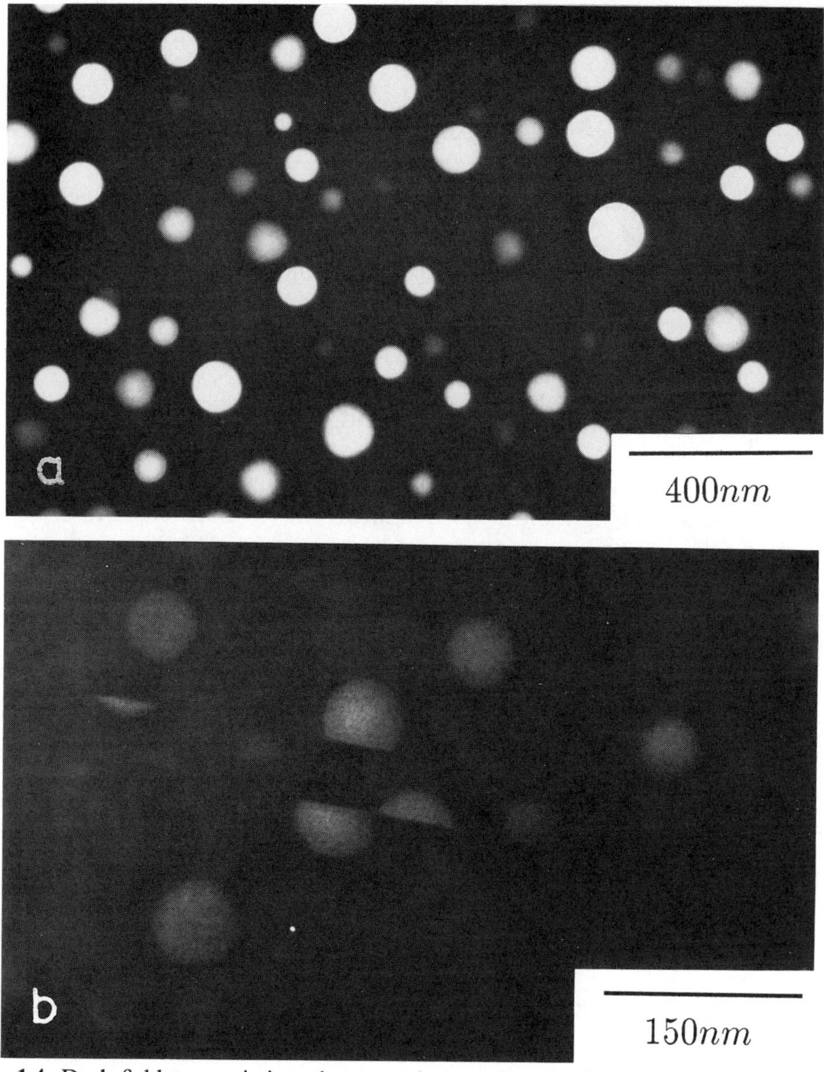

Fig. 1.4. Dark-field transmission electron micrographs of coherent γ'-particles in the nickel–base superalloy NIMONIC PE16. (*a*) Undeformed specimen. (Courtesy of C. Schlesier, Münster.) (*b*) Deformed specimen. Some γ'-particles have been sheared. (Courtesy of W. Mangen, Münster.)

simultaneously pinning the dislocation. Consequently, the CRSS is expected to increase with the interaction force more rapidly than linearly. The two solutions to this statistical problem which have become the bases for all later developments are Friedel's [1.22] (Section 4.2.1.3) analytical solution and Foreman and Makin's [1.23] (Section 4.2.2.1) computer simulations.

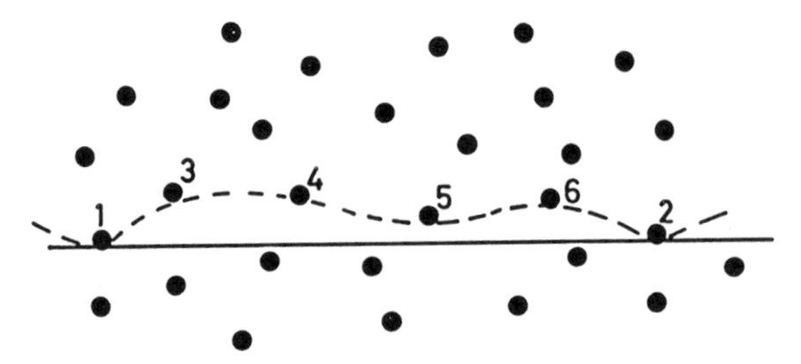

Fig. 1.5. Variation of the particle spacing along a dislocation with its flexibility, shown schematically. Solid line: rigid dislocation; dashed line: flexible dislocation.

1.4 UNDERAGED, PEAK-AGED, AND OVERAGED MATERIALS

In many materials it is possible to let individual coherent particles grow in size with the duration t of the heat treatment and to keep their total volume constant; that is, they become larger but less numerous (Section 3.2.1.2). If the particles are sheared by dislocations, the CRSS increases with t until a maximum is reached at $t = t_m$. For $t < t_m$ the material is called underaged; at t_m it is in the peak-aged state. Further heat treating lowers the CRSS. The material becomes overaged: though the particles may still be coherent, they have become so widely spaced that it is easier for dislocations to circumvent them than to shear them (Sections 4.2.1.3.2 and Chapter 6). Figure 1.6 illustrates this for copper strengthened by cobalt-rich particles. They as well as the matrix have the face-centered cubic (fcc) crystal structure. Since the lattice constants differ by about 1.5%, the particles are surrounded by a stress field via which they interact with dislocations (Section 5.4.1). The peak-aged state is reached when the particle radius is around 10 nm.

1.5 SOME TECHNICALLY IMPORTANT PARTICLE-STRENGTHENED MATERIALS

Of the myriad of technically important particle-strengthened materials, a selection of five groups of them is presented here; two systems have already been mentioned above: NIMONIC PE16 and aluminum–copper–magnesium alloys. The basic metal is always quoted first.

1. *Aluminum–copper–magnesium alloys* [1.26]
   Particles: Coherent, copper–magnesium-rich thin plates.
   Particle–dislocation interaction mechanism: The particles are sheared; the interaction is via their strain field, the increase in particle–matrix

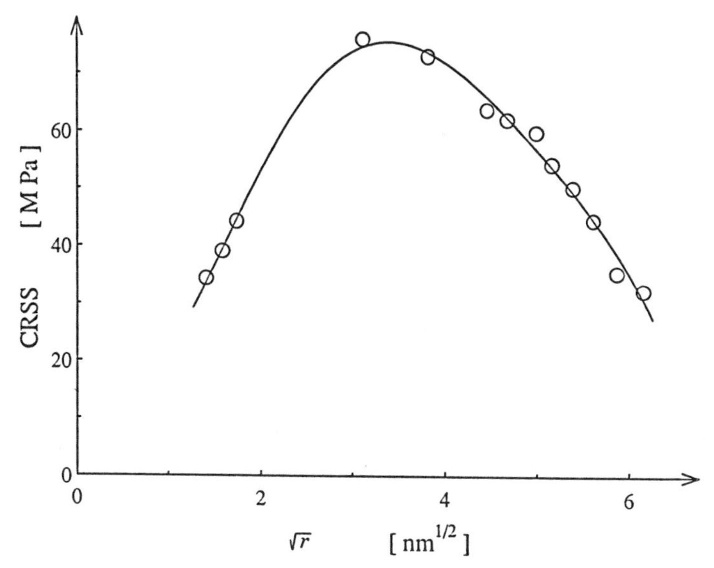

Fig. 1.6. CRSS of copper strengthened by cobalt-rich particles versus the square root of their radius r; their volume fraction is 0.02. (After Refs. 1.24 and 1.25.)

interface also contributes to the CRSS. At later stages of aging, incoherent particles grow; they are circumvented.

2. *Copper–beryllium alloys* [1.2, 1.27]
   Similar to 1.
3. *Nickel-base superalloys* [1.2, 1.6, 1.28–1.30]
   Particles: Coherent γ'-precipitates whose composition is approximately $Ni_3(Al, Ti)$.
   Particle–dislocation interaction mechanism: The particles are sheared; they are long-range-ordered, and dislocations create antiphase boundaries (Sections 4.1 and 5.5). In addition to the coherent γ'-precipitates, incoherent oxide particles can be introduced; they are circumvented.
4. *Maraging steels* [1.31–1.34]
   Particles: Precipitates of various intermetallic phases, similar to 3.
   Particle–dislocation interaction mechanism: The particles are long-range-ordered. If they are coherent and small, they are sheared; otherwise they are circumvented, similar to 3.
5. *Titanium–RE oxides* [1.35–1.38]
   Particles: Incoherent RE oxides; RE stands for rare earth (cerium, erbium, etc. or yttrium).
   Particle–dislocation interaction mechanism: The particles are circumvented.

BIBLIOGRAPHY

General Reading

1.1 A. J. Ardell, *Met. Trans. A*, **16A**, 2131, 1985.

1.2 L. M. Brown and R. K. Ham, in *Strengthening Methods in Crystals* (edited by A. Kelly and R. B. Nicholson), p. 9, Applied Science Publishers, London, 1971.

1.3 V. Gerold, in *Dislocations in Solids*, Vol. 4 (edited by F. R. N. Nabarro), p. 219, North-Holland, Amsterdam, 1979.

1.4 A. Kelly and R. B. Nicholson, *Prog. Mater. Sci.*, **10**, 149, 1963.

1.5 J. W. Martin, *Micromechanisms in Particle-Hardened Alloys*, Cambridge University Press, Cambridge, 1980.

1.6 E. Nembach and G. Neite, *Prog. Mater. Sci.*, **29**, 177, 1985.

1.7 B. Reppich, in *Materials Science and Technology*, Vol. 6 (edited by R. W. Cahn, P. Haasen, and E. J. Kramer), p. 311, VCH Verlagsgesellschaft, Weinheim, 1993.

1.8 J.-L. Strudel, in *Physical Metallurgy*, Part II (edited by R. W. Cahn and P. Haasen), p. 1411, North-Holland, Amsterdam, 1983.

Film

1.9 E. Nembach, V. Ruth, S. Takeuchi, K. Suzuki, and M. Ichihara, *Dislocation Glide in Precipitation Hardened Materials—Tensile Tests in HVEM*, Film D 1681, Institut für den Wissenschaftlichen Film, Göttingen, 1988.

REFERENCES

1.10 A. Wilm, *Metallurgie*, **8**, 225, 1911.

1.11 E. Orowan, *Z. Phys.*, **89**, 634, 1934.

1.12 M. Polanyi, *Z. Phys.*, **89**, 660, 1934.

1.13 G. I. Taylor, *Proc. R. Soc. London A*, **145**, 362, 1934.

1.14 J. M. Raynal, J. Schelten, and W. Schmatz, *J. Appl. Cryst.*, **4**, 511, 1971.

1.15 G. Kostorz, in *Treatise on Materials Science and Technology*, Vol. 15, *Neutron Scattering* (edited by G. Kostorz), p. 227, Academic Press, New York, 1979.

1.16 T. Sato and A. Kamio, *Mater. Sci. Eng.*, **A146**, 161, 1991.

1.17 K. Hono, T. Sakurai, and H. W. Pickering, *Met. Trans. A*, **20A**, 1585, 1989.

1.18 A. Guinier, *Nature*, **142**, 569, 1938.

1.19 G. D. Preston, *Nature*, **142**, 570, 1938.

1.20 T. Sato and T. Takahashi, *Trans. JIM*, **24**, 386, 1983.

1.21 E. Orowan, *Symposium on Internal Stresses in Metals and Alloys*, p. 451, Institute of Metals, London, 1948.

1.22 J. Friedel, *Les Dislocations*, Gauthier-Villars, Paris, 1956.

1.23 A. J. E. Foreman and M. J. Makin, *Philos. Mag.*, **14**, 911, 1966.

1.24 E. Nembach and M. Martin, *Acta Metall.*, **28**, 1069, 1980.

1.25 N. Büttner, K.-D. Fusenig, and E. Nembach, *Acta Metall.*, **35**, 845, 1987.

1.26 J. T. Staley, R. F. Ashton, I. Broverman, and P. R. Sperry, in *Aluminum, Properties and Physical Metallurgy* (edited by J. E. Hatch), 3rd printing, p. 134, American Society for Metals, Metals Park, Ohio, 1988.

1.27 R. J. Price and A. Kelly, *Acta Metall.*, **11**, 915, 1963.

1.28 N. S. Stoloff, in *Alloying* (edited by J. L. Walter, M. R. Jackson, and C. T. Sims), p. 371, American Society for Metals, Metals Park, Ohio, 1988.

1.29 W. Betteridge and J. Heslop, editors, *The Nimonic Alloys*, 2nd edition, Edward Arnold, London, 1974.

1.30 C. T. Sims, N. S. Stoloff, and W. C. Hagel, editors, *Superalloys II*, John Wiley & Sons, New York, 1987.

1.31 L. K. Singhal and J. W. Martin, *Acta Metall.*, **16**, 947, 1968.

1.32 S. Floreen, *Met. Rev.*, **13**, 115, 1968.

1.33 J. W. Christian, in *Strengthening Methods in Crystals* (edited by A. Kelly and R. B. Nicholson), p. 261, Applied Science Publishers, London, 1971.

1.34 R. B. Nicholson, in *Strengthening Methods in Crystals* (edited by A. Kelly and R. B. Nicholson), p. 535, Applied Science Publishers, London, 1971.

1.35 E. W. Collings, in *Alloying* (edited by J. L. Walter, M. R. Jackson, and C. T. Sims), p. 257, American Society for Metals, Metals Park, Ohio, 1988.

1.36 S. M. L. Sastry, P. J. Meschter, and J. E. O'Neal, *Met. Trans. A*, **15A**, 1451, 1984.

1.37 S. M. L. Sastry, T. C. Peng, and L. P. Beckerman, *Met. Trans. A*, **15A**, 1465, 1984.

1.38 D. G. Konitzer, B. C. Muddle, and H. L. Fraser, *Met. Trans. A*, **14A**, 1979, 1983.

CHAPTER 2

Dislocations

Because the theory of dislocations is covered in many books [2.1–2.14], only a short compilation of the relevant definitions and equations is presented here.

2.1 GEOMETRY AND CRYSTALLOGRAPHY

Dislocations are characterized by the following geometric parameters:

$\mathbf{b}$ = Burgers vector

b = $|\mathbf{b}|$ = length of the Burgers vector

$\mathbf{s}$ = vector of unit length parallel to the local direction of the dislocation line

$d\mathbf{s}$ = infinitesimal segment of a dislocation, $d\mathbf{s}$ is parallel to $\mathbf{s}$

ds = $|d\mathbf{s}|$

θ_d = angle between $\mathbf{s}$ and $\mathbf{b}$

$\mathbf{n}$ = normal of the glide plane, $|\mathbf{n}| = 1$

A dislocation whose Burgers vector is a translational vector of the crystal is called perfect. It may, however, dissociate into two or more partial dislocations. In fcc cystals, perfect dislocations have Burgers vectors of the type $(a/2)\langle 110 \rangle$, where a is the lattice constant. Such a dislocation may split into two Shockley partial dislocations, whose Burgers vectors are of the type $(a/6)\langle 211 \rangle$:

$$(a/2)[10\bar{1}] = (a/6)[11\bar{2}] + (a/6)[2\bar{1}\bar{1}] \tag{2.1}$$

In body-centered cubic (bcc) crystals, $\mathbf{b}$ of perfect dislocations is of the type $(a/2)\langle 111 \rangle$; many different dissociations are possible in this system.

2.2 DISLOCATION LINE ENERGY AND TENSION

The energy E_l of a perfect straight dislocation of unit length is written in the form

$$E_l(\theta_d) = K_E(\theta_d) b^2 \ln(R_o/R_i) \tag{2.2}$$

where R_o is the outer and R_i the inner cutoff radius. Linear isotropic theory of elasticity yields the following for $K_E(\theta_d)$:

$$K_E(\theta_d) = \frac{\mu}{4\pi}[\cos^2\theta_d + \sin^2\theta_d/(1-\nu)] \tag{2.3a}$$

or

$$K_E(\theta_d) = \frac{\mu}{4\pi(1-\nu)}[1 - \nu\cos^2\theta_d] \tag{2.3b}$$

where μ is the shear modulus and ν is Poisson's ratio. In the core of the dislocation, atomistic calculations are required. Let the radius of the core and its energy per unit length be R_c and E_c, respectively. Then R_i follows from

$$K_E b^2 \ln(R_o/R_i) = E_c(R_c) + K_E b^2 \ln(R_o/R_c).$$

R_i turns out to be close to b. Evidently, R_c is larger than R_i. Four useful empirical core models will be introduced in Section 5.2.1.

The energy E_{loop} stored in a circular dislocation loop of radius R_{loop} is given by

$$E_{\text{loop}} = 2\pi R_{\text{loop}} \frac{2-\nu}{2(1-\nu)} \frac{\mu b^2}{4\pi}\left[\ln\left(\frac{4R_{\text{loop}}}{R_i}\right) - 2\right] \tag{2.4}$$

b lies in the plane of the loop. Equation (2.4), too, is based on linear isotropic theory of elasticity. Equation (2.4) correctly allows for the mutual elastic interaction between each segment of the loop with the rest of it; that is, the self-interaction is fully taken into account. In Section 4.2.1.4.3 it will be shown that the actual equilibrium shape of such a loop is not a circle, but it is close to an ellipse.

The flexibility of dislocations can be described by their line tension S introduced by deWit and Koehler [2.15]. Here only dislocation bow-outs in the glide plane are considered. S differs from E_l because dislocations change their character as they bow out: from edge to screw character and vice versa. $S(\theta_d)$ is related to $E_l(\theta_d)$ by

$$S(\theta_d) = E_l(\theta_d) + \partial^2 E_l(\theta_d)/\partial\theta_d^2 \tag{2.5}$$

In the derivation of this equation the elastic interaction between different parts of the bent dislocation—that is, the change in self-interaction energy—has been disregarded. Actually the dislocation half-arc AB shown in Figure 2.1 interacts with the half-arc BC and vice versa. This interaction depends on the strength of the dislocation's bow-out.

In analogy to Eq. (2.2), $S(\theta_d)$ is written as

$$S(\theta_d) = K_S(\theta_d) b^2 \ln(R_o/R_i) \tag{2.6}$$

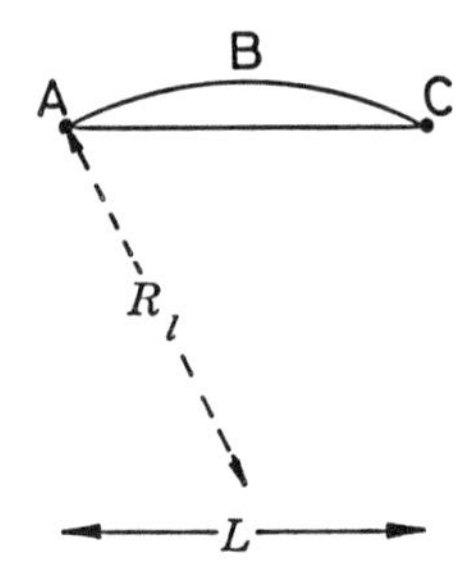

Fig. 2.1. A dislocation bows out between the two pinning centers A and C; the point B lies on the arc. R_l is its radius of curvature.

$K_S(\theta_d)$ of isotropic materials follows from Eqs. (2.2), (2.3), and (2.5):

$$K_S(\theta_d) = \frac{\mu}{4\pi(1-\nu)}[1 - \nu\cos^2\theta_d + 2\nu\cos(2\theta_d)] \qquad (2.7a)$$

or

$$K_S(\theta_d) = \frac{\mu}{4\pi(1-\nu)}[(1+\nu)\cos^2\theta_d + (1-2\nu)\sin^2\theta_d] \qquad (2.7b)$$

Barnett et al. [2.16] treated the case of arbitrary elastic anisotropy. These authors related $K_E(\theta_d)$ and $K_S(\theta_d)$ to the stiffness constants C_{ik} of the material. Bacon and Scattergood [2.17, 2.18] carried out the numerical integrations involved. They presented their results for $K_E(\theta_d)$ and $K_S(\theta_d)$ in the form of Fourier series and tabulated the respective coefficients. $K_E(\theta_d)$ and $K_S(\theta_d)$ of nickel are plotted in Figure 2.2. Evidently, isotropic and anisotropic calculations yield similar results for K_E. The effects of the anisotropy on K_S tend to exceed those on K_E. Edge dislocations have higher line energies, but lower line tensions, than screw dislocations. If the anisotropy is allowed for, the ratios $K_E(\theta_d = 90°)/K_E(\theta_d = 0°)$ and $K_S(\theta_d = 90°)/K_S(\theta_d = 0°)$ of nickel are 1.6 and 0.24, respectively.

Because of the term $\partial^2 E_l/\partial\theta_d^2$ appearing in Eq. (2.5), the line energy and the line tension of a dislocation differ from each other. The ratios $K_E(\theta_d = 0°)/K_S(\theta_d = 0°)$ and $K_E(\theta_d = 90°)/K_S(\theta_d = 90°)$ of nickel are 0.40 and 2.65, respectively. Again, the anisotropy has been taken into account.

If a dislocation bows out, θ_d varies along the arc. This gives rise to a variation of S. If the dislocation bows out only slightly, S may be considered as constant along the arc and $S(\theta_{d0})$ may be inserted. θ_{d0} is the angle before the dislocation bows out. In Section 4.1.2 a quantitative limit will be set for "slight bowing." Further details will be given in Section 4.2.1.4.3.

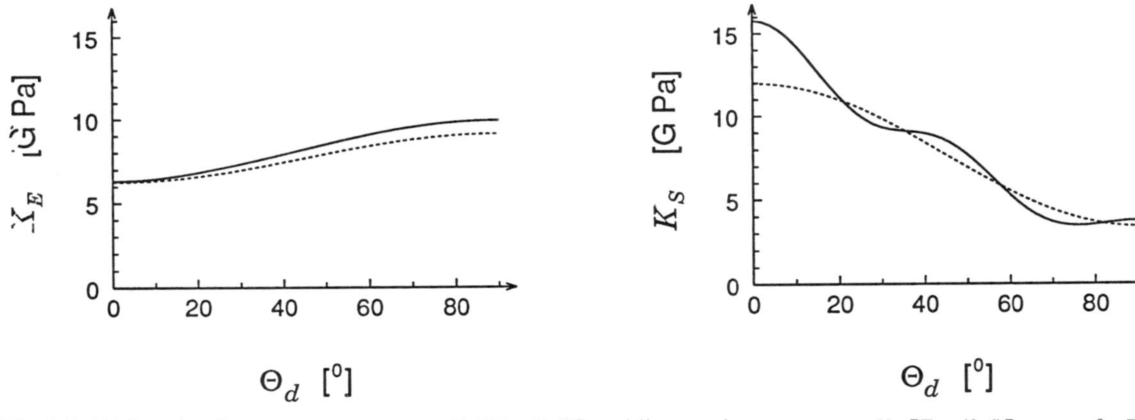

Fig. 2.2. Dislocation line energy parameter K_E[Eq. (2.2)] and line tension parameter K_S [Eq. (2.6)] versus θ_d. The material is nickel at 280 K: $C_{11} = 252$ GPa, $C_{12} = 150$ GPa, $C_{44} = 124$ GPa [2.19], C_{ik} = stiffnesses. Solid line: The elastic anisotropy is fully allowed for [2.17]. Dashed line: Elastic isotropy is assumed, Eqs. (2.3) and (2.7); $\mu = [C_{44}(C_{11} - C_{12})/2]^{1/2}$, $\nu = 0.31$ [2.20].

The choice of the ratio R_o/R_i in the line tension [Eq. (2.6)] of a bent dislocation is somewhat arbitrary. In the following, one of two alternatives will be inserted; the latter one has been suggested by Rönnpagel [2.21, 2.22]:

$$R_o/R_i = L/b \tag{2.8a}$$

$$R_o/R_i = 1.47R_l/b \tag{2.8b}$$

The length L is indicated in Figure 2.1; it is the length of the dislocation before it bows out between the pinning points A and C. R_l is the radius of the local curvature of the dislocation. Equation (2.8b) will be justified in Section 2.3. The length b of the Burgers vector has been inserted for R_i. R_o will be discussed further in Section 4.2.1.4.3.

2.3 FORCES ACTING ON A STATIONARY DISLOCATION

Let σ be the stress tensor describing the local stress acting on the dislocation segment $d\mathbf{s}$. σ may be due to external stresses, particles, other dislocations, and so on. The force $d\mathbf{F}$ experienced by the dislocation segment $d\mathbf{s}$ is given by the famous Peach and Koehler [2.23] equation:

$$d\mathbf{F} = (\mathbf{b}\sigma) \times d\mathbf{s} \tag{2.9a}$$

In most cases only the component dF acting in the glide plane in the glide direction is relevant. Evidently dF cannot be expected to be equal to $|d\mathbf{F}|$. In order to obtain dF, one resolves σ to the glide system; this resolved shear stress will be denoted by τ:

$$\tau = \mathbf{b}\sigma\mathbf{n}/b \tag{2.10}$$

where $\mathbf{n}$ is the normal of the glide plane. Thus one finds the following for dF:

$$dF = \tau b\, ds \tag{2.9b}$$

dF is normal to $d\mathbf{s}$ and $\mathbf{n}$. The total force acting on the dislocation arc sketched in Figure 2.1 is obtained by integrating Eq. (2.9b). This force has the two components F_{arc} and F'_{arc}. F_{arc} is normal to the base line AC, and F'_{arc} is parallel to it. The subscript "arc" of F_{arc} and of F'_{arc} stands for dislocation "arc." If τ is spacially constant, the result for F_{arc} is

$$F_{\text{arc}} = \tau b L \tag{2.9c}$$

L is the spacing of the pinning centers A and C (Figure 2.1). As long as τ is below the critical resolved shear stress (CRSS), the centers A and C can balance F_{arc}. In reality the dislocation does not end at A and C, but instead it forms bow-outs to the left of A and to the right of C. If the pinning centers are collinear and equidistantly spaced, the forces F'_{arc} compensate at each center.

Due to the line tension S, the segment $d\mathbf{s}$ of the bent dislocation shown in Figure 2.1 experiences the resolved shear stress τ_{back}, which tends to pull the dislocation back into the straight original configuration. τ_{back} depends on the radius R_l of the local curvature of the dislocation arc and on S:

$$\tau_{\text{back}} = \frac{S}{bR_l} \tag{2.11a}$$

This equation is analogous to the one describing the bowing of an elastic string under the action of delocalized forces—for example, the sagging of power transmission lines in the earth's field of gravity. In most cases, R_l and S of a dislocation bowing out slightly between two particles can be assumed to be constant; then τ_{back} too is constant along the dislocation arc. L is inserted for R_o of S. Solving Eq. (2.11a) for R_l yields the radius of the local curvature of a dislocation bowing out under an applied stress τ_{ext}:

$$R_l = \frac{S}{\tau_{\text{ext}} b} \tag{2.11b}$$

If and only if τ_{ext} and S are constant, the dislocation assumes the shape of a circular arc. Actually S is not constant because it varies with θ_d.

Rönnpagel [2.21, 2.22] derived S in Eq. (2.11a) not from Eq. (2.5) but from Eq. (2.4). This author considered the energy change dE_{loop} if the radius of a dislocation loop is extended from R_{loop} to $(R_{\text{loop}} + dR_{\text{loop}})$:

$$dE_{\text{loop}} = (\partial E_{\text{loop}}/\partial R_{\text{loop}})\, dR_{\text{loop}}$$

This energy equals the work done against τ'_{back}:

$$dE_{\text{loop}} = \tau'_{\text{back}} b 2\pi R_{\text{loop}}\, dR_{\text{loop}}$$

which leads to

$$\tau'_{\text{back}} b = \frac{2-\nu}{2(1-\nu)} \frac{\mu b^2}{4\pi R_{\text{loop}}} \ln\left(\frac{4R_{\text{loop}}}{eR_i}\right) \tag{2.12}$$

where e equals 2.7183. Rönnpagel's further steps were (i) to drop the factor $[(2-\nu)/2] \approx 0.83$, (ii) to multiply the right-hand side of Eq. (2.12) by the function of θ_d which appears in Eq. (2.7b) in square brackets, and (iii) to identify R_{loop} with R_l and τ'_{back} with τ_{back}. The result is

$$\tau_{\text{back}} = \frac{1}{bR_l} \cdot \frac{\mu b^2}{4\pi(1-\nu)} [(1+\nu)\cos^2\theta_d + (1-2\nu)\sin^2\theta_d] \cdot \ln\left(\frac{4R_l}{eb}\right) \qquad (2.11c)$$

For comparison, Eq. (2.11a) is repeated here for isotropic materials:

$$\tau_{\text{back}} = \frac{1}{bR_l} \cdot \frac{\mu b^2}{4\pi(1-\nu)} [(1+\nu)\cos^2\theta_d + (1-2\nu)\sin^2\theta_d] \cdot \ln\left(\frac{L}{b}\right) \qquad (2.11d)$$

Except for the outer cutoff radii R_o, Eqs. (2.11c) and (2.11d) are the same. R_o equals $4R_l/e = 1.47R_l$ and L, respectively. The former choice is probably the better one. Both alternatives for R_o have already been quoted in Eqs. (2.8). In Eq. (2.11c) as well as in Eq. (2.11d), b has been inserted for the inner cutoff radius R_i. Though the dislocation's self-interaction has been correctly allowed for in the derivation of Eq. (2.4), this is not really true for Eq. (2.11c) because it is applied to a dislocation arc instead of to a full circle. Moreover, it must be stressed that (i) the equilibrium shape of a dislocation is actually not a circle but is close to an ellipse (Section 4.2.1.4.3) and (ii) Eq. (2.11c) is based on linear isotropic elasticity.

2.4 FORCES ACTING ON A MOVING DISLOCATION

If a dislocation moves, it experiences two more forces: One is due to its inertia and the other one is due to viscous drag. Following Rönnpagel [2.21], these forces are represented by the resolved stresses τ_{inert} and τ_{drag}, respectively:

$$b\tau_{\text{inert}} = m_{\text{eff}}(\partial v_n/\partial t) \qquad (2.13)$$

and

$$b\tau_{\text{drag}} = Bv_n \qquad (2.14)$$

where m_{eff} is the effective mass per unit length of the dislocation, v_n is its velocity normal to $d\mathbf{s}$, and B is the drag coefficient [2.24, 2.25]. Equation (2.13) is only an approximation. Often one inserts the product of the density ρ_m of the material and b^2 for m_{eff} [2.24]:

$$m_{\text{eff}} = \rho_m b^2 \qquad (2.15)$$

Dislocation effects caused by τ_{inert} and τ_{drag} are called *dynamic dislocation effects*.

2.5 EQUATION OF MOTION OF A DISLOCATION

The dislocation glides in the plane $z = 0$. The position of the segment $d\mathbf{s}$ at time t is described by the function $y(x, t)$. The radius R_l of the local curvature and the velocity v_n follow from standard geometric considerations:

$$R_l = -\frac{[1 + (\partial y/\partial x)^2]^{3/2}}{\partial^2 y/\partial x^2} \tag{2.16a}$$

$$v_n = \frac{\partial y/\partial t}{[1 + (\partial y/\partial x)^2]^{1/2}} \tag{2.17a}$$

Tacitly, it has so far been assumed that R_l and τ_{back} are positive. Since dislocations may be bent forward or backward (Figures 1.5 and 4.6), R_l and τ_{back} may actually be positive or negative. To allow for this in a consistent way, minus signs have been introduced on the right-hand sides of Eqs. (2.16a) and (2.16b). Therefore these definitions of R_l differ from the standard ones.

The stress tensor σ is separated into two parts: that due to external macroscopic forces (σ_{ext}) and that due to the particles (σ_{obst}). Other stress contributions—for example, those of other dislocations or solved atoms—are disregarded for the time being. In principle, any inhomogeneity of the material may act as an obstacle and thus contribute to σ. Though in most cases the term "obstacle" and the subscript "obst" will refer to particles, the more general expression "obstacle" is kept. The effect of the external forces and that of the obstacles is described by the resolved stresses τ_{ext} and τ_{obst}, respectively.

There is a local balance of stresses [2.21, 2.22]:

$$\tau_{\text{ext}} + \tau_{\text{obst}} = \tau_{\text{inert}} + \tau_{\text{drag}} + \tau_{\text{back}} \tag{2.18}$$

σ_{ext} and τ_{ext} are assumed to be constant throughout the entire specimen. Some authors put τ_{obst} on the right-hand side of Eq. (2.18), but then its sign has to be reversed. With the aid of Eq. (2.9b), τ_{obst} can be derived from the force dF_{obst} exerted by the obstacles on the dislocation segment of length ds. dF_{obst} is parallel to the glide direction of the segment:

$$\tau_{\text{obst}} = \frac{1}{b}(\partial F_{\text{obst}}/\partial s) \tag{2.19}$$

dF_{obst} and τ_{obst} represent the sum over the forces and stresses, respectively, exerted by all obstacles interacting simultaneously with the dislocation segment. If the particle–dislocation interaction is long-ranged—for example, if the particles are surrounded by an elastic stress field—many particles contribute to dF_{obst} and τ_{obst}.

Even if the local curvature of the dislocation is only slight, the change of its direction from one end to the other one may be large. Therefore one cannot assume that the entire dislocation stays approximately parallel to a straight line. A relatively short length of it, which still exceeds the interparticle spacing severalfold, may, however, be supposed to be parallel to the x-axis of the coordinate system and to move towards $+y$. The coordinate system has to be defined accordingly. Then $|\partial y/\partial x|$ is much smaller than unity, and Eqs. (2.16a) and (2.17a) take more simple forms:

$$R_l = -1/(\partial^2 y/\partial x^2) \tag{2.16b}$$

$$v_n = \partial y/\partial t \tag{2.17b}$$

Introducing these relations into Eq. (2.18) results in the equation of motion [2.24–2.27]:

$$b(\tau_{ext} + \tau_{obst}) = m_{eff}(\partial^2 y/\partial t^2) + B(\partial y/\partial t) - S(\partial^2 y/\partial x^2) \tag{2.20}$$

It must be remembered that this equation is based on the assumption that the dislocation is approximately parallel to the x-axis and bows out only slightly. Therefore S is constant. In Section 4.1.2 a quantitative limit will be given for "slight bowing."

In the derivations of Eqs. (2.18) and (2.20), thermal activation has been disregarded. Therefore these equations are correct only at 0 K. Up to moderately high temperatures the effects of thermal activation on particle hardening are believed to be negligible, provided that the dislocations do not leave their slip planes by cross-slip or climb. The neglect can be justified on the ground that the activation energies involved in particle shearing are high. Thermal activation can be allowed for by adding the stochastic stress τ_{therm} to the left-hand sides of Eqs. (2.18) and (2.20). τ_{therm} fluctuates in time and space [2.28, 2.29].

REFERENCES

2.1 H. G. van Bueren, *Imperfections in Crystals*, North-Holland, Amsterdam, 1960.

2.2 A. H. Cottrell, *Dislocations and Plastic Flow in Crystals*, Clarendon Press, Oxford, 1961.

2.3 J. Friedel, *Les Dislocations*, Gauthier-Villars, Paris, 1956.

2.4 J. Friedel, *Dislocations*, Pergamon Press, Oxford, 1964.

2.5 J. P. Hirth and J. Lothe, *Theory of Dislocations*, 2nd edition, John Wiley & Sons, New York, 1982.

2.6 D. Hull and D. J. Bacon, *Introduction to Dislocations*, 3rd edition, Butterworth-Heinemann, Oxford, 1984.

2.7 I. Kovács and L. Zsoldos, *Dislocations and Plastic Deformation*, Pergamon Press, Oxford, 1973.

2.8 F. R. N. Nabarro, *Theory of Crystal Dislocations*, Oxford University Press, London, 1967.

2.9 F. R. N. Nabarro (editor), *Dislocations in Solids*, Vols. 1–3, North-Holland, Amsterdam, 1979–1980.

2.10 W. T. Read, Jr., *Dislocations in Crystals*, McGraw-Hill, New York, 1953.

2.11 A. Seeger, in *Handbuch der Physik*, Band VII, Teil 1, *Kristallphysik I* (edited by S. Flügge), p. 383, Springer-Verlag, Berlin, 1955.

2.12 J. W. Steeds, *Introduction to Anisotropic Elasticity Theory of Dislocations*, Clarendon Press, Oxford, 1973.

2.13 C. Teodosiu, *Elastic Models of Crystal Defects*, Springer-Verlag, Berlin, 1982.

2.14 J. Weertmann and J. R. Weertmann, *Elementary Dislocation Theory*, 2nd printing, Macmillan, New York, 1965.

2.15 G. deWit and J. S. Koehler, *Phys. Rev.*, **116**, 1113, 1959.

2.16 D. M. Barnett, R. J. Asaro, S. D. Gavazza, D. J. Bacon, and R. O. Scattergood, *J. Phys. F: Metal Phys.*, **2**, 854, 1972.

2.17 D. J. Bacon and R. O. Scattergood, *J. Phys. F: Metal Phys.*, **4**, 2126, 1974.

2.18 D. J. Bacon, D. M. Barnett, and R. O. Scattergood, *Prog. Mater. Sci.*, **23**, 51, 1979.

2.19 G. A. Alers, J. R. Neighbours, and H. Sato, *J. Phys. Chem. Sol.*, **13**, 40, 1960.

2.20 G. Simmons and H. Wang, *Single Crystal Elastic Constants and Calculated Aggregate Properties: A Handbook*, 2nd edition, The M.I.T. Press, Cambridge, Massachusetts, 1971.

2.21 D. Rönnpagel, *Proceedings of the 8th Risø International Symposium on Metallurgy and Materials Science, Constitutive Relations and Their Physical Basis* (edited by S. I. Andersen, J. B. Bilde-Sørensen, N. Hansen, T. Leffers, H. Lilholt, O. B. Pedersen, and B. Ralph), p. 503, Risø National Laboratory, Roskilde, Denmark, 1987.

2.22 T. Pretorius and D. Rönnpagel, *Proceedings of the 10th International Conference of the Strength of Materials* (edited by H. Oikawa, K. Maruyama, S. Takeuchi, and M. Yamaguchi), p. 689, JIM, 1994.

2.23 M. Peach and J. S. Koehler, *Phys. Rev.*, **80**, 436, 1950.

2.24 R. W. Balluffi and A. V. Granato, in *Dislocations in Solids*, Vol. 4 (edited by F. R. N. Nabarro), p. 1, North-Holland, Amsterdam, 1979.

2.25 E. Nadgornyi, *Prog. Mater. Sci.*, **31**, 1, 1988.

2.26 R. B. Schwarz and R. Labusch, *J. Appl. Phys.*, **49**, 5174, 1978.

2.27 K.-D. Fusenig and E. Nembach, *Acta Metall. Mater.*, **41**, 3181, 1993.

2.28 R. Labusch and R. B. Schwarz, in *Proceedings of the 9th International Conference on the Strength of Metals and Alloys* (edited by D. G. Brandon, R. Chaim, and A. Rosen), p. 47, Freund Publishing House, London, 1991.

2.29 D. Rönnpagel, Th. Streit, and Th. Pretorius, *phys. stat. sol. (a)*, **135**, 445, 1993.

CHAPTER 3

Particles

3.1 PARTICLE STATISTICS

Particles will be assumed to be spherical. This is no serious restriction because they are spherical anyway in most underaged and peak-aged (Section 1.4) specimens. In overaged ones, however, they may be cuboidal, but for such specimens many of the relations derived in this section are not needed. If nonspherical particles are referred to, this will be explicitly stated.

The following characteristic lengths and parameters are defined [3.1–3.8]; means over particle characteristics refer to the entire specimen. The word "parallel" appears in the definitions of r_r, r_d, q, L_{max}, L_{min}, and n_s because the results of the averaging procedures to be detailed below depend on whether the planes and lines are parallel or not.

ρ = radius of an individual spherical particle

r = mean radius of the spherical particles

r_r = mean radius of the intersections of parallel planes with spherical particles

$2r_d$ = mean linear traverse length of parallel straight lines over spherical particles

q = mean area of the intersections of parallel planes with spherical particles

L_{max} = mean particle spacing along parallel straight lines

L_{min} = center-to-center nearest-neighbor particle spacing in parallel planes if the particles form square arrays in the planes = square lattice spacing

n_s = number of particles intersecting unit area of parallel planes

n_v = number of particles per unit volume

f = volume fraction of particles

The above parameters are related to each other by

$$L_{min}^2 n_s = 1.0 \tag{3.1a}$$

$$f = n_v \frac{4\pi}{3} \overline{\rho^3} = n_v \omega_3 \frac{4\pi}{3} r^3 \tag{3.2}$$

$$f = q/L_{\text{min}}^2 \tag{3.3a}$$

$$f = n_s q = n_s \omega_q \pi r^2 \tag{3.3b}$$

$$L_{\text{min}} = r(\pi \omega_q / f)^{1/2} \tag{3.3c}$$

$$f = 2r_d/L_{\text{max}} = 2\omega_d r/L_{\text{max}} \tag{3.4}$$

with

$$\omega_r = r_r/r \tag{3.5a}$$

$$\omega_n = \overline{\rho^n}/r^n \tag{3.5b}$$

$$\omega_d = r_d/r \tag{3.5c}$$

$$\omega_q = q/(\pi r^2) \tag{3.5d}$$

where $\overline{\rho^n}$ is the mean of (ρ^n). Equations (3.1a), (3.3a), and (3.3c) hold only for square lattice arrangements of particles. If they form a hexagonal arrangement instead of a square one, Eq. (3.1a) is replaced by

$$L_6^2 n_s = 2/\sqrt{3} \tag{3.1b}$$

where L_6 is the nearest-neighbor spacing. Below, n_s, which is well-defined for any particle arrangement, will be represented by $1/L_{\text{min}}^2$ given by Eq. (3.3c).

The statistical parameters ω_r, ω_n, ω_d, and ω_q are governed by the spectrum of the specimen's particle radii. Their distribution is described by the function $g(\rho/r)$. Its argument is the reduced particle radius ρ/r. g is normalized to unity:

$$\int_0^\infty g(\rho/r)\, d(\rho/r) = 1.0$$

Of the many distributions possible, only two will be analyzed in detail:

1. g_0: All particles have the same radius r; that is,

$$g_0(\rho/r) = \delta(\rho/r - 1) \tag{3.6}$$

   where δ is Dirac's delta function.

2. g_{WLS}: This function has been introduced by Wagner [3.9] and Lifshitz and Slyozov [3.10], it will be discussed in Section 3.2.1.2:

$$
\begin{aligned}
g_{\text{WLS}}(\rho/r) &= \frac{4}{9}(\rho/r)^2 \left(\frac{3}{3+\rho/r}\right)^{7/3} \left(\frac{1.5}{1.5-\rho/r}\right)^{11/3} \\
&\quad \cdot \exp\left(\frac{\rho/r}{\rho/r - 1.5}\right) \qquad \text{for } \rho < 1.5r \\
g_{\text{WLS}}(\rho/r) &= 0 \qquad\qquad\qquad\qquad\qquad\quad \text{for } \rho \geqslant 1.5r
\end{aligned}
\tag{3.7}
$$

g_{WLS} is plotted in Figure 3.1. The full width at half-maximum of $g_{\text{WLS}}(\rho/r)$ is 0.42.

The evaluation of ω_n is straightforward:

$$
\omega_n = \int_0^{\infty} (\rho/r)^n \, g(\rho/r) \, d(\rho/r) \tag{3.8}
$$

In the case of g_0—that is, for particles of uniform size—ω_n equals unity for all n. The results for g_{WLS} are quoted in Table 3.1 [3.11].

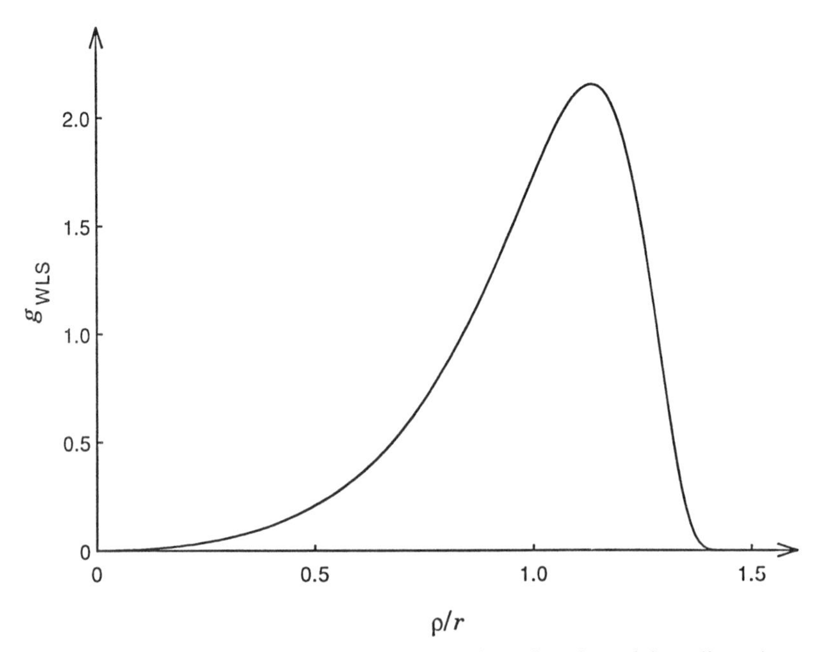

Fig. 3.1. Distribution function g_{WLS} versus the reduced particle radius ρ/r.

TABLE 3.1. ω_n Defined in Eq. (3.5b) for the Distribution Function g_{WLS} of Eq. (3.7)[a]

n	ω_n
1	1.000
2	1.046
3	1.130
4	1.249
5	1.406
6	1.609

[a]After Ref. 3.11.

Before deriving ω_r, ω_d, and ω_q, the two-dimensional problem sketched in Figure 3.2a is analyzed. The probability that a straight horizontal line cuts the circle of radius ρ at a level between x and $(x + dx)$ is dx/ρ. This is also the probability that the semichord has the length $v = (\rho^2 - x^2)^{1/2}$. Thus one obtains for the average $\bar{v}$:

$$\bar{v} = \int_0^\rho (\rho^2 - x^2)^{1/2}\, dx/\rho \tag{3.9a}$$

$$\bar{v} = \frac{1}{2\rho}\{x(\rho^2 - x^2)^{1/2} + \rho^2 \arcsin(x/\rho)\}|_0^\rho$$

$$\bar{v} = (\pi/4)\rho \tag{3.9b}$$

Now the three-dimensional problem of Figure 3.2b is addressed for g_0, for which ρ is identical with r. Evidently the mean radius r_r equals $\bar{v}$ of Eq. (3.9b):

$$r_r = (\pi/4)\rho \tag{3.9c}$$

That is, ω_r equals $\pi/4$ for g_0.

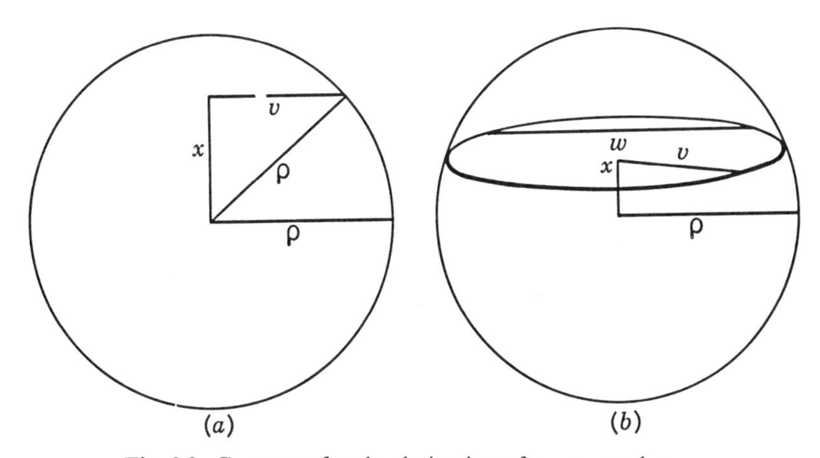

Fig. 3.2. Geometry for the derivation of ω_r, ω_d, and ω_q.

For the derivation of q, the term $(\rho^2 - x^2)^{1/2}$ in Eq. (3.9a) is replaced by $\pi(\rho^2 - x^2)$:

$$q = \pi \int_0^\rho (\rho^2 - x^2)\, dx/\rho$$

$$q = \tfrac{2}{3}\pi\rho^2 \tag{3.10}$$

That is, ω_q equals 2/3 for g_0.

To obtain the average chord $\bar{w}$ of a spherical particle, one has to insert different weights into Eq. (3.9a). The average w within the plane at height x is $2\frac{\pi}{4}(\rho^2 - x^2)^{1/2}$ [Eq. (3.9b)]. The probability dp that a straight horizontal line hits the sphere at heights between x and $(x + dx)$ is proportional to the extension of this circle normal to the image plane of Figure 3.2b. This yields the following for dp:

$$dp = \frac{[2(\rho^2 - x^2)^{1/2}]dx}{\pi\rho^2/2} \tag{3.11a}$$

and thus one obtains for $\bar{w} = 2r_d$

$$2r_d = \bar{w} = \int_0^\rho 2\frac{\pi}{4}(\rho^2 - x^2)^{1/2}\,[2(\rho^2 - x^2)^{1/2}]dx/(\pi\rho^2/2)$$

$$r_d = \tfrac{2}{3}\rho \tag{3.11b}$$

That is, ω_d is 2/3 for g_0.

It must be stressed that dp will be different when r_d is defined for randomly oriented straight lines instead of for parallel straight lines. Since ω_r, ω_d, and ω_q will be used in connection with the glide of dislocations in one glide system, it is appropriate to stay with the definitions given at the beginning of this section: r_r, ω_r, q, ω_q, r_d, and ω_d refer to parallel planes and lines.

In the case of a general distribution function $g(\rho/r)$, one starts with the results for each ρ quoted above and weights them with the product of $g(\rho/r)$ and the probability that parallel planes or parallel straight lines actually hit particles with radii between ρ and $(\rho + d\rho)$. These weights are as follows:

Planes: $$\frac{(\rho g)d(\rho/r)}{\int_0^\infty (\rho g)d(\rho/r)} = \frac{1}{\omega_1 r}[\rho g]d(\rho/r)$$

Straight lines: $$\frac{(\pi\rho^2 g)d(\rho/r)}{\int_0^\infty (\pi\rho^2 g)d(\rho/r)} = \frac{1}{\omega_2 r^2}[\rho^2 g]d(\rho/r)$$

ω_1 equals unity for all functions g [Eq. (3.8)]. Thus one obtains (the terms in square brackets reflect the weights):

$$r_r = \frac{1}{\omega_1 r}\int_0^\infty \frac{\pi}{4}\rho[\rho g]\, d(\rho/r)$$

$$r_r = \frac{\pi}{4}\omega_2 r \tag{3.12}$$

$$\omega_r = \frac{\pi}{4}\omega_2$$

TABLE 3.2. Parameters ω_r, ω_q, and ω_d of Eqs. (3.5) for the Distribution Functions g (General), g_0 [Eq. (3.6)], and g_{WLS} [Eq. (3.7)]

	g	g_0	g_{WLS}
ω_r	$(\pi/4)\omega_2$	$\pi/4$	0.82
ω_q	$(2/3)\omega_3$	2/3	0.75
ω_d	$(2/3)(\omega_3/\omega_2)$	2/3	0.72

$$q = \frac{1}{\omega_1 r} \int_0^\infty \frac{2\pi}{3} \rho^2 [\rho g] \, d(\rho/r)$$

$$q = \frac{2\pi}{3} \omega_3 r^2 \tag{3.13}$$

$$\omega_q = \tfrac{2}{3} \omega_3$$

$$r_d = \frac{1}{\omega_2 r^2} \int_0^\infty \tfrac{2}{3} \rho [\rho^2 g] \, d(\rho/r)$$

$$r_d = \tfrac{2}{3} \omega_3 r / \omega_2 \tag{3.14}$$

$$\omega_d = \tfrac{2}{3} \omega_3 / \omega_2$$

The results for ω_r, ω_q, and ω_d are compiled in Table 3.2.

3.2 PRODUCTION OF SECOND-PHASE PARTICLES

There are at least three different methods to produce fine particles of secondary phases in materials:

- Precipitation from a supersaturated solid solution
- Mechanical alloying
- Internal oxidation

The processes have been quoted in the order of decreasing technical relevance. Particles useful for hardening have diameters below about 0.5 μm.

3.2.1 Precipitation from a Supersaturated Solid Solution

This is by far the most important process. It is described here with reference to the nickel-rich part of the nickel–aluminum phase diagram, shown in Figure 3.3. Nickel with an atomic fraction $c = 0.15$ of aluminum is a homogeneous solid solution at 1500 K; at 1000 K, however, precipitates of the γ'-phase form.

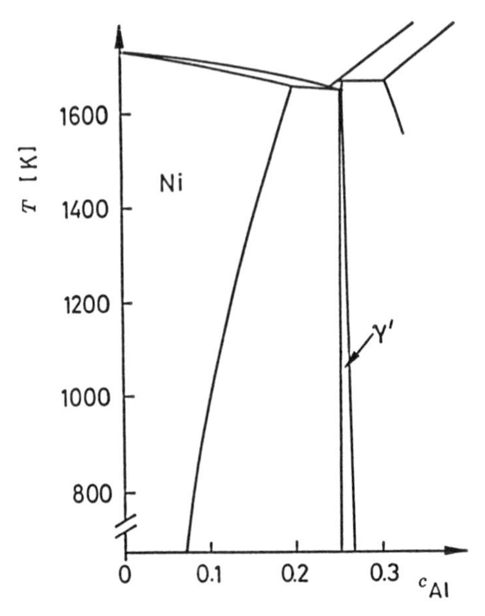

Fig. 3.3. Nickel-rich part of the nickel–aluminum phase diagram. c_{Al} = atomic fraction of aluminum. (After Ref. 3.12.)

This intermetallic compound has the approximate composition Ni_3Al. Three stages can be distinguished in the classical precipitation process:

1. Nucleation of the γ'-phase
2. Growth of the γ'-particles and accompanying solute depletion of the matrix
3. Coagulation of the particles

Though all three steps may occur concurrently, there are many systems in which stages 1 and 2 are completed after relatively short time, whereas stage 3 continues indefinitely. Examples are γ'-precipitates in nickel-base superalloys [3.13, 3.14] (Section 5.5.4.1) and cobalt-rich particles in copper–cobalt alloys [3.15] (Section 5.4.2.1.8).

3.2.1.1 Nucleation and Growth of the Nuclei of the New Phase. For particle strengthening it is of overriding importance that nucleation is homogeneous throughout the specimen. Inhomogeneities caused by heterogeneous nucleation lead to soft spots in the material and are therefore intolerable. Examples will be given in Section 3.2.1.3.

The classical description of homogeneous nucleation [3.16, 3.17] based on Volmer and Weber's [3.18] and Becker and Döring's [3.19] original ideas suffices for the present purpose. Later developments have been reviewed by Wagner and Kampmann [3.20] and by Cohen [3.21]. The nucleus is assumed to be spherical of radius ρ. The Gibbs free energies per unit volume of the supersaturated matrix and of the nucleus will be referred to by g_s and by g_p, respectively. The subscripts s and p of g stand for "solid solution" and "particle," respectively. In the above example (Figure 3.3) the nucleus of the γ'-phase forms at 1000 K from the supersaturated, thermodynamically unstable nickel-rich matrix. g_s is larger than g_p. The total free energy difference ΔG connected with the formation of the nucleus involves two more terms: the misfit strain energy E_{St} [Eq. (3.15a)] and the energy E_Γ [Eq. (3.15b)] stored in the phase boundary.

E_{St} of a coherent nucleus is related to ρ, the shear modulus μ, and the unconstrained lattice mismatch δ by [3.22–3.24]:

$$E_{St} = \left(\frac{4\pi}{3}\rho^3\right)4\mu\delta^2 \tag{3.15a}$$

Three assumptions have been made: Matrix and nucleus are elastically isotropic, they have the same elastic parameters, and Poisson's ratio ν equals 1/3. δ describes the relative difference of the lattice constants a_s and a_p of the two phases involved:

$$\delta = \frac{a_s - a_p}{(a_s + a_p)/2} \tag{3.16}$$

a_s and a_p are to be measured on the *unconstrained* single-phased materials; this designation is carried over to δ: unconstrained mismatch. The *constrained* lattice mismatch ε will be defined in Section 3.3.4. The above definition of δ is meant for cases where the matrix as well as the particles have the same cubic crystal structure. In all other cases one must carefully check whether δ defined in Eq. (3.16) can be used in Eq. (3.15a). Many authors chose $(a_p - a_s)$ as the numerator of Eq. (3.16); that is, they gave δ the opposite sign.

Incoherent nuclei are more difficult to treat because in general their elastic properties differ drastically from those of the matrix: In most cases they are harder. A possible approximation is to consider them as incompressible. The misfit energy of incoherent spherical nuclei also is proportional to ρ^3, and it tends to be smaller than that of coherent ones.

The derivation of E_Γ is straightforward; Γ is the specific energy of the phase boundary:

$$E_\Gamma = 4\pi\rho^2\Gamma \tag{3.15b}$$

Possible anisotropies of Γ are disregarded. Thus one obtains the following for

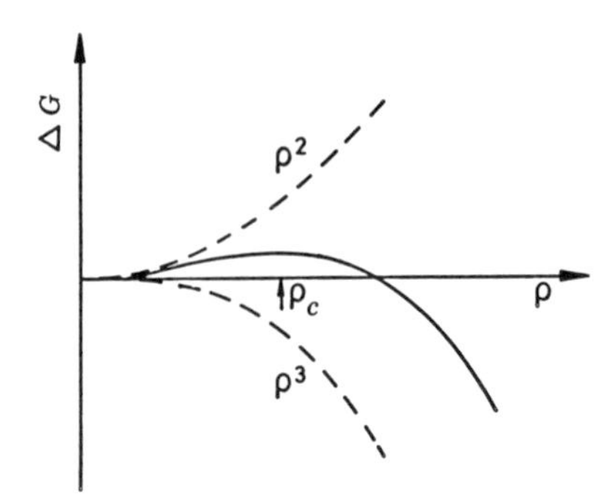

Fig. 3.4. Schematic sketch of ΔG versus ρ(—). The two terms appearing on the right-hand side of Eq. (3.17) are shown separately (---). ΔG is the free energy of the nucleus and ρ is its radius.

ΔG of coherent nuclei:

$$\Delta G = \frac{4\pi}{3}\rho^3(g_p - g_s + 4\mu\delta^2) + 4\pi\rho^2\Gamma \tag{3.17}$$

The sign of ΔG is such that negative values of ΔG indicate that the nucleus is stable. The variation of the term proportional to ρ^3, the variation of that proportional to ρ^2, and the variation of ΔG with ρ are sketched in Figure 3.4. ΔG has a maximum at the critical radius ρ_c. If a nucleus with $\rho < \rho_c$ is formed by a thermal fluctuation, it cannot grow because this would raise ΔG and thus destabilize the nucleus further; hence it dissolves. A nucleus with $\rho \geqslant \rho_c$, however, lowers ΔG by its growth; consequently it will grow and deplete the matrix of solute. ρ_c follows from the condition $\partial \Delta G/\partial \rho|_{\rho_c} = 0$:

$$\rho_c = \frac{2\Gamma}{(g_s - g_p - 4\mu\delta^2)} \tag{3.18a}$$

Inserting ρ_c into Eq. (3.17) yields

$$\Delta G(\rho_c) = \frac{16\pi\Gamma^3}{3(g_s - g_p - 4\mu\delta^2)^2} \tag{3.18b}$$

These two equations are meant for coherent nuclei; for incoherent ones the term $(4\mu\delta^2)$ has to be replaced by another one. The important point is that it will be rather small.

The rate of formation of nuclei with ρ_c is proportional to $\exp[-\Delta G(\rho_c)/(k_B T_H)]$, where k_B is the Boltzmann constant and T_H is the temperature of the heat treatment. Only the difference $(g_s - g_p)$ varies strongly with T_H; it increases as T_H is lowered. The same holds for the rate of formation of nuclei with $\rho \geqslant \rho_c$—that is, for the actual nucleation rate. Since diffusion is involved in this process, this rate decreases again at very low temperatures.

Generally, Γ of coherent phase boundaries is below $0.2\,\mathrm{J/m^2}$ (often it is even much smaller), whereas Γ of incoherent boundaries is higher: up to about $1\,\mathrm{J/m^2}$ [3.17, 3.20]. This explains why the precipitation of particles of a coherent metastable phase often precedes that of the incoherent, thermodynamic equilibrium phase. The following relations hold: $\Gamma(\text{coherent}) < \Gamma(\text{incoherent})$ and $g_p(\text{coherent}) > g_p(\text{incoherent})$. If the effect of Γ on ρ_c outweighs those of g_p and δ [Eq. (3.18a)], the coherent particles nucleate first; they are metastable. Examples are the Guinier–Preston zones in the aluminum–copper alloys mentioned in Section 1.1. In the final stage of aging, the zones are replaced by incoherent particles of the stable θ-phase. Its composition is Al_2Cu.

Semicoherent phase boundaries may be considered as being derived from coherent ones. Part of the strain associated with the latter ones is relieved by introducing dislocations into the boundaries, which then resemble small-angle boundaries. The spacing of the dislocations is directly related to δ. Measuring this spacing in quenched specimens yields δ at the temperature of the heat treatment. This has been done by Lasalmonie and Strudel [3.25] for γ'-particles in a nickel-base alloy.

So far the supersaturated matrix has been assumed to have been homogenized in the solid state. Consequently the solute content is below the solubility limit at the homogenization temperature. In the technical rapid solidification process [3.26] the melt is quenched at rates exceeding $10^4\,\mathrm{K/s}$. Thus high supersaturations can be achieved; during the subsequent solid state processing they lead to high particle volume fractions. There are also other benefits of this technology: Grain size and dendrite arm spacings are reduced [3.27]. Rapid solidification processing has been applied to aluminum [3.27] and titanium alloys [3.28–3.30].

3.2.1.2 Coagulation of the Particles.

Nuclei with $\rho \geqslant \rho_c$ will grow and deplete the matrix of solute. Because nucleation is a statistical process, there will be a distribution of the particle radii. The energy per unit volume stored in a particle's interface equals $[4\pi\rho^2\Gamma/(4\pi\rho^3/3)] = 3\Gamma/\rho$. Thus small particles can lower their energy by coagulation. This process has been called *Ostwald ripening* [3.31].

The Gibbs–Thomson equation relates the equilibrium solute concentration $c_s(\rho)$ in the matrix next to the interface of a particle to its radius ρ:

$$c_s(\rho) = c_s(\rho \to \infty)\exp\left(\frac{2\Gamma V_a}{k_B T_H \rho}\right) \tag{3.19a}$$

After linearization one obtains

$$c_s(\rho) = c_s(\rho \to \infty)\left(1 + \frac{2\Gamma V_a}{k_B T_H \rho}\right) \tag{3.19b}$$

where V_a is the atomic volume, k_B the Boltzmann constant, T_H the temperature of the heat treatment, and $c_s(\rho \to \infty)$ the solute concentration in equilibrium with a plane interface. Evidently, small particles are surrounded by a dense cloud of solute, whereas the cloud surrounding larger particles is less dense. Therefore solute diffuses to the larger particles and makes them grow at the expense of the smaller ones, which dissolve. Consequently the mean particle radius r increases with time t. This diffusion problem has been treated by Wagner [3.9] and Lifshitz and Slyozov [3.10] under the following restrictive assumptions:

1. The precipitated volume fraction f is very small; that is, $f \ll 1.0$.
2. The matrix has virtually reached its equilibrium solute concentration; this implies that f also has its final value; no new particles nucleate.
3. The coagulation rate is governed by volume diffusion.

The results were as follows:

$$r^3(t) - r^3(t_0) = k_{\mathrm{WLS}}(t - t_0) \tag{3.20a}$$

with

$$k_{\mathrm{WLS}} = \frac{8\Gamma V_a D}{9 k_B T_H} \frac{c_s(\rho \to \infty)}{c_p - c_s(\rho \to \infty)} \tag{3.20b}$$

where D is the diffusion coefficient of the solute and c_p is its concentration in the particles. The distribution function of particle radii is g_{WLS} of Eq. (3.7). If $r(t_0)$ is much smaller than $r(t)$, Eq. (3.20a) can be replaced by

$$r(t) = k'_{\mathrm{WLS}} t^{1/3} + r_0 \tag{3.20c}$$

where r_0 is a constant.

Equations (3.20a) and (3.20c) describe experimental data very well, even for f up to 0.5 [3.14]. Figure 3.5 shows $r(t)$ of γ'-particles in the nickel-base superalloy NIMONIC PE16 heat-treated at various temperatures T_H. f of NIMONIC PE16 is at most 0.1. The different slopes k'_{WLS} mainly reflect the temperature dependence of D and of $c_s(\rho \to \infty)$ [Eq. (3.20b)]. As to be expected on the basis of the binary Ni–Al phase diagram shown in Figure 3.3, f decreases as T_H is raised. Equations (3.20) do not hold at the beginning of the precipitation process [3.14, 3.20].

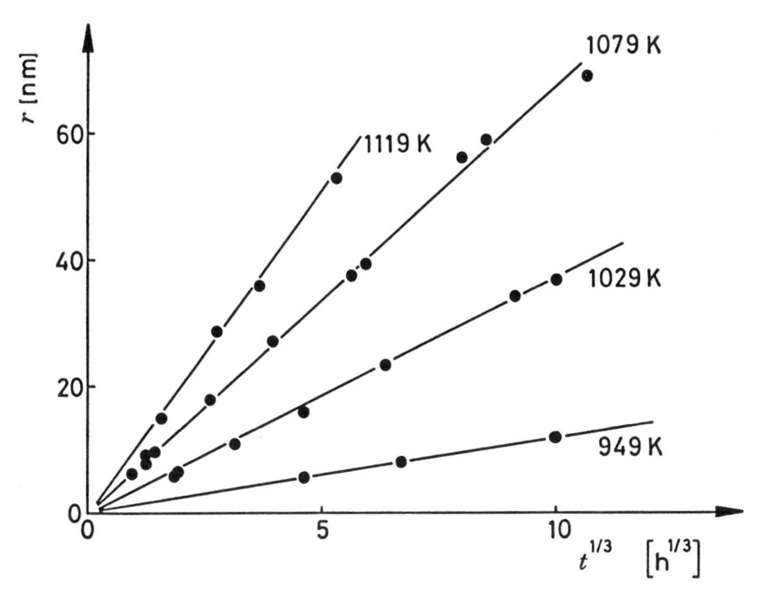

Fig. 3.5. Mean particle radius r of γ'-particles in the superalloy NIMONIC PE16 versus $t^{1/3}$. t is the duration of the heat treatment at the indicated temperatures. (After Ref. 3.32.)

The observed distribution functions of particle radii are in reasonable agreement with $g_{\mathrm{WLS}}(\rho/r)$; this will be discussed further in Section 3.3.1.5. There also the question of the spatial distribution of the particles will be addressed. Here it suffices to state that a particle of radius ρ depletes a "sphere of interest" of solute atoms. The radius of this sphere is proportional to $\rho/f^{1/3}$ [Eq. (3.29)]. This leads to the correct volume fraction of the particles. The "spheres of interest" form a random, densely packed arrangement. There are no "white areas" in such an arrangement and consequently no soft spots (Figures 3.10a and 3.10b). Instead of the expression "sphere of interest," often the term "hard core" is used in the literature; see, for example, Ref 3.33. Both terms are meant to indicate that each precipitate is in the center of a sphere which is free of other precipitates. In most cases, the statistical parameters ω_r, ω_d, and ω_q defined in Eqs. (3.5) will be assumed to be those of g_{WLS}.

Later authors primarily aimed at eliminating Wagner's [3.9] and Lifshitz and Slyozov's [3.10] supposition that f must be much smaller than unity: They treated finite volume fractions [3.20]; Davies et al. [3.34] specifically allowed for the encounter of growing particles. The principal results of these developments are as follows: Equation (3.20a) is maintained, but k_{WLS} increases with f, $g(\rho/r)$ broadens and extends beyond $\rho/r = 1.5$. There are many systems in which after relatively short times of aging, f stays constant, whereas r increases continuously. If the particles consist of many constituents, whose diffusion coefficients are not identical, the particles are expected to be chemically inhomogeneous.

3.2.1.3 Spatially Inhomogeneous Precipitate Distributions. Heterogeneous nucleation leads to spatially inhomogeneous precipitate distributions, which are detrimental to the strength of materials. In polycrystals, grain boundaries often disturb the precipitation process: They lead to precipitate-free zones (PFZs) along the grain boundaries. At the beginning of an isothermal aging treatment the precipitates may form even rather close to the grain boundaries, but during prolonged aging one or more of the precipitating elements migrate into the grain boundaries and the precipitates dissolve again. The results are PFZs. Two examples are mentioned here: precipitates of the δ'-phase in aluminum–lithium alloys and precipitates of the γ'-phase in the nickel-base superalloy NIMONIC PE16 (Sections 1.2 and 5.5). Aluminum-base aluminum–lithium alloys are a new class of materials for aerospace applications [3.35] (Section 5.5.4.1).

δ'-Precipitates: Their approximate composition is Al_3Li, they are metastable. The stable δ-phase AlLi grows in grain boundaries [3.36, 3.37]. This depletes a layer next to them of lithium and thus gives rise to PFZs [3.38].

γ'-Precipitates: This case is somewhat more involved. The approximate γ'-composition in NIMONIC PE16 is $Ni_3(Al, Ti)$. In the grain boundaries two different types of carbides grow: titanium-rich and chromium-rich ones. Figure 3.6 shows both of them at a rather early stage: Titanium carbides form a thin layer along the grain boundary, whereas the chromium carbide is blockish. At a later stage, titanium carbides are difficult to recognize because they are masked by the bigger chromium carbides. The formation of the titanium-rich carbides in the grain boundaries leads to a titanium depletion along them and—since titanium is a constituent of the γ'-precipitates—to PFZs [3.39].

Such PFZs are soft. Their detrimental effect has been demonstrated by Mangen and Nembach [3.40], who found that γ'-PFZs lowered the yield strength of NIMONIC PE16 polycrystals by 10–15%.

Practically, dislocations are far less likely to become sites of heterogeneous nucleation [3.41] because most dislocations are eliminated during the homogenization, which precedes the precipitation treatment. Ngi et al. [3.42] reported the precipitation of γ'-particles on dislocations in a nickel-base alloy which had been slightly deformed after its homogenization. Cassada et al. [3.43] observed heterogeneous, coherent nucleation of δ'-precipitates on dislocations in an aluminum–lithium alloy. The nucleation occurred on the compressive side of edge dislocations.

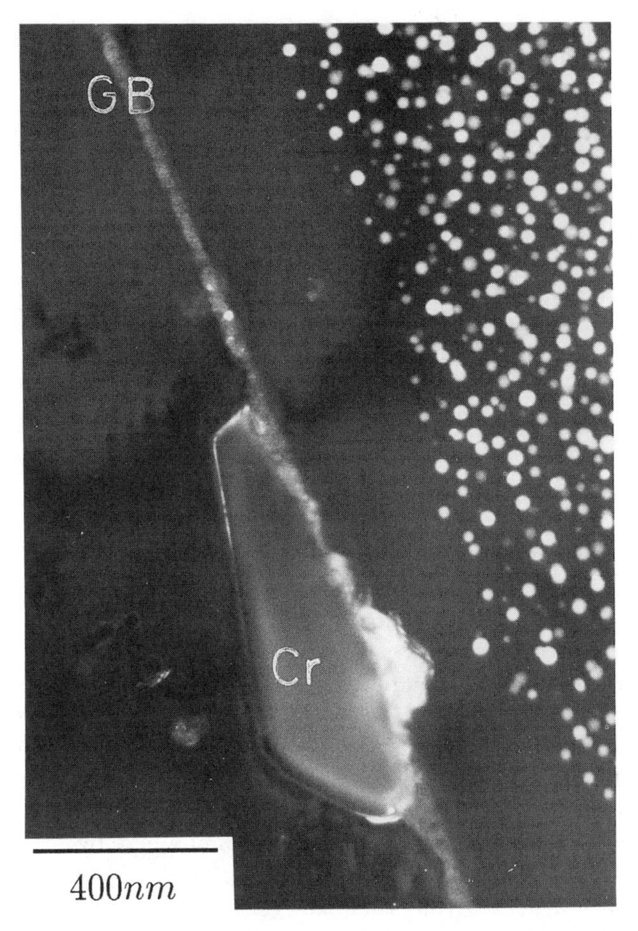

Fig. 3.6. A γ'-precipitate-free zone along a grain boundary (GB) in the nickel-base superalloy NIMONIC PE16 (dark-field TEM image). The left grain and its zone are out of contrast. The grain boundary is decorated by titanium carbides, and a chromium carbide is marked by Cr. (Courtesy of R. Maldonado, Münster.)

3.2.2 Mechanical Alloying

In the 1960s Benjamin [3.44, 3.45] developed mechanical alloying as a method for uniformly dispersing incoherent oxide particles in nickel-base superalloys. The results are oxide-dispersion-strengthened (ODS) superalloys. Typically yttria (Y_2O_3) is the oxide. The process involves several steps [3.46]:

1. Mixing of powders of pure metals or of crushed master alloys and of the oxides

2. Ball milling of the mixture
3. Compacting the powder drained from the mill by extrusion or by hot isostatic pressing
4. Directional recrystallization
5. Precipitation of γ'-particles by a final heat treatment

Whittenberger [3.47] characterized the particles in the commercial ODS superalloy MA 6000 as follows; MA 6000 is produced by Inco Alloys International:

γ'-particles:	Volume fraction, about 0.5
Mixed (Y_2O_3–Al_2O_3) particles:	Volume fraction, 0.025; average radius, around 15 nm; planar particle spacing, around 150 nm

The most important production step is the high-energy ball milling process, during which the grains of the powders are repeatedly welded and fractured. The further steps—extrusion and directional recrystallization—yield large, elongated grains.

The free energies of formation of yttria and alumina exceed that of the γ'-phase several fold [3.48]. This guarantees the high thermal stability of the oxides, whereas γ'-precipitates dissolve as the solvus temperature is approached. By introducing oxide particles by mechanical alloying, many severe thermodynamic limitations are overcome.

It should be mentioned that also ODS–aluminum alloys can be produced by mechanical alloying [3.45, 3.46, 3.48, 3.49]. They derive their high strength from alumina particles. Al_2O_3 is already present on the surface of the original aluminum powder grains; in addition, Al_2O_3 may develop during the milling process. Moreover, aluminum carbides are formed from the organic process control agent applied during mechanical alloying.

3.2.3 Internal Oxidation

The objective of this process is the same as that of mechanical alloying (Section 3.2.2): to obtain materials containing homogeneously distributed, incoherent oxide particles [3.44, 3.50, 3.51]. The approach is, however, different. The starting materials are not powders, but dilute, compact solid solutions. During a heat treatment in an oxidizing atmosphere, oxygen diffuses into the material and oxidizes only the highly reactive solute. The result is an ODS material. The main problems with this procedure are as follows:

1. Since the rate of internal oxidation is diffusion-controlled, only thin specimens can be treated within reasonable times.

2. The size distribution of the oxide particles is spatially inhomogeneous: They are small near the surface and relatively large deeper inside [3.52, 3.53].
3. In more complex systems, it is impossible to limit the oxidation process to those elements which had been meant to be oxidized; in nickel-base superalloys the γ'-forming elements aluminum and titanium also may be oxidized [3.44] (Section 5.5.1).

For the above reasons, internal oxidation has not become a technical process for the production of ODS materials; it has, however, been used in basic research. In contrast to mechanical alloying, which yields polycrystalline ODS materials, internal oxidation can be used to obtain ODS single crystals. Ebeling and Ashby [3.53] investigated the strenghtening effect of SiO_2 particles in copper single crystals (Section 6.2.1). These authors procured a spatially rather uniform size distribution of SiO_2 particles throughout the specimens by removing their outer 0.3 mm-thick layers by a chemical polish. The final diameter of the cylindrical crystals was 1.5 mm.

3.3 EXPERIMENTAL CHARACTERIZATION OF PARTICLE DISTRIBUTIONS

Two lengths are of prime importance for the strengthening effects of a distribution of spherical particles: their mean radius r and their mean nearest-neighbor spacing l in parallel planes (Section 3.1). To be more precise: l is the mean center-to-center nearest-neighbor spacing of the intersections of glide planes with particles. Since dislocation glide is mainly planar, l is defined for planes and not for the volume. This agrees with the analogous definitions given for other parameters in Section 3.1. Some more advanced models of particle hardening involve also the distribution functions $g(\rho/r)$ (Sections 4.2.2.1 and 7.1) and $g^*(\lambda/l)$ (Sections 4.2.1.4 and 4.2.2). $g^*(\lambda/l)$ is the analogue to $g(\rho/r)$: λ is an individual nearest-neighbor spacing (center-to-center) in the glide plane. g enters models of strengthening via the statistical parameters ω_r, ω_n, ω_d, and ω_q introduced in Section 3.1.

Particle radii relevant to hardening are normally below 500 nm; in most cases they are actually below 50 nm. r can be directly measured, whereas l is calculated from the number of particles per unit volume (n_v) or per unit area (n_s) or from their volume fraction f. It is standard practice to set l equal to the square lattice spacing L_{min} defined in Section 3.1. Equations (3.1)–(3.3) connect r, n_v, n_s, f, and L_{min} with each other.

The determination of f does not necessarily involve particle counting. There are two methods that directly yield f: (1) chemical analyses (Section 3.3.4) and (2) magnetic measurements (Sections 3.3.4 and 3.3.5). The latter method can only be applied if the particles or the matrix are ferromagnetic.

Particle distributions can be characterized by the following experimental methods; the parameters that they yield directly are also listed:

1. TEM (transmission electron microscope, microscopy): r, n_v
2. SEM (scanning electron microscope, microscopy): q, n_s [Eqs. (3.3)]
3. SANS (small-angle neutron scattering): r, n_v
4. Chemical analyses: f
5. Superparamagnetism: r, f

These five methods have been listed in the order of decreasing practical relevance. They will be described below. Field ion microscopy [3.54] is normally only used for particles that are smaller than those relevant to particle hardening.

3.3.1 Transmission Electron Microscopy

Transmission electron microscopy (TEM) is by far the most widely applied technique for measuring r and n_v. Radii above about 2 nm can be measured with satisfactory accuracy by conventional TEM. For smaller ones, high-resolution TEM is required; but such small particles are rarely relevant strengtheners. Exceptions are disc-shaped particles—for example, Guinier–Preston zones (Section 1.1). If r exceeds about 5 nm and the corrections described in Section 3.3.1.4 are performed, the relative error limits of r are 5% or even less. Those of f are larger: around 20% [3.13]. The various TEM techniques available nowadays are far too diversified and complex to be detailed here. Only the bases of the most important methods will be discussed; beyond that, reference is made to some of the excellent books on TEM [3.55–3.65].

In the TEM a beam of monoenergetic, coherent electrons with wave vector **k** is transmitted through a thin foil of the material under investigation. The final image is a magnified parallel projection of the foil into the plane of the photographic film. The electrons are scattered by the specimen's atoms. Normally only elastically scattered electrons are used for imaging. Let the wave vector of scattered electrons be $\mathbf{k}'$. The differential Rutherford [3.66] scattering cross section is proportional to $Z^2/\sin^4\theta_k$, where Z is the material's atomic number and $2\theta_k$ is the angle between $\mathbf{k}$ and $\mathbf{k}'$. Evidently, heavy elements scatter electrons strongly. Particles are expected to appear in the image either brighter or darker than the matrix; that is, they must produce some contrast. Below three effects that lead to the desired contrast will be discussed [3.55–3.59, 3.61, 3.63]; the respective sections are given in parentheses:

1. Structure factor contrast (Section 3.3.1.1): Matrix and particles differ in their structure factors.

2. Atomic number contrast (Section 3.3.1.2): Matrix and particles differ in their atomic numbers.
3. Strain contrast (Section 3.3.1.3): The particles distort the matrix surrounding them.

Structure factor contrast, as well as strain contrast, are governed by the interference of electron waves scattered at the specimen's atoms; that is, both are diffraction contrasts. Atomic number contrast, also known as Z-contrast, can be observed even in amorphous materials. The results obtained by any one of these three methods must be subjected to the correction procedures outlined in Section 3.3.1.4. Contrast mechanisms not listed above are described in Refs. 3.55–3.59, 3.61, and 3.63.

The TEM can be operated either in the bright-field mode or in the dark-field mode. In the bright-field mode, the objective aperture intercepts all electrons except those that have not been scattered; that is, only those that kept their original wave vector $\mathbf{k}$ pass the aperture. Scatterers appear in dark. In the dark-field mode, only scattered electrons with a chosen wave vector $\mathbf{k}'$, $\mathbf{k}' \neq \mathbf{k}$ can pass the aperture; that is, if no electrons are scattered, the entire image is dark.

The number n_A of particle images per unit area of the thin foil is related to its thickness z and to the number n_v of particles per unit volume:

$$n_A = n_v z \tag{3.21}$$

The most reliable method for the determination of z is convergent beam electron diffraction [3.13, 3.67]. The acceleration voltage of today's standard TEMs is 200 keV. For them z is usually between 50 and 100 nm.

3.3.1.1 Structure Factor Contrast. The structure factor F_{hkl} of the hkl reflection is governed by the atomic numbers of the atoms in the elementary cell and by their positions in it. The diffracted intensity is proportional to $|F_{hkl}|^2$. If the types of elements contained in the elementary cell and/or if their positions in it are not the same for the matrix and for the particles, their structure factors differ from each other. The same holds for the diffracted intensities. The following case is of special interest: Matrix and particles have the same crystal structure, nearly the same lattice constants, and constituents of similar atomic numbers, but the particles have a superlattice, whereas the matrix is disordered. Examples are the $L1_2$-long-range-ordered γ'-precipitates in the disordered matrices of nickel-base superalloys. F_{100} vanishes for the matrices, but not for the γ'-particles. They appear bright in 100 dark-field images. This is illustrated in Figure 3.7a. The contrast in the corresponding bright-field image (Figure 3.7b) is much weaker.

There is no exact 1:1 correspondence between the two micrographs. Some particles (e.g., A in Figure 3.7a) are only visible in the dark-field image, whereas others (e.g., B in Figure 3.7b) appear only in the bright-field image. A likely explanation for this discrepancy is the following: Particle A is only a flat cap as indicated for particle No. 3 in Figure 3.9. Particle A's bright-field contrast is just too weak to be noticeable in Figure 3.7b. Particle B has fallen out of the foil during its electropolish. Later, particle B got stuck loosely to the surface of the foil. Since particle B no longer has the same orientation and diffraction conditions as those particles that are still embedded in the foil, particle B cannot be dark-field imaged together with them: It is invisible in Figure 3.7a. Evidently, dark-field images taken with a superlattice reflection are most reliable. If possible, one chooses them.

3.3.1.2 Atomic Number Contrast. It has been stated in Section 3.3.1 that the Rutherford [3.66] scattering cross section is proportional to $Z^2/\sin^4 \theta_k$, where $2\theta_k$ is the scattering angle of the electrons. The variation with Z^2 explains the term Z-contrast.

Before the advent of the scanning electron microscope (Section 3.3.2), Z-contrast was widely exploited in replica techniques [3.55, 3.56, 3.68–3.70].

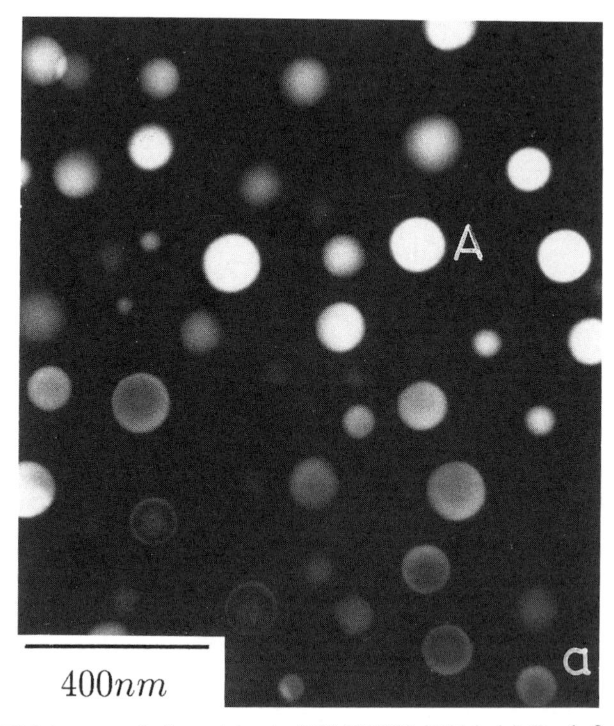

Fig. 3.7. TEM images of γ'-particles in NIMONIC PE16. (*a*) Dark-field, superlattice reflection. (*b*) Bright-field. Both micrographs show the same area of the thin foil. The γ'-particles A and B are visible only in (*a*) and in (*b*), respectively. (Courtesy of R. Maldonado, Münster.)

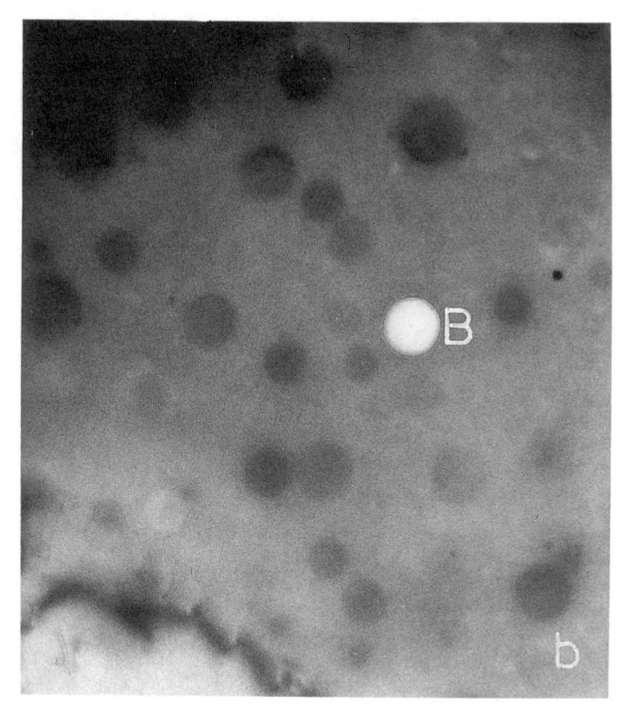

Fig. 3.7. (*Continued*).

One of them was to evaporate a thin carbon film normal to the etched surface of the specimen and to shadow this film by the oblique evaporation of a heavy element (e.g., platinum). After stripping the film from the specimen, the film was observed in the TEM. Particles created a surface relief. The shadows cast by them were bright-field imaged. Since the evaporated metal was very finely grained, there was predominately Z-contrast. Figure 3.8a gives an example.

Extraction replicas yielded direct particle images. First the polished surface of the compact specimen was etched. The applied etchant was supposed to attack the matrix preferentially. Therefore the particles stuck out of it and were enveloped by an evaporated carbon film. A second etch through the film dissolved the matrix further and released the film together with the particles from the specimen. The film with the particles sticking to it was imaged in the TEM [3.70, 3.71]. In this way, Ashby and Ebeling [3.8, 3.53] studied silica particles in internally oxidized copper–silicon (Section 3.2.3) and Randle and Ralph [3.71] studied γ'-precipitates in the superalloy NIMONIC PE16.

Nowadays Z-contrast is used to dark-field image particles in scanning transmission electron microscopes [3.72–3.74]. An electron probe of about 1 nm diameter is scanned over a thin foil of the material to be studied. The electrons scattered through angles of about 100 mrad are collected by an

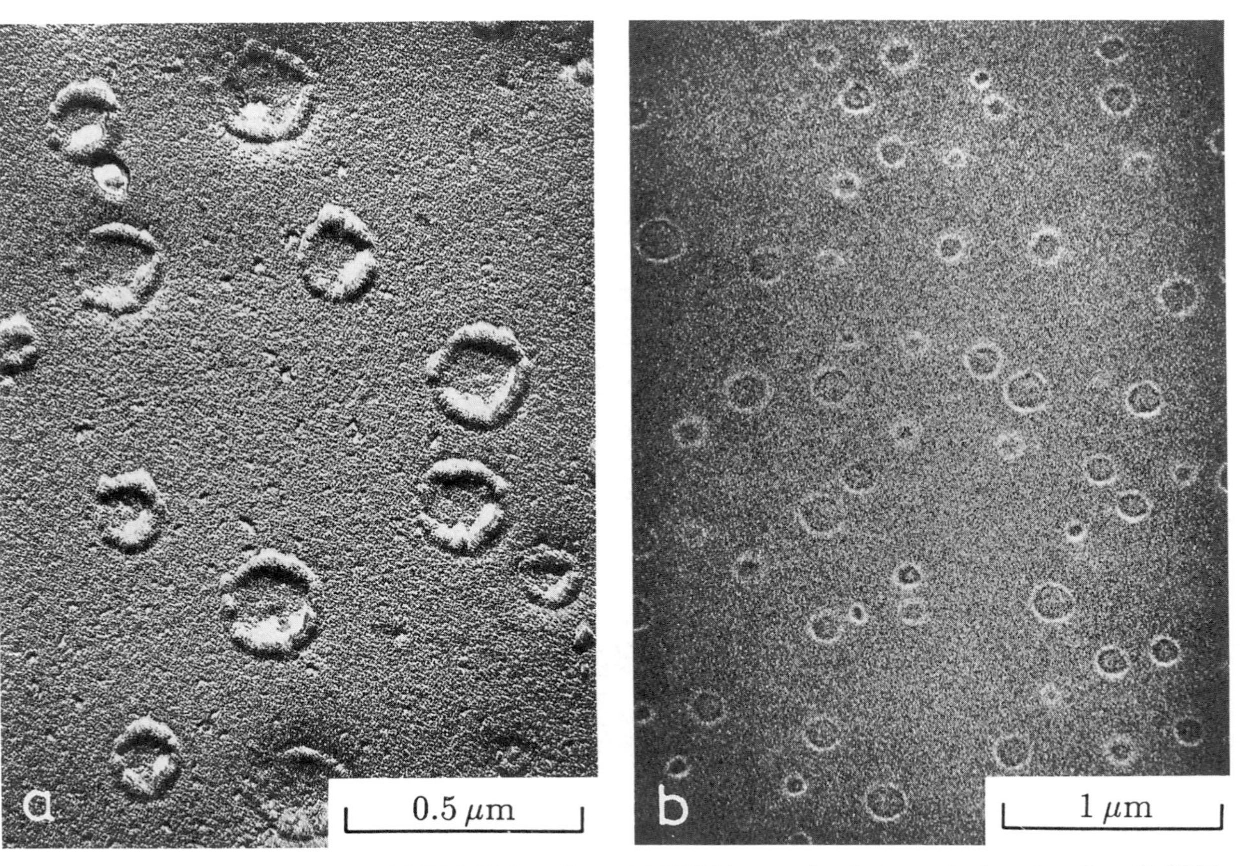

Fig. 3.8. γ'-particles in an etched NIMONIC PE16 specimen: (*a*) TEM image of a platinum-shadowed replica. (*b*) SEM image of the same specimen. (Courtesy of W. Mangen, Münster.)

annular detector mounted below the foil and the lenses. At these large angles, scattering is mostly thermal diffuse; that is, there is hardly any diffraction contrast. Since the scattering cross section increases with Z, the majority of the detected electrons has been scattered by high-Z atoms. Instead of the described arrangement, one may scan a hollow cone of electrons over the foil and collect the electrons scattered in the axial direction. In this way, Hattenhauer et al. [3.75] imaged precipitates in an aluminum-rich aluminum–silver alloy. Provided that the mean Z of the particles and that of the matrix differ sufficiently from each other, this method is to be preferred to that based on strain contrast (Section 3.3.1.3).

3.3.1.3 Strain Contrast. Particles having a lattice mismatch distort the lattice planes next to them and thus modify the diffraction conditions locally. If the specimen is orientated such that the incident electron beam is strongly diffracted by the undisturbed matrix, it will not be diffracted by the material next to the particles. This yields a contrast that has a symmetry plane even though the distortion due to the particles may have spherical symmetry [3.76]. The reason is that the spherical symmetry is broken by the plane extended by the incident and the diffracted electron beams.

Ashby and Brown [3.76, 3.77] calculated the contrast produced by a misfitting particle and found that it depends on the radius ρ, the constrained lattice mismatch ε [Eq. (3.32)], the exact diffraction conditions, and the particle's position within the thin foil. Therefore it is rather difficult to derive precise values of ρ from strain contrast images.

3.3.1.4 Correction Procedures. Even the best transmission electron micrograph suffers from various deficiencies. The three most important ones are [3.11, 3.13]:

1. **Finite resolution of the TEM, $i = 1$.** The smallest particles present in the thin foil may not be resolved by the TEM.
2. **Overlap of images, $i = 2$.** A small particle may be invisible if its image is overlapped by that of a larger particle. This is the case for particle 1 in Figure 3.9.
3. **Truncation, $i = 3$.** Particles that intersect the surface of the foil AND (logic AND) whose centers lie outside of it may not be imaged with their correct radius. Particle 3 in Figure 3.9 is an example.

In the following, these three deficiencies will be referred to by their respective number i, $1 \leqslant i \leqslant 3$. Deficiency number i changes the actual mean radius r to r_i and changes the actual number of particle images per unit area of the thin foil, n_A, to n_{Ai}. Three correction factors u_i and v_i, $1 \leqslant i \leqslant 3$, are defined as follows:

$$u_i = r_i/r \tag{3.22a}$$

$$v_i = n_{Ai}/n_A \tag{3.22b}$$

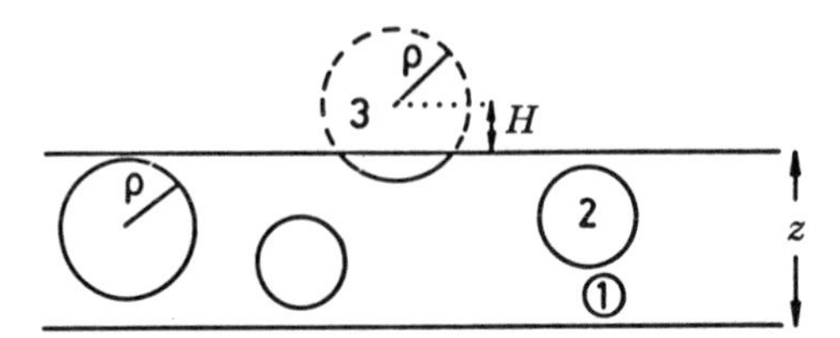

Fig. 3.9. Schematic diagram of particles in a thin foil of thickness z. The incident electron beam is normal to the foil. In a TEM image, particle 1 may not be visible because its image is overlapped by that of the larger particle 2. Particle 3 is truncated; it may not be imaged with its true radius.

If all three deficiencies have to be allowed for, the uncorrected mean radius r_M derived directly from the micrographs is related to r by

$$r_M/r = u = u_1 u_2 u_3 \tag{3.23a}$$

An analogous equation holds for n_{AM} and n_A:

$$n_{AM}/n_A = v = v_1 v_2 v_3 \tag{3.23b}$$

The subscript M of r_M and n_{AM} stands for "measurement."

The volume fraction f has to be corrected correspondingly; Eqs. (3.2) and (3.21) are used:

$$f = \frac{4\pi\omega_3 n_{AM} r_M^3}{3u^3 vz} \tag{3.24}$$

where z is the thickness of the thin foil; the statistical parameter ω_3 has been defined in Eq. (3.5b).

The coefficients u_i and v_i depend on the distribution of particle radii. Schlesier and Nembach [3.13] calculated u_i and v_i for g_0 [Eq. (3.6)], for g_{WLS} [Eq. (3.7)], and for g_{exp} defined by

$$g_{exp}(\rho/r) = 2.2472 \exp\{-[(\rho/r) - 1.0014]^2/0.0635\} \tag{3.25}$$

This Gaussian has the same full width at half-maximum as g_{WLS}. The three quoted deficiencies will be analyzed in succession.

***i* = 1: Finite resolution.** Let the radius of the smallest particle which is still visible in the TEM, be ρ_0. Hence when the number of *visible* particles, n_{A1}, is calculated, the integration over the distribution function g does not start at $\rho/r = 0.0$ but at $\rho/r = \rho_0/r$.

This yields the following for v_1:

$$v_1 = \int_{\rho 0/r}^{\infty} g(\rho/r)\, d(\rho/r) \tag{3.26a}$$

The result for u_1 is analogous:

$$u_1 = \int_{\rho 0/r}^{\infty} (\rho/r)\, g(\rho/r)\, d(\rho/r)/v_1 \tag{3.26b}$$

The quoted authors [3.13] expressed their results in the following form:

$$u_1 = \sum_{j=0}^{n} a_j(\rho_0/r)^j \tag{3.26c}$$

$$v_1 = \sum_{j=0}^{n} b_j(\rho_0/r)^j \tag{3.26d}$$

The coefficients a_j and b_j, $0 \leqslant j \leqslant 6$, are listed in Table 3.3 for g_{WLS} and g_{exp}. u_1 and v_1 of g_0 equal unity.

***i* = 2: Overlap.** A particle is assumed to be invisible due to overlap if the image of its center is overlapped by the image of a larger particle. The centers of both particles are supposed to lie inside of the foil. The probability that a particle of radius ρ is invisible, $p(\rho/r)$, equals the total cross-sectional area of all particles that lie in unit area of the thin foil AND (logic AND) whose radius is larger than ρ; p must not exceed unity:

$$p(\rho/r) = n_v z \int_{\rho/r}^{\infty} [\pi\rho'^2 g(\rho'/r)]\, d(\rho'/r)$$

n_v is the number of particles per unit volume and z is the thickness of the foil. Hence the product $(n_v z)$ equals the total number of particles per unit area of the foil [Eq. (3.21)]. The integral gives the average particle cross section available for overlap. With the aid of Eq. (3.2), n_v can be replaced by f and r:

$$\begin{aligned} p(\rho/r) &= \left[\int_{\rho/r}^{\infty} (\rho'/r)^2 g(\rho'/r)\, d(\rho'/r)\right] z\pi r^2 f/(4\pi\omega_3 r^3/3) \\ &= A\alpha(\rho/r) \end{aligned}$$

where A and $\alpha(\rho/r)$ are given by

$$A = fz/r \tag{3.27a}$$

$$\alpha(\rho/r) = \frac{3}{4\omega_3} \int_{\rho/r}^{\infty} (\rho'/r)^2\, g(\rho'/r)\, d(\rho'/r) \tag{3.27b}$$

Overlap modifies the distribution function $g(\rho/r)$ to $\{[1 - p(\rho/r)]g(\rho/r)/v_2\}$ or $\{[1 - A\alpha(\rho/r)]g(\rho/r)/v_2\}$. Hence the parameters v_2 and u_2, which are related to the number of visible particles and to their radii, respectively, are given by

$$v_2 = \int_0^\infty \{[1 - A\alpha(\rho/r)]g(\rho/r)\}\ d(\rho/r) \tag{3.27c}$$

$$u_2 = \int_0^\infty \{[1 - A\alpha(\rho/r)](\rho/r)g(\rho/r)\}\ d(\rho/r)/v_2 \tag{3.27d}$$

These two parameters are also written as polynomials, and their coefficients c_j and d_j, $0 \leqslant j \leqslant 2$, are quoted in Table 3.3 for g_{WLS} and g_{exp}; c_j and d_j vanish for $j > 2$:

$$u_2 = \sum_{j=0}^{n} c_j A^j \tag{3.27e}$$

$$v_2 = \sum_{j=0}^{n} d_j A^j \tag{3.27f}$$

In this derivation, double and multiple overlap have been disregarded. Therefore the effect of particle overlap is slightly overestimated for A larger than 0.5 [3.11]. Since in the case of g_0, all particles have the same radius, there is no overlap by larger ones. Hence u_2 and v_2 of g_0 equal unity.

***i* = 3: Truncation.** The derivation of u_3 and v_3 involves two more parameters: H and W. H is the elevation of the center of the truncated particle above the surface of the foil; H is indicated in Figure 3.9. The surface is assumed to be plane. Consequently, particle 3 is imaged with the radius $(\rho^2 - H^2)^{1/2}$ instead of with ρ. To be visible in the TEM, the particle must reach a minimum depth W into the foil; that is, $\rho - H$ must exceed W. At elevation H, only those particles are visible whose radii ρ are not smaller than H + W. This is reflected in the lower limits of the inner integrals in Eqs. (3.28). A reasonable choice for W is ρ_0. One finds the following for n_{A3} and v_3; the factors 2 before the integrals represent the two surfaces of the foil:

$$n_{A3} = n_v \left\{ z + 2\int_0^\infty \int_{(H+W)/r}^\infty g(\rho/r)\ d(\rho/r)dH \right\} \tag{3.28a}$$

and with Eq. (3.21) one obtains

$$v_3 = 1 + \frac{2}{z}\int_0^\infty \int_{(H+W)/r}^\infty g(\rho/r)\ d(\rho/r)\,dH \tag{3.28b}$$

The particles whose centers lie inside of the foil are imaged with their correct radii; they yield the term zr in Eq. (3.28c). The truncated particles are represented by the double integral. Thus for r_3 and u_3 one obtains

$$r_3 = n_v \left\{ zr + 2\int_0^\infty \int_{(H+W)/r}^\infty r[(\rho/r)^2 - (H/r)^2]^{1/2} g(\rho/r)\ d(\rho/r)dH \right\} \Big/ n_{A3} \tag{3.28c}$$

TABLE 3.3. Coefficients a_j, b_j, c_j, d_j, e_j, and f_j of the Polynomials Defined in Eqs. (3.26c), (3.26d), (3.27e), (3.27f), (3.28e), and (3.28f)[a]

	Finite Resolution				Overlap				Truncation			
	g_{WLS}		g_{exp}		g_{WLS}		g_{exp}		g_{WLS}		g_{exp}	
j	a_j	b_j	a_j	b_j	c_j	d_j	c_j	d_j	e_j	f_j	e_j	f_j
0	1.0000	0.9992	1.0011	1.0038	1.0009	1.0000	1.0008	1.0011	0.8219	0.9995	0.8107	1.0005
1	0.0000	0.0258	−0.0462	−0.1571	0.0304	−0.4221	0.0262	−0.4321	−0.1599	−0.9947	−0.1563	−0.9919
2	−0.0002	−0.1889	0.4104	2.0321	0.0371	0.0000	0.0321	0.0000	−1.5692	−0.0438	−1.6385	−0.1017
3	0.1482	0.3561	−1.3097	−10.184					2.0068	0.1395	2.2991	0.4240
4	0.0000	−0.6160	1.5971	23.295					−1.6331	−0.1694	−2.1568	−0.7602
5	0.0000	0.0000	−0.5110	−24.029					0.6350	0.1539	0.9269	0.4998
6	0.0000	0.0000	0.0000	8.539								

[a]After Ref. 3.13.

and

$$u_3 = \frac{1}{v_3} + \frac{2}{zv_3} \int_0^{\infty} \int_{(H+W)/r}^{\infty} [(\rho/r)^2 - (H/r)^2]^{1/2} \, g(\rho/r) \, d(\rho/r) dH \tag{3.28d}$$

The double integrals in Eqs. (3.28b) and (3.28d) will be referred to as $I_2(W/r)$ and $I_1(W/r)$, respectively. They too are written as polynomials for g_{WLS} and g_{exp}:

$$I_1(W/r) = \frac{1}{r} \int_0^{\infty} \int_{(H+W)/r}^{\infty} [(\rho/r)^2 - (H/r)^2]^{1/2} \, g(\rho/r) \, d(\rho/r) dH$$

$$= \sum_{j=0}^{n} e_j (W/r)^j \tag{3.28e}$$

$$I_2(W/r) = \frac{1}{r} \int_0^{\infty} \int_{(H+W)/r}^{\infty} g(\rho/r) \, d(\rho/r) dH = \sum_{j=0}^{n} f_j \, (W/r)^j \tag{3.28f}$$

The coefficients e_j and f_j, $0 \leqslant j \leqslant 5$, are given in Table 3.3 for g_{WLS} and g_{exp}. u_3 and v_3 of g_0 are analytical functions of W/r:

$$u_3 = 1/v_3 + \{(1 - W/r)[1 - (1 - W/r)^2]^{1/2} + \arcsin(1 - W/r)\}/(v_3 z/r) \tag{3.28g}$$

$$v_3 = 1 + 2(1 - W/r)/(z/r) \tag{3.28h}$$

The most important correction is that for truncation. This is evident in Table 3.4. There the results of the correction procedures are compiled for two γ'-particle distributions in NIMONIC PE16 [3.13]. The term $[1/(u^3 v)]$ governs the correction of f [Eq. (3.24)]. The coefficients were determined in the following order: v_3, u_3, v_2, u_2, v_1, and u_1. With the aid of Eqs. (3.21) and (3.24), n_v and f can be derived from n_{AM} and r_M.

The parameters u_1 (finite resolution) and u_2 (overlap) are larger than unity; that is, these deficiencies raise r_M above the true mean radius r; for u_3 (truncation) the opposite holds. The importance of overlap increases as z increases, and for truncation the opposite is true.

The three deficiencies discussed above make it rather impractical to derive actual distribution functions $g(\rho/r)$ from TEM images. Though there have been attempts to do this (e.g., Refs. 3.34 and 3.78), the approach which has been chosen by Rönnpagel and associates [3.79, 3.80] and which will be described in the next section, is to be preferred. Its basis are computer simulations.

3.3.1.5 Computer Simulations of TEM Images. Rönnpagel and coworkers [3.79, 3.80] simulated TEM images of thin foils on a computer. These images are compared with real ones. The simulation runs as follows. f, r, and $g(\rho/r)$ are selected. A cube of volume V is traced out in space. $N = Vn_v = [(3Vf)/(4\pi r^3 \omega_3)]$ particles are generated. Their radii are distributed according to g and they are given random numbers j, $1 \leqslant j \leqslant N$. Particle j, $1 \leqslant j \leqslant N$, of radius ρ_j is assigned a concentric "sphere of interest" (Section

TABLE 3.4. Correction Factors u_i, u, v_i, v, and $1/(u^3v)$ [Eqs. (3.22)–(3.24), and (3.26)–(3.28)] for Two TEM Micrographs[a]

	g	W/r	Correction Factors								
			u_1	u_2	u_3	u	v_1	v_2	v_3	v	$1/(u^3v)$
$r=7.8$ nm	g_{WLS}	0.0	1.003	1.014	0.953	0.969	0.997	0.873	1.357	1.180	0.933
$f=0.054$	g_{WLS}	0.3	1.003	1.014	0.993	1.009	0.997	0.873	1.250	1.087	0.896
$z=44$ nm	g_{exp}	0.0	1.001	1.012	0.950	0.962	1.002	0.873	1.357	1.187	0.947
$\rho_0=2.0$ nm	g_0	0.0	1.000	1.000	0.944	0.944	1.000	1.000	1.357	1.357	0.877
$r=44$ nm	g_{WLS}	0.0	1.002	1.009	0.928	0.938	0.997	0.914	1.686	1.536	0.790
$f=0.070$	g_{WLS}	0.3	1.002	1.009	0.989	0.999	0.997	0.914	1.481	1.349	0.743
$z=130$ nm	g_{exp}	0.0	1.001	1.008	0.923	0.930	1.002	0.914	1.687	1.546	0.804
$\rho_0=11$ nm	g_0	0.0	1.000	1.000	0.913	0.913	1.000	1.000	1.687	1.687	0.780

[a]After Ref. 3.13.

3.2.1.2) of radius P_j:

$$P_j = \xi \rho_j / f^{1/3} \tag{3.29}$$

The factor ξ is close to unity, and its exact value is determined later. The idea behind these "spheres of interest" is that in order to produce particle j of radius ρ_j, a sphere of radius P_j has to be depleted of solute. These spheres are placed into the volume V in the order 1, 2,..., N. They are "shaken" such that they form a densely packed arrangement. Then truncations at the cube's upper boundary are allowed for and ξ is adjusted to yield the input value of f. Finally a foil of thickness z is cut out of V, and the particles, not the "spheres of interest" surrounding them, are projected into a plane parallel to the foil. Particles whose radii are below the chosen minimum radius ρ_0 or whose caps are less high than the chosen minimum value W (Section 3.3.1.4) are eliminated. Similarly, particle overlap can be allowed for. This simulated micrograph is compared with real TEM pictures.

Later Rönnpagel and coworkers devised an alternative algorithm for arranging the spheres of radius P_j in the cube: They distribute them at random over the sites of a cubic primitive lattice. In the beginning its lattice constant is so large that the spheres do not touch each other. Then this arrangement is "shaken" to make it fill up the space densely; at the same time the volume V of the cube is reduced. In the end the spheres take up about 60% of V. For comparison, the corresponding percentage of hexagonal close-packed spheres of uniform size is 74%.

The results are presented in Figure 3.10; the second algorithm of arranging the spheres in space has been applied. Figure 3.10a is a digitized actual TEM image of γ'-particles in the nickel-base superalloy NIMONIC PE16. Figure 3.10b is the image simulated for the distribution function g_{WLS} of particle radii. g_{WLS} has been defined in Eq. (3.7). Both figures look very similar: The particle images are rather uniformly distributed. In the simulation of Figure 3.10c the coordinates of the particle centers have been determined by a random number generator. In contrast to Figures 3.10a and 3.10b, the particle images are nonuniformly distributed: There are closely spaced images and there are "white" areas. The latter ones are unrealistic. If there were one in a real specimen, a new particle would nucleate there. In the beginning it would grow fast because the matrix is still supersaturated (Section 3.2.1.1). In spite of this the distribution functions $g^{*\prime}(\lambda')$ of all three figures are indistinguishable. λ' is the center-to-center nearest-neighbor spacing of particle images. The reason for this surprising result is that since a particle image on the border of a "white" area may have a close neighbor on the opposite side, such areas do not affect $g^{*\prime}(\lambda')$ very strongly. Therefore an alternative measure for the spatial uniformity of particle image distributions has been introduced. A square-shaped mask of $141 \times 141(\text{nm})^2$ has been placed at random over Figures 3.10a–3.10c. φ is the fraction of the mask covered by particle images. Figures 3.10d–3.10f give the φ-histograms for Figures 3.10a–3.10c, respectively. There were 500 random

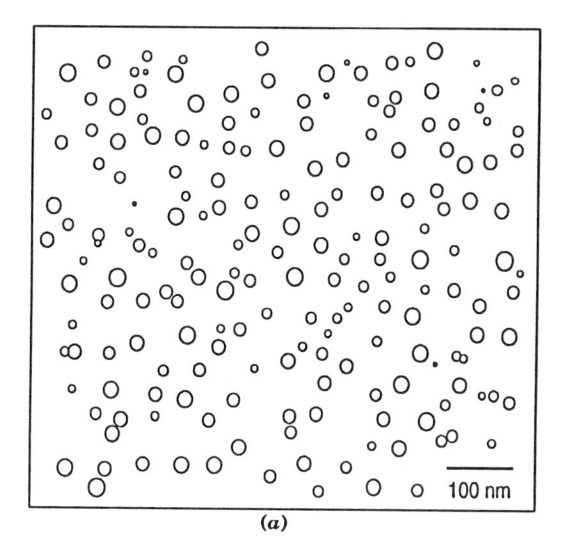

(a)

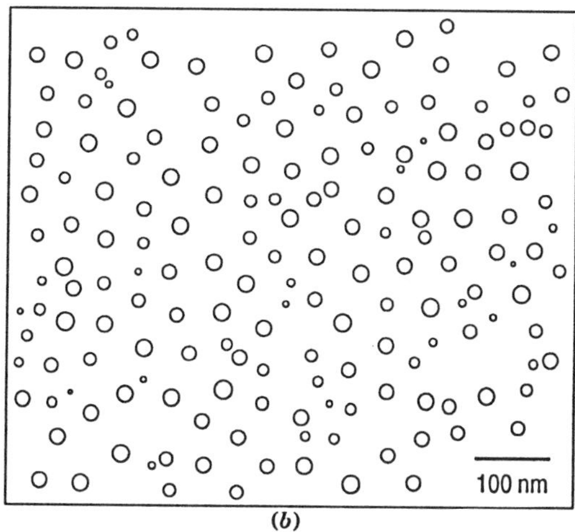

(b)

Fig. 3.10. Computer simulation of TEM images of γ'-particles in NIMONIC PE16. $r = 8.0\,\text{nm}$, $f = 0.024$, $z = 38\,\text{nm}$, $\rho_0 = 2.3\,\text{nm}$; (*a*) Digitized actual dark-field TEM image. (*b*) Simulated TEM image for g_{WLS}, $W = 0\,\text{nm}$. Densely packed "spheres of interest" of radius P_j [Eq. (3.29)]. Particles whose image center is overlapped by the image of a larger particle are not shown. (*c*) Simulated TEM image for g_{WLS} and $W = 0\,\text{nm}$. The coordinates of the particle centers are determined by a random number generator. No allowance has been made for particle overlap: The real particles may intersect each other, and/or their images may overlap. (*d*) Histogram $h(\varphi)$ of (*a*) (digitized actual TEM image). φ is the fraction of the mask covered by particle images. The vertical arrow marks the average $\bar{\varphi}$ of φ, and the horizontal double arrow indicates the standard deviation of $\bar{\varphi}$. (*e*) As (*d*) for the simulated [(*b*), spheres of interest] TEM image. (*f*) As (*d*) for the simulated [(*c*), random] TEM image. (Courtesy of D. Rönnpagel, Braunschweig.)

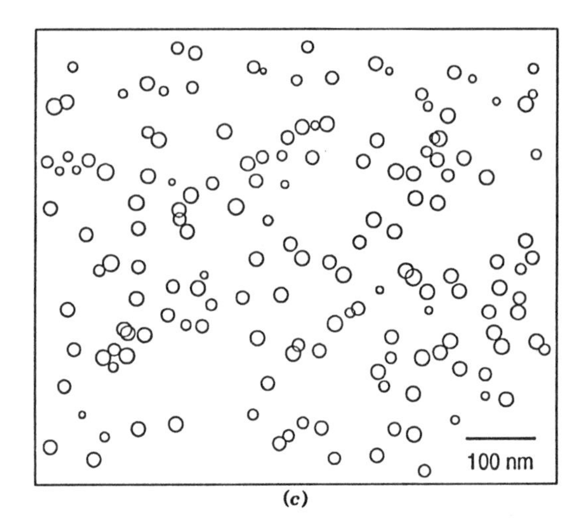

(*c*)

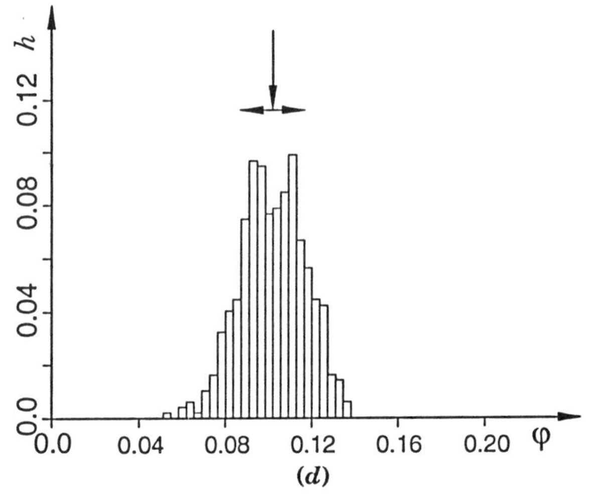

(*d*)

Fig. 3.10. (*Continued*)

positions of the mask. Evidently the histogram shown in Figure 3.10f differs drastically from the two other ones. This difference can be quantified as follows: $\bar{\varphi}$ is the average over 500 random positions of the mask, and ψ is the standard deviation of $\bar{\varphi}$. The results are as follows:

Figure	$\psi/\bar{\varphi}$
3.10a, actual TEM image	0.15
3.10b, simulated TEM image, "spheres of interest"	0.13
3.10c, simulated TEM image, random	0.27

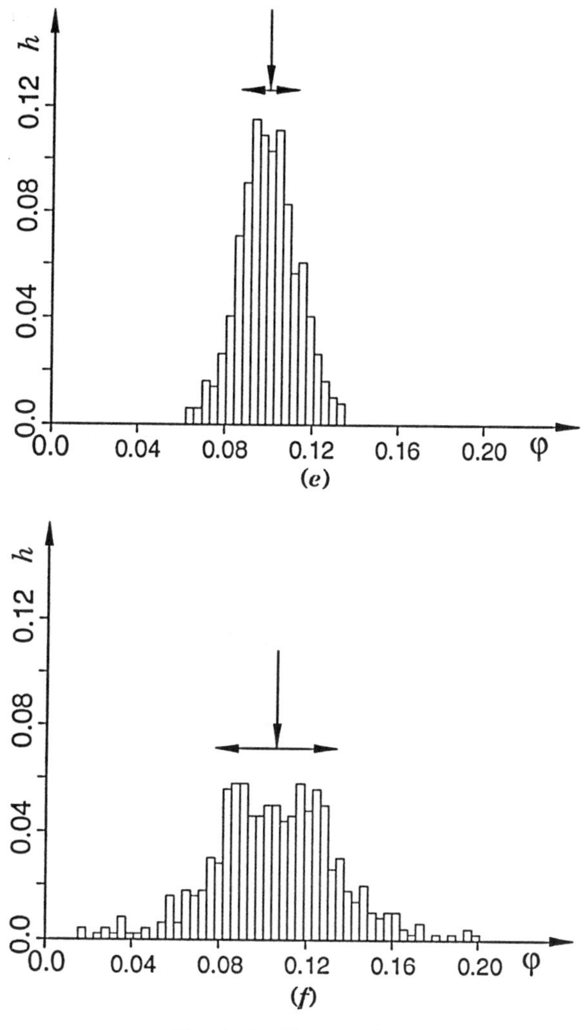

Fig. 3.10. (*Continued*).

The ratios $\psi/\bar{\varphi}$ of Figures 3.10a and 3.10b are nearly the same, whereas that of Figure 3.10c differs strongly from them. This reflects the nonuniform distribution of the particle images apparent in Figure 3.10c. In Chapter 4 it will be shown that a uniform particle distribution is favorable for particle strengthening. The result of these comparisons is that the "spheres of interest" model describes the spatial distribution of actual particles well.

The spatial distribution of the particles may be truly random only at the very beginning of a precipitation process when it is mainly governed by random nucleation. If a specimen has been subjected to an Ostwald ripening

(Section 3.2.1.2) treatment, each particle has its "sphere of interest," which is free of other particles. The radius of this sphere is given by Eq. (3.29). The images of the γ'-particles shown in Figures 4.35a, 4.36a, and 4.37 have been generated as described above.

Each of the Figures 3.10a–3.10c shows only about 180 particles. This is not sufficient for the derivations of the distribution function $g'(\rho')$ of the radii ρ' of particle images with convincing accuracy. Normally, measurements of the average radii and of the volume fraction of the particles are based on about 1000 particle images [3.13]. Within the large limits of error, $g'(\rho')$ of Figures 3.10a and 3.10b are compatible with each other. Hence it can be concluded that the particle radii in actual NIMONIC PE16 specimens are distributed according to g_{WLS} (Figure 3.10b).

3.3.2 Scanning Electron Microscopy

In the scanning electron microscope (SEM), a fine electron probe is raster-scanned over the surface of the specimen. The interaction of the electrons with the specimen's atoms leads to three different signals. The intensity of either of them can be used to modulate the brightness of a cathode ray tube which is synchronized with the electron probe [3.81–3.85]. The three signals are: secondary electrons (SEs), back-scattered electrons (BSEs), and characteristic X-rays. Electrons leaving the specimen with energies below 50 eV are called SEs. All others are considered to be BSEs. SEs have been generated not deeper than about 5 nm below the surface. The intensity of BSEs increases with the atomic numbers Z of the material; for SEs, this dependence appears to be much weaker. Because the SE and BSE intensities depend also on the angle between the incident electron beam and the specimen's surface, excellent topographical contrast is obtained. Therefore one often etches the specimen slightly after polishing. Figure 3.8b illustrates this for NIMONIC PE16. The γ'-particles are clearly visible. During etching very small particles may be lost, however.

SEM pictures correspond to optical micrographs in that both are images of a surface, whereas TEM images are projections of a foil of thickness z into a plane.

3.3.3 Small-Angle Neutron Scattering

The coherent neutron scattering cross section $\partial\sigma/\partial\Omega$ per atom is related to the following: n_v, the number of particles per unit volume; V_p, the particle volume; N, the number of atoms per unit volume; ρ_{bp} and ρ_{bs}, the average scattering length densities of the particles and of the matrix, respectively; and $F_p(\mathbf{k} - \mathbf{k}')$, the single-particle form factor, where $\mathbf{k}$ and $\mathbf{k}'$ are the wave vectors of the incident and of the scattered neutrons, respectively [3.86–3.88]:

$$\partial\sigma/\partial\Omega = \frac{V_p^2 n_v}{N} (\rho_{bp} - \rho_{bs})^2 |F_p|^2 \tag{3.30a}$$

If the product $(|\mathbf{k} - \mathbf{k}'|r)$ is small, $|F_p|^2$ can be approximated by

$$|F_p|^2 = \exp(-|\mathbf{k} - \mathbf{k}'|^2 r_G^2/3) \tag{3.30b}$$

where r_G is the radius of gyration [3.86, 3.87]. r_G of spherical particles of uniform size equals $(3/5)^{1/2}r$. In a plot of $\log(\partial\sigma/\partial\Omega)$ versus $|\mathbf{k} - \mathbf{k}'|^2$, the slope and the ordinate intercept yield r_G and $V_p^2 n_v/N$, respectively. Equations (3.30) apply to particles of uniform size. In general there is a distribution of particle radii; then these equations have to be weighted by the distribution function g.

The advantages of small-angle neutrons scattering (SANS) are that compact specimens can be studied nondestructively and that very small radii can easily be measured. The drawbacks are that neutron scattering facilities are not readily available and that in the case of g, the evaluation of r requires involved deconvolution procedures. Wagner et al. [3.89] measured cobalt-rich particles in copper; Beddoe et al. [3.90], Windsor et al. [3.91], Ali et al. [3.92], and Degischer et al. [3.93] measured γ'-particles in NIMONIC PE16 by SANS.

3.3.4 Chemical Analysis

The volume fraction f can be directly derived from the overall composition c of the alloy, that of the matrix (c_s), and that of the particles (c_p):

$$f = 1 \bigg/ \left[1 + \frac{1 - f_A}{f_A(1 - \varepsilon)^3}\right] \tag{3.31a}$$

with

$$f_A = \frac{c - c_s}{c_p - c_s} \tag{3.31b}$$

Equation (3.31b) expresses the lever rule. f_A is the fraction of atoms contained in the particles. Since they and the matrix may differ in their atomic volumina, f_A may differ from the volume fraction f. In case of multicomponent systems, f_A, c, c_S, and c_p may refer to any of the constituents. Equation (3.31a) applies only to coherent particles, and it compensates for the mentioned difference. Equation (3.31b) can be used for coherent as well as for incoherent particles. ε stands for the constrained lattice mismatch. "Constrained" means that ε describes the relative difference in lattice constant when the particles are constrained by the surrounding matrix. Both phases adjust their lattice constants somewhat to each other. Mott and Nabarro [3.22] related ε to the unconstrained lattice mismatch δ defined in Eq. (3.16):

$$\varepsilon = \frac{\delta}{1 + 4\mu_s/(3\chi_p)} \tag{3.32}$$

where μ_s is the shear modulus of the matrix and χ_p is the particles' bulk modulus. In the derivation of Eq. (3.32), elastic isotropy has been assumed. Normally ε is close to $2\delta/3$. If the particles are coherent but do not have the same crystal structure as the matrix, the applicability of Eq. (3.31a) has to be carefully checked.

c is normally known from the preparation of the alloy. c_s and c_p of binary systems can often be read from the phase diagram. Figure 3.3 is an example. c_s and c_p can also be determined by energy dispersive X-ray analysis in the TEM [3.60, 3.62, 3.65]; this is possible even for multicomponent systems. Because the diameter of the electron probe is normally not less than 2 nm, the dimensions of the two phases must be sufficiently large. Moreover, the particle to be analyzed must reach from the upper surface of the thin foil to the lower one; that is, the particle radius should be equal to or exceed the thickness of the foil. Such chemical analyses have been performed for the matrix and for the γ'-particles of superalloys [3.13, 3.14, 3.94].

The matrix of binary nickel-rich nickel–aluminum alloys is ferromagnetic. The Curie temperature θ_C varies linearly with the aluminum concentration c_s [3.95, 3.96]:

$$c_s = 0.20(1 - \theta_C/631\,\text{K}) \tag{3.33}$$

θ_C, which can easily be measured, yields c_s.

Equations (3.31) lead to accurate f-values only if the difference $(c - c_s)$ is not too small; that is, if f is relatively large.

3.3.5 Superparamagnetism

Copper can be strengthened by coherent spherical cobalt-rich precipitates. Because this system serves as a model for lattice mismatch hardening, it has been extensively investigated [3.15, 3.97–3.101]; it will be discussed in Sections 5.4.2.1.8, 5.4.2.2.2, and 5.4.3. The particles are ferromagnetic. Their individual magnetic moment m_p, however, is so small that the magnetization behavior of the macroscopic copper–cobalt specimen resembles that of a paramagnetic material. This explains the term *superparamagnetism* [3.102–3.104]. "super" indicates that m_p is still much larger than the moment of a single atom. Let M be the overall magnetization of the macroscopic superparamagnetic specimen. Langevin's [3.105] equation relates M to the number of particles per unit volume, n_v, to m_p, to the applied magnetic field H, and to the temperature T:

$$M = n_v m_p[\coth(x) - 1/x] \tag{3.34a}$$

with

$$x = \frac{m_p \mu_0 H}{k_B T} \tag{3.34b}$$

μ_0 and k_B are the permeability of vacuum [$4\pi \times 10^{-7}$ Vs/(Am)] and the Boltzmann constant, respectively. First spherical particles of uniform radius r are considered. Their moment m_p is given by

$$m_p = \frac{4\pi}{3} r^3 M_{ps} \tag{3.34c}$$

M_{ps} is the saturation magnetization of an individual particle. There are two approximations for Eq. (3.34a): one for $x \gg 1$ and another one for $x \ll 1$—that is, one for high and one for low fields:

$$\text{For } x \gg 1: \qquad M = n_v m_p \tag{3.35a}$$

or

$$M = n_v \frac{4\pi}{3} r^3 M_{ps} = f M_{ps} \tag{3.35b}$$

and

$$\text{For } x \ll 1: \qquad M = n_v m_p \frac{m_p \mu_0 H}{3k_B T} \tag{3.35c}$$

or

$$M = n_v \left(\frac{4\pi}{3}\right)^2 r^6 \frac{M_{ps}^2 \mu_0 H}{3k_B T} \tag{3.35d}$$

In Eq. (3.35b), the particles' volume fraction f has been inserted for $n_v 4\pi r^3/3$ [Eq. (3.2)].

For a general distribution function g of particle radii, Eqs. (3.35) have to be weighted with g. This yields

$$\text{For } x \gg 1: \qquad M = f M_{ps} \tag{3.36a}$$

and

$$\text{For } x \ll 1: \qquad M = \frac{4\pi}{9} \frac{\omega_6}{\omega_3} f r^3 \frac{M_{ps}^2 \mu_0 H}{k_B T} \tag{3.36b}$$

Here n_v has been replaced by f and r [Eq. (3.2)]. The statistical parameters ω_3 and ω_6 have been defined in Eq. (3.5b). By measuring M in high and low fields, f and r are obtained, respectively. Equation (3.36a) yields precise values for f.

There is a limit to the range of particle radii which can be derived from Eqs. (3.35d) and (3.36b). Large particles show a hysteresis because their crystal anisotropy comes into play. Therefore these equations can only be used if the product of the particle volume and the anisotropy constant is less than [$25k_B T$] [3.102–3.104].

Breu et al. [3.106] determined f and r of cobalt-rich particles in copper as described above. Their results were in satisfactory agreement with those gained by TEM, SANS, and field ion miroscopy. Because strain contrast was used in the TEM, the respective results were of relatively low precision (Section 3.3.1.3). The problems associated with SANS have been detailed in Section 3.3.3.

The advantage of the magnetic method is that it is nondestructive. Its main drawback is of course that the particles have to be superparamagnetic. As mentioned above, this condition is fulfilled by cobalt-rich particles in copper.

REFERENCES

3.1 E. E. Underwood, *Quantitative Stereology*, Addison-Wesley, Reading, Massachusetts, 1970.

3.2 H. J. Rack and R. W. Newman, in *Physical Metallurgy* (edited by R. W. Cahn), 3rd printing, p. 705, North-Holland, Amsterdam, 1977.

3.3 J. W. Martin, *Micromechanisms in Particle-Hardened Alloys*, Cambridge University Press, Cambridge, 1980.

3.4 E. R. Weibel, *Stereological Methods*, Vol. 2, Academic Press, London, 1980.

3.5 H. E. Exner and H. P. Hougardy, *Quantitative Image Analysis of Microstructures*, DGM Informationsgesellschaft Verlag, Oberursel, 1988.

3.6 R. L. Fullman, *Trans. AIME J. Metals*, **5**, 447, 1953.

3.7 J. E. Hilliard, *Trans. AIME*, **224**, 906, 1962.

3.8 M. F. Ashby and R. Ebeling, *Trans. Met. Soc. AIME*, **236**, 1396, 1966.

3.9 C. Wagner, *Z. Elektrochem.*, **65**, 581, 1961.

3.10 I. M. Lifshitz and V. V. Slyozov, *J. Phys. Chem. Solids*, **19**, 35, 1961.

3.11 E. Nembach, *Z. Metallkde.*, **62**, 291, 1971.

3.12 T. B. Massalski, editor, *Binary Alloy Phase Diagrams*, Vol. 1, p. 142, American Society for Metals, Metals Park, Ohio, 1986.

3.13 C. Schlesier and E. Nembach, *Mater. Sci. Eng.*, **A119**, 199, 1989.

3.14 K. Trinckauf, J. Pesicka, C. Schlesier, and E. Nembach, *phys. stat. sol. (a)*, **131**, 345, 1992.

3.15 E. Nembach and M. Martin, *Acta Metall.*, **28**, 1069, 1980.

3.16 J. W. Christian, *The Theory of Transformations in Metals and Alloys*, Pergamon Press, Oxford, 1965.

3.17 D. A. Porter and K. E. Easterling, *Phase Transformations in Metals and Alloys*, 2nd edition, Chapman and Hall, London, reprinted 1993.

3.18 M. Volmer and A. Weber, *Z. Phys. Chem.*, **119**, 277, 1926.

3.19 R. Becker and W. Döring, *Annalen der Physik*, **24**, 719, 1935.

3.20 R. Wagner and R. Kampmann, in *Material Science and Technology*, Vol. 5 (edited by R. W. Cahn, P. Haasen, and E. J. Kramer), p. 213, VCH Verlagsgesellschaft mbH, Weinheim, 1991.

3.21 J. B. Cohen, *Met. Trans. A*, **23A**, 2685, 1992.

3.22 N. F. Mott and F. R. N. Nabarro, *Proc. Phys. Soc.*, **52**, 86, 1940.

3.23 F. R. N. Nabarro, *Proc. R. Soc., Series A, London*, **175**, 519, 1940.

3.24 J. D. Eshelby, *Proc. R. Soc., Series A, London*, **241**, 376, 1957.

3.25 A. Lasalmonie and J. L. Strudel, *Philos. Mag.*, **32**, 937, 1975.

3.26 L. A. Jacobson and J. McKittrick, *Mater. Sci. Eng.*, **R11**, 355, 1994.

3.27 E. M. Dunn, A. P. Davidson, J. P. Faunce, C. G. Levi, S. Maitra, and R. Mehrabian, in *Aluminum, Properties and Physical Metallurgy* (edited by J. E. Hatch), 3rd printing, p. 25, American Society for Metals, Metals Park, Ohio, 1988.

3.28 S. M. L. Sastry, P. J. Meschter, and J. E. O'Neal, *Met. Trans. A*, **15A**, 1451, 1984.

3.29 S. M. L. Sastry, T. C. Peng, and L. P. Beckerman, *Met. Trans. A*, **15A**, 1465, 1984.

3.30 E. W. Collings, in *Alloying* (edited by J. L. Walker, M. R. Jackson, and C. T. Sims), p. 257, American Society for Metals, Metals Park, Ohio, 1988.

3.31 W. Ostwald, *Zeitschr. Phys. Chem.*, **34**, 495, 1900.

3.32 E. Nembach and G. Neite, *Prog. Mater. Sci.*, **29**, 177, 1985.

3.33 R. Hattenhauer and P. Haasen, *Philos. Mag. A*, **68**, 1195, 1993.

3.34 C. K. L. Davies, P. Nash, and R. N. Stevens, *Acta Metall.*, **28**, 179, 1980.

3.35 E. A. Starke, Jr., in *Alloying* (edited by J. L. Walter, M. R. Jackson, and C. T. Sims), p. 165, American Society for Metals, Metals Park, Ohio, 1988.

3.36 C. Sigli and J. M. Sanchez, *Acta Metall.*, **34**, 1021, 1986.

3.37 D. B. Williams, in *Proceedings of the 5th International Al–Li Conference*, Vol. II, (edited by T. H. Sanders, Jr. and E. A. Starke, Jr.), p. 551, Materials and Component Engineering Publishers, Birmingham, England, 1989.

3.38 S. C. Jha, T. H. Sanders, Jr., and M. A. Dayananda, *Acta Metall.*, **35**, 473, 1987.

3.39 R. Maldonado and E. Nembach, *Acta Mater.*, in press.

3.40 W. Mangen and E. Nembach, *Acta Metall.*, **37**, 1451, 1989.

3.41 F. C. Larché, in *Dislocations in Solids*, Vol. 4 (edited by F. R. N. Nabarro), p. 135, North-Holland, Amsterdam, 1979.

3.42 N. H. Ngi, A. I. Novikov, Y. A. Skakov, and V. A. Solov'Yev, *Phys. Met. Metall.*, **64**, No. 5, 75, 1987.

3.43 W. A. Cassada, G. J. Shiflet, and W. A. Jesser, *Acta Metall. Mater.*, **40**, 2101, 1992.

3.44 J. S. Benjamin, *Met. Trans.*, **1**, 2943, 1970.

3.45 J. S. Benjamin, in *New Materials by Mechanical Alloying Techniques* (edited by E. Arzt and L. Schultz), p. 3, DGM Informationsgesellschaft Verlag, Oberursel, 1989.

3.46 G. A. J. Hack, in *Advanced Materials and Processing Techniques for Structural Applications* (edited by T. Khan and A. Lasalmonie), p. 310, Office National d'Etudes et de Recherches Aerospatiales, Chatillon, 1987.

3.47 J. D. Whittenberger, *Met. Trans. A*, **15A**, 1753, 1984.

3.48 R. C. Benn and P. K. Mirchandani, in *New Materials by Mechanical Alloying Techniques* (edited by E. Arzt and L. Schultz), p. 19, DGM Informationsgesellschaft Verlag, Oberursel, 1989.

3.49 J. S. Benjamin and M. J. Bomford, *Met. Trans. A*, **8A**, 1301, 1977.

3.50 J. C. Chaston, *J. Inst. Metals*, **71**, 23, 1945.

3.51 J. L. Meijering, *Report of a Conference on Strength of Solids*, p. 140, Bristol, 1947, The Physical Society, London, 1948.

3.52 G. Böhm and M. Kahlweit, *Acta Metall.*, **12**, 641, 1964.

3.53 R. Ebeling and M. F. Ashby, *Philos. Mag.*, **13**, 805, 1966.

3.54 R. Wagner, *Field-Ion Microscopy*, Springer-Verlag, Berlin, 1982.

3.55 G. Thomas, *Transmission Electron Microscopy of Metals*, John Wiley & Sons, New York, 1962.

3.56 L. Reimer, *Elektronenmikroskopische Untersuchungs- und Präparationsmethoden*, 2nd edition, Springer-Verlag, Berlin, 1967.

3.57 M. von Heimendahl, *Einführung in die Elektronenmikroskopie*, Vieweg-Verlag, Braunschweig, 1970.

3.58 E. Hornbogen, *Durchstrahlungs-Elektronenmikroskopie fester Stoffe*, Verlag Chemie GmbH, Weinheim, 1971.

3.59 P. Hirsch, A. Howie, R. B. Nicholson, D. W. Pashley, and M. J. Whelan, *Electron Microscopy of Thin Crystals*, Krieger Publishing Company, Malabar, Florida, 1977.

3.60 J. J. Hren, J. I. Goldstein, and D. C. Joy, editors, *Introduction to Analytical Electron Microscopy*, Plenum Press, New York, 1979.

3.61 S. Amelinckx, R. Gevers, and J. Van Landuyt, editors, *Diffraction and Imaging Techniques in Materials Science*, Vol. I: *Electron Microscopy*, 2nd edition, North-Holland, Amsterdam, reprinted 1979.

3.62 M. H. Loretto, *Electron Beam Analysis of Materials*, Chapman & Hall, London, 1984.

3.63 L. Reimer, *Transmission Electron Microscopy*, Springer-Verlag, Berlin, 1984.

3.64 P. R. Buseck, J. M. Cowley, and L. Eyring, editors, *High-Resolution Transmission Electron Microscopy and Associated Techniques*, Oxford University Press, New York, 1988.

3.65 D. C. Joy, A. D. Romig, Jr., and J. I. Goldstein, editors, *Principles of Analytical Electron Microscopy*, 2nd printing, Plenum Press, New York, 1989.

3.66 E. Rutherford, *Philos. Mag.*, **21**, 669, 1911.

3.67 P. M. Kelly, A. Jostsons, R. G. Blake, and J. G. Napier, *phys. stat. sol. (a)*, **31**, 771, 1975.

3.68 D. E. Bradley, in *Techniques for Electron Microscopy* (edited by D. H. Kay), 2nd edition, p. 96, Blackwell Scientific Publications, Oxford, 1965.

3.69 H. Bethge, M. Krohn, and H. Stenzel, in *Electron Microscopy in Solid State Physics* (edited by H. Bethge and J. Heydenreich), p. 202, Elsevier, Amsterdam, 1987.

3.70 R. M. Fisher, *J. Appl. Phys.*, **24**, 113, 1953.

3.71 V. Randle and B. Ralph, *Metallography*, **19**, 49, 1986.

3.72 M. M. J. Treacy, A. Howie, and C. J. Wilson, *Philos. Mag. A*, **38**, 569, 1978.

3.73 M. Isaacson, M. Ohtsuki, and M. Utlaut, in *Introduction to Analytical Electron Microscopy* (edited by J. J. Hren, J. I. Goldstein, and D. C. Joy), p. 343, Plenum Press, New York, 1979.

3.74 S. J. Pennycook and D. E. Jesson, *Phys. Rev. Lett.*, **64**, 938, 1990.

3.75 R. Hattenhauer, G. Schmitz, P. J. Wilbrandt, and P. Haasen, *phys. stat. sol. (a)*, **137**, 429, 1993.

3.76 M. F. Ashby and L. M. Brown, *Philos. Mag.*, **8**, 1083, 1963.

3.77 M. F. Ashby and L. M. Brown, *Philos. Mag.*, **8**, 1649, 1963.

3.78 J. M. G. Crompton, R. M. Waghorne, and G. B. Brook, *Br. J. Appl. Phys.*, **17**, 1301, 1966.

3.79 A. Fuchs and D. Rönnpagel, *Mater. Sci. Eng.*, **A164**, 340, 1993.

3.80 T. Pretorius and D. Rönnpagel, *Proceedings of the 10th International Conference on the Strength of Materials*, (edited by H. Oikawa, K. Maruyama, S. Takeuchi, and M. Yamaguchi), p. 689, The Japan Institute of Metals, 1994.

3.81 L. Reimer, *Scanning Electron Microscopy*, Springer-Verlag, Berlin, 1985.

3.82 U. Werner and H. Johansen, in *Electron Microscopy in Solid State Physics* (edited by H. Bethge and J. Heydenreich), p. 170, Elsevier, Amsterdam, 1987.

3.83 D. E. Newbury, D. C. Joy, P. Echlin, C. E. Fiori, and J. I. Goldstein, *Advanced Scanning Electron Microscopy and X-Ray Microanalysis*, 2nd printing, Plenum Press, New York, 1987.

3.84 J. I. Goldstein, D. E. Newbury, P. Echlin, D. C. Joy, A. D. Romig, Jr., C. E. Lyman, C. Fiori, and E. Lifshin, *Scanning Electron Microscopy and X-Ray Microanalysis*, Plenum Press, New York, 1992.

3.85 J. B. Bindell, *Adv. Mater. Proc.*, **143**, No. 3, 20, 1993.

3.86 A. Guinier and G. Fournet, *Small-Angle Scattering of X-Rays*, John Wiley & Sons, New York, Chapman & Hall, London, 1955.

3.87 G. Kostorz, in *Treatise on Materials Science and Technology*, Vol. 15, *Neutron Scattering* (edited by G. Kostorz), p. 227, Academic Press, New York, 1979.

3.88 G. Kostorz, *Phys. Scripta*, **T49**, 636, 1993.

3.89 W. Wagner, J. Piller, H.-P. Degischer, and H. Wollenberger, *Z. Metallkde.*, **76**, 693, 1985.

3.90 R. E. Beddoe, P. Haasen, and G. Kostorz, *Z. Metallkde.*, **75**, 213, 1984.

3.91 C. G. Windsor, V. S. Rainey, P. K. Rose, and V. M. Callen, *J. Phys. F: Met. Phys.*, **14**, 1771, 1984.

3.92 K. Ali, S. Messoloras, and R. J. Stewart, *J. Phys. F Met. Phys.*, **15**, 487, 1985.

3.93 H. P. Degischer, W. Hein, H. Strecker, W. Wagner, and R. P. Wahi, *Z. Metallkde.*, **78**, 237, 1987.

3.94 W. Mangen, E. Nembach, and H. Schäfer, *Mater. Sci. Eng.*, **70**, 205, 1985.

3.95 A. J. Ardell, *Acta Metall*, **16**, 511, 1968.

3.96 W. O. Gentry and M. E. Fine, *Acta Metall.*, **20**, 181, 1972.

3.97 J. D. Livingston and J. J. Becker, *Trans. Met. Soc. AIME*, **212**, 316, 1958.

3.98 J. D. Livingston, *Trans. Met. Soc. AIME*, **215**, 566, 1959.

3.99 V. Gerold and H. Haberkorn, *phys. stat. sol.*, **16**, 675, 1966.

3.100 M. Witt and V. Gerold, *Z. Metallkde.*, **60**, 482, 1969.

3.101 E. Nembach, *Scripta Metall.*, **18**, 105, 1984.

3.102 J. J. Becker, *Trans. Met. Soc. AIME*, **212**, 138, 1958.

3.103 C. P. Bean and J. D. Livingston, *J. Appl. Phys.*, **30**, 120S, 1959.

3.104 E. Kneller, in *Magnetism and Metallurgy*, Vol. 1 (edited by A. E. Berkowitz and E. Kneller), p. 365, Academic Press, New York, 1969.

3.105 M. P. Langevin, *Ann. Chim. Phys.*, **5**, 70, 1905.

3.106 M. Breu, W. Gust, B. Predel, and E. Wachtel, *Z. Metallkde.*, **79**, 1, 1988.

CHAPTER 4

The Critical Resolved Shear Stress of Particle-Hardened Materials. Particle Shearing: General Considerations

The critical resolved shear stress (CRSS) is defined as the resolved shear stress necessary to make dislocations glide over macroscopic distances. In particle-hardened materials the external stress must be high enough to overcome the interaction forces between particles and dislocations. In a real material the total CRSS τ_t comprises different contributions: that due to the particles, referred to as τ_p; that due to atoms solved in the matrix, referred to as τ_s; and so on. Provided that the particles are the only obstacles to the glide of dislocations, the CRSS equals τ_p. In Chapters 4–6, only τ_p will be dealt with; the question of how τ_p, τ_s, and so on, superimpose to τ_t will be addressed in Chapter 7. The obstacles are supposed to be immobile.

It has already been stated in Section 1.3 that in relating τ_p to the properties of the particles one proceeds in two steps. With reference to Eqs. (2.18) and (2.20), which describe the motion of a dislocation through a field of obstacles, the two steps are formulated as follows:

1. Derivation of τ_{obst} at the position of the dislocation segment $d\mathbf{s}$. τ_{obst} is the resolved shear stress exerted by all particles which interact simultaneously with $d\mathbf{s}$.
2. Solving Eqs. (2.18) or (2.20).

In the present and in the following chapter, particle shearing will be treated. Only coherent particles can be sheared by dislocations. This process is sketched in Figure 4.1a. Figure 1.4b shows sheared γ'-particles in a deformed specimen of the nickel-base superalloy NIMONIC PE16. The term "coherent" implies that there are no dislocations in the particle–matrix interface. Circumventing of particles will be discussed in Chapter 6. If the interaction between a particle and a dislocation is long-ranged, there is a noticeable interaction force even if the particle does not intersect the glide plane. This is demonstrated in Figure 4.1b for a particle that is surrounded by a stress field, via which it interacts with the passing edge dislocation. Such hardening mechanisms also will be discussed in Chapters 4 and 5. Therefore the conventional term "particle

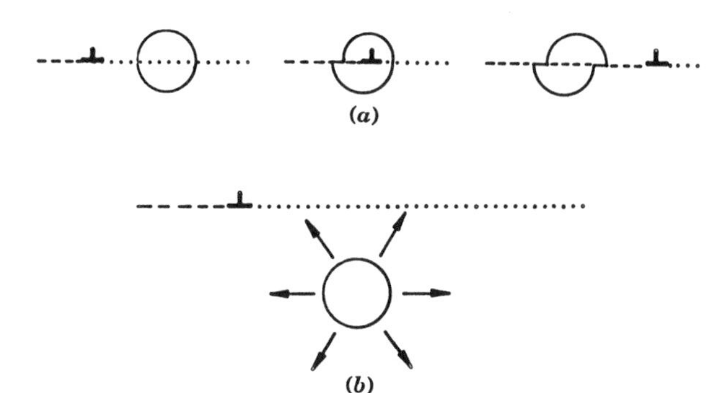

Fig. 4.1. An edge dislocation (*a*) shears a particle and (*b*) passes a particle having a lattice mismatch. The arrows surrounding the particle indicate its stress field.

shearing," which appears in their titles, is meant to refer to all processes except for only the Orowan process.

In the present chapter the two problems—establishing τ_{obst} and solving Eqs. (2.18) or (2.20)—will be discussed from a general point of view. In Chapter 5 the different mechanisms of particle hardening will be analyzed in detail.

Thermal activation will be disregarded throughout; this has been justified in Section 2.5. Many of the models presented in Chapter 4 can also be applied to solid solution hardening at 0 K. The latter limitation is necessary because thermal activation has strong effects on solid solution hardening.

4.1 DERIVATION OF τ_{obst}

The geometry is similar to the one in Section 2.5: The dislocation segment $d\mathbf{s}$ has the coordinates $(x, y, 0)$, the glide plane is the plane $z = 0$, and the segment glides towards $+y$. The coordinates of the center of a spherical particle of radius ρ are (h_x, h_y, h_z). This is sketched in Figure 4.2. As the dislocation segment passes the particle, y assumes the values $-\infty, h_y, +\infty$ in succession. It has already been mentioned in Section 2.5 that though in most cases the term "obstacle" and the subscript "obst" will refer to particles, the more general expression "obstacle" is often used.

Equation (2.19) relates the resolved shear stress τ_{obst} to $\partial F_{obst}/\partial s$. dF_{obst} is the sum of the forces exerted by all obstacles which simultaneously interact with $d\mathbf{s}$. For the time being, it will be assumed that only the particle centered at (h_x, h_y, h_z) interacts with $d\mathbf{s}$. If the tensor $\boldsymbol{\sigma}_{obst}(x - h_x, y - h_y, -h_z)$, which represents the stress produced by the particle, is known, τ_{obst} follows from Eq. (2.10):

$$\tau_{obst}(x - h_x, y - h_y, -h_z) = \sigma_{obst\,xz}(x - h_x, y - h_y, -h_z)\cos\alpha + \sigma_{obst\,yz}(x - h_x, y - h_y, -h_z)\sin\alpha \qquad (4.1)$$

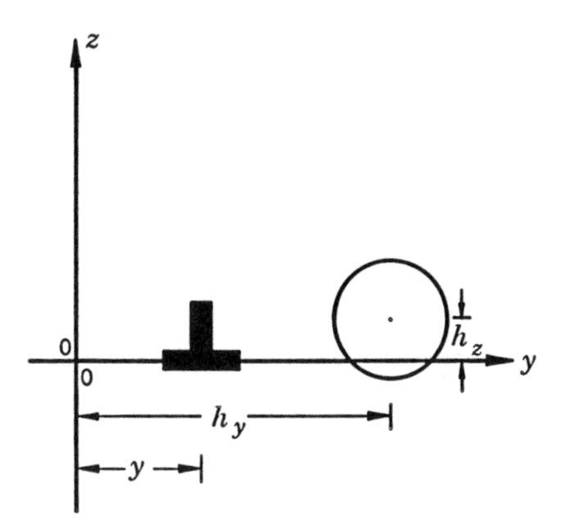

Fig. 4.2. Schematic diagram showing the position of the particle relative to the dislocation segment.

where α is the angle between the x-axis and the Burgers vector **b**. $\sigma_{obst\,xz}$ and $\sigma_{obst\,yz}$ are the shear stress components in the glide plane. Since the dislocation segment glides in the plane $z = 0$, only $-h_z$ appears in the argument of τ_{obst} and the z-component of **b** vanishes. Evidently, τ_{obst} is independent of $d\mathbf{s}$. This is a direct consequence of Eq. (2.10), which does not involve the direction of the dislocation. τ_{obst} is governed by $\boldsymbol{\sigma}_{obst}$, by the direction of **b**, and by the normal to the glide plane. If the dislocation segment is parallel to the x-axis, α equals θ_d defined in Section 2.1. θ_d is the angle between $d\mathbf{s}$ and **b**.

In Figures 4.3a and 4.3b, $\tau_{obst}(x - h_x, y - h_y, -h_z)$ is plotted for a particle that has the constrained lattice mismatch ε [Eq. (3.32)]. The coordinates of its center are ($h_x = 0$, $h_y = 0$, $h_z = 1.5\rho$). ρ is the radius. ε is positive; that is, the lattice constant of the matrix is larger than that of the particle. The dislocation is supposed to be undissociated. Linear isotropic elasticity has been assumed. Because h_z exceeds ρ, the glide plane does not intersect the particle. α equals 90° in Figure 4.3a; that is, the dislocation segment is of edge character. Figure 4.3b gives the corresponding stress for $\alpha = 0°$—that is, for a screw segment. Because h_z and ε are positive, the edge dislocation element is attracted by the particle. The derivation of $\boldsymbol{\sigma}_{obst}$ as a function of ε and of the shear modulus μ_s of the matrix will be detailed in Section 5.4.1. ε and μ_s of the copper–cobalt system (Sections 3.3.5, 5.4.2.1.8, and 5.4.2.2.2) have been inserted into Eqs. (5.40)–(5.42). Inside of the particle, $\boldsymbol{\sigma}_{obst}$ has only hydrostatic components; therefore the shear stress τ_{obst} is zero for

$$[(x - h_x)^2 + (y - h_y)^2 + (-h_z)^2] \leqslant \rho^2 \quad \text{[Eq. (5.40a)]}.$$

The energy of the system (particle plus dislocation segment) vanishes for the two extreme situations: (1) The segment has not yet approached the particle ($y = -\infty$), and (2) the particle has been left far behind ($y = +\infty$). The interaction with such obstacles is purely elastic.

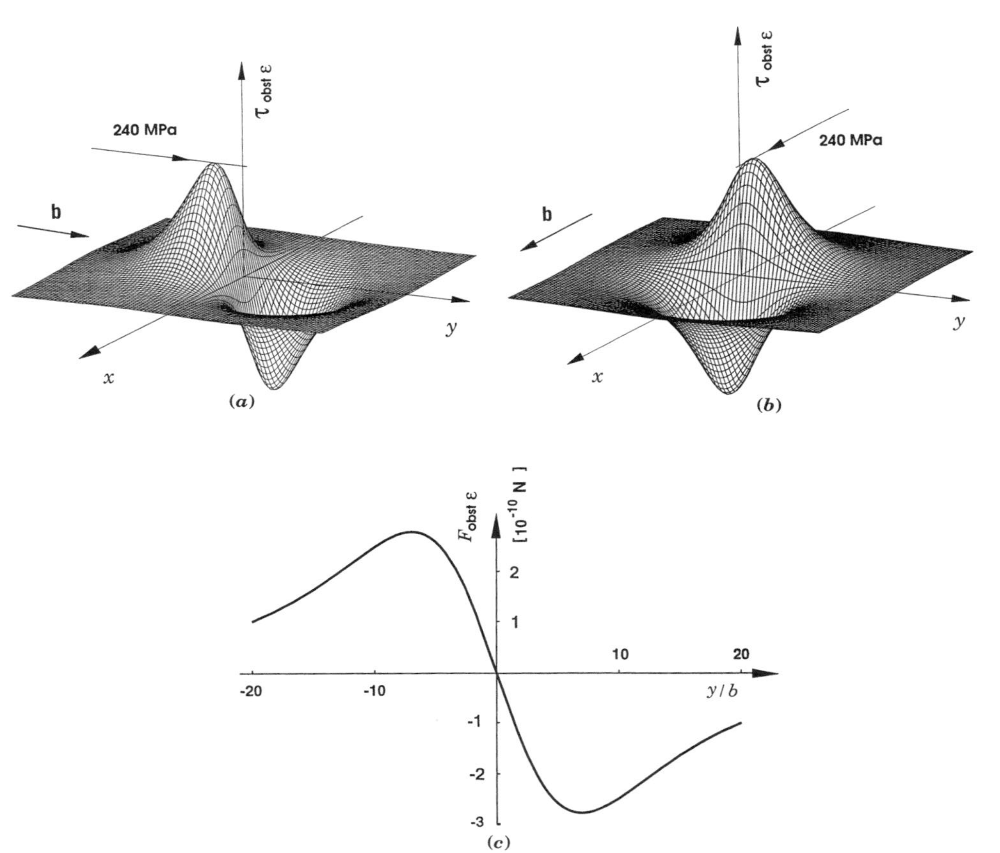

Fig. 4.3. Resolved shear stress $\tau_{\text{obst}\,\varepsilon}(x, y, -h_z = -1.5\rho)$ and force $F_{\text{obst}\,\varepsilon}(y, -h_z = -1.5\rho)$ [Eq. (4.2)] produced by a spherical particle having a lattice mismatch. Its center is at $(0, 0, \text{h}_z = 1.5\rho)$. Constrained lattice mismatch $\varepsilon = +0.015$ $(a_s > a_p)$, $b = 0.25$ nm, shear modulus of the matrix = 31 GPa, radius $\rho = 8b$. (*a*) Edge dislocation: $d\mathbf{s} = [ds, 0, 0]$, $\mathbf{b} = [0, b, 0]$, $\tau_{\text{obst}\,\varepsilon}$ versus (x, y). (*b*) Screw dislocation: $d\mathbf{s} = [ds, 0, 0]$, $\mathbf{b} = [\text{b}, 0, 0]$, $\tau_{\text{obst}\,\varepsilon}$ versus (x, y). (*c*) $F_{\text{obst}\,\varepsilon}$ of the edge dislocation versus y/b. (Courtesy of D. Rönnpagel, Braunschweig.)

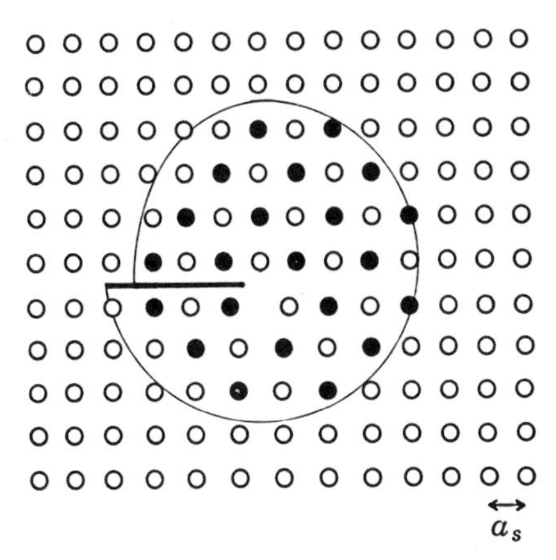

Fig. 4.4. Schematic diagram showing a matrix dislocation with a Burgers vector of length a_s that has entered the long-range-ordered particle. The antiphase boundary is indicated by the thick line.

The other type of obstacles is of the energy storing type: The energy at $y = -\infty$ differs from that at $y = +\infty$. In this case the particle is permanently damaged by the passage of the dislocation segment. This is illustrated in Figure 4.4 for a particle which—in contrast to the matrix—is long-range-ordered. In this example the length of the Burgers vector of a perfect *matrix* dislocation equals the lattice constant a_s of the matrix, but in the *particle* the length of the Burgers vector of a perfect dislocation is $2a_s$. Evidently a perfect matrix dislocation is only a partial dislocation in the particle. Along its glide plane, the matrix dislocation creates an antiphase boundary [4.1]. Its energy is stored in the particle. This is the relevant hardening mechanism in nickel-base superalloys strengthened by coherent, shearable γ'-precipitates. They have the long-range-ordered $L1_2$ structure. This will be detailed in Section 5.5. Figure 4.5a shows τ_{obst} for such an ordered particle. A second dislocation with $b = a_s$ following the first one in the same glide plane eliminates the antiphase boundary. The trailing dislocation is attracted by the particle. It is for this reason that dislocations glide in pairs in materials that are strengthened by long-range-ordered particles. This is demonstrated in Figure 4.6. A thin foil of the nickel-base superalloy NIMONIC PE16 (Section 5.5.1) has been tensile tested inside of a TEM. This micrograph has been taken while the specimen was under full load.

Besides being long-range-ordered, a particle may differ from the matrix in the shear modulus. This alters the energy density stored in the dislocation's stress field and thus gives rise to an interaction force (Section 5.2). dF_{obst} is the total force exerted by the particle on the dislocation segment $d\mathbf{s}$. The analogue holds for τ_{obst}: τ_{obst} includes the stresses due to both interaction mechanisms [4.3–4.7]. This has to be generalized to more than two mechanisms. In the

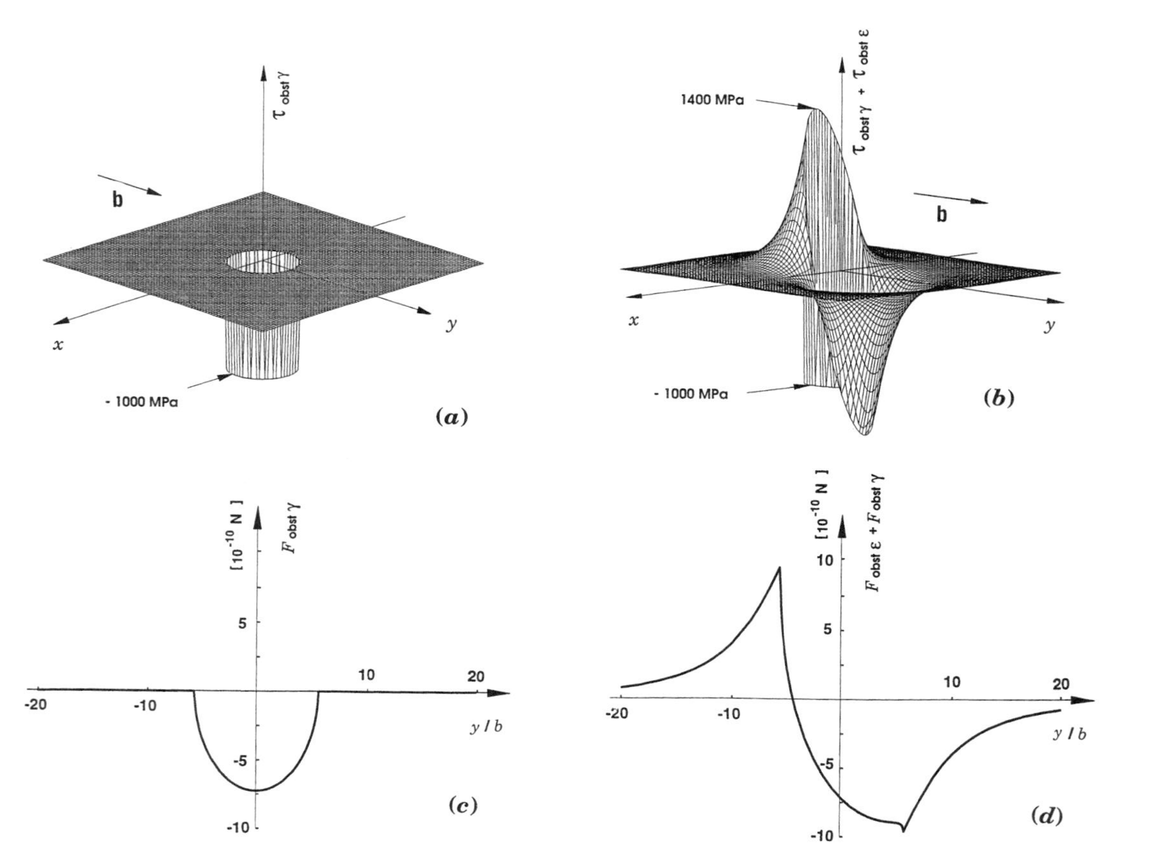

Fig. 4.5. Resolved shear stress $\tau_{\text{obst}}(x, y, -h_z = -\rho/\sqrt{2})$ and force $F_{\text{obst}}(y, -h_z = -\rho/\sqrt{2})$ [Eq. (4.2)] due to a long-range-ordered spherical particle centered at $(0, 0, h_z = \rho/\sqrt{2})$, edge dislocation: $d\mathbf{s} = [ds, 0, 0]$, $\mathbf{b} = [0, b, 0]$, $b = 0.25$ nm, radius $\rho = 8b$, specific antiphase boundary energy $\gamma = 0.25$ J/m, lattice mismatch and elastic parameters are the same as in Figure 4.3. The subscripts γ and ε of τ_{obst} and F_{obst} refer to the effect of long-range order and lattice mismatch, respectively. (*a*) $\tau_{\text{obst}\,\gamma}$ versus (x, y), (*b*) $(\tau_{\text{obst}\,\gamma} + \tau_{\text{obst}\,\varepsilon})$ versus (x, y), (*c*) $F_{\text{obst}\,\gamma}$ versus y/b, (*d*) $(F_{\text{obst}\,\gamma} + F_{\text{obst}\,\varepsilon})$ versus y/b. (Courtesy of D. Rönnpagel, Braunschweig.)

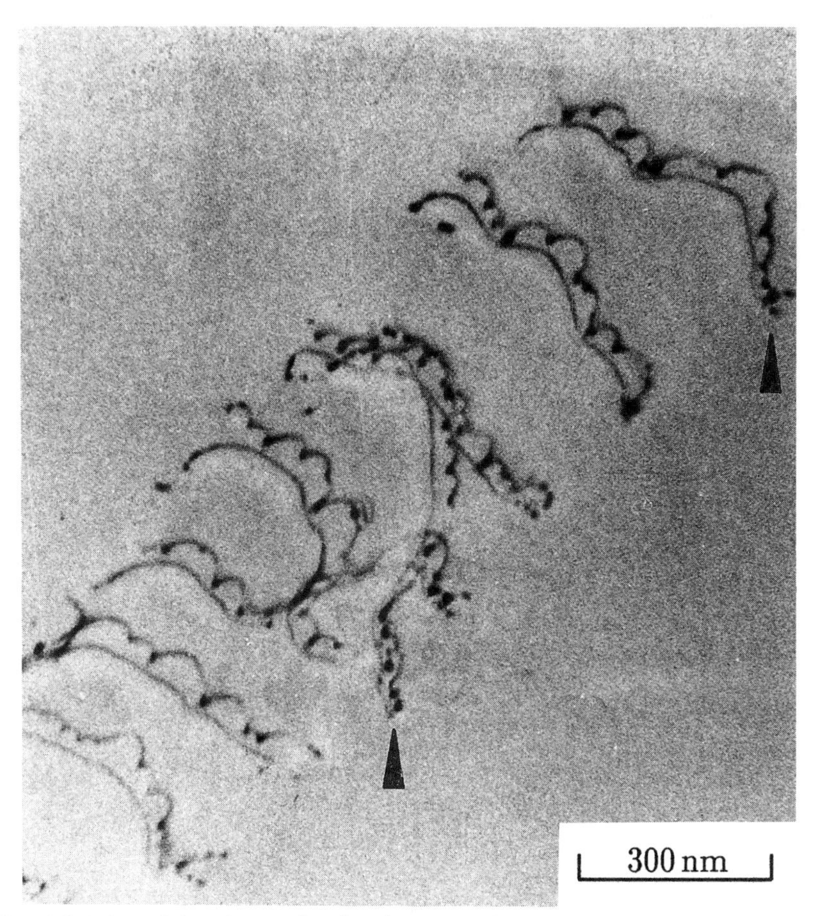

Fig. 4.6. Edge dislocation pairs in the superalloy NIMONIC PE16, which is strengthened by long-range-ordered γ'-particles. Their volume fraction is 0.09 and their mean radius is 8 nm. A thin foil has been plastically deformed inside of the transmission electron microscope. The micrograph shows the fully loaded specimen. The two dislocation pairs marked by arrows do not lie in the same slip plane as the other pairs, but in an adjacent, parallel one. (After Ref. 4.2.)

most general case, many obstacles simultaneously interact with the dislocation segment $d\mathbf{s}$. In such cases, dF_{obst} and $\tau_{obst}(x, y)$ are the respective sums and no obstacle coordinates are quoted in the argument of τ_{obst}. dF_{obst} and τ_{obst} are meant to be resolved to the glide system. Figure 4.5b shows τ_{obst} of a long-range-ordered particle that has a lattice mismatch.

The two quoted types of shearable obstacles will be referred to as

- Elastically interacting obstacles
- Energy storing obstacles

Most authors did not insert $\tau_{obst}(x, y)$ into Eqs. (2.18) and (2.20), but made one of the two approximations detailed in the next two sections.

4.1.1 Straight-Line Approximation

In the following the subscript "obst" of F_{obst} is dropped. It is, however, kept in τ_{obst}. The total force $F(y - h_y, -h_z)$ exerted by a single shearable particle positioned at (h_x, h_y, h_z) on a *straight* dislocation of infinite length is obtained by integrating $\tau_{obst}(x - h_x, y - h_y, -h_z)$ produced by this particle:

$$F(y - h_y, -h_z) = b \int_{-\infty}^{\infty} \tau_{obst}(x - h_x, y - h_y, -h_z)\, dx \tag{4.2}$$

Because the dislocation lies in the plane $z = 0$, z does not appear as argument of F and τ_{obst}.

τ_{obst} is now approximated by

$$\tau_{obst}(x - h_x, y - h_y, -h_z) = \delta(x - h_x)\, F(y - h_y, -h_z)/b \tag{4.3}$$

δ is Dirac's δ-function. In this "straight-line approximation," the dislocation segment $d\mathbf{s}$ experiences a force only at $x = h_x$: The range w_x (in the x-direction) of $\tau_{obst}(x, y)$ has been eliminated by the integration involved in Eq. (4.2). Because the range w_y (in the y-direction) is still allowed for, Eq. (4.3) describes a one-dimensional stress profile τ_{obst}. It is stressed that Eqs. (4.2) and (4.3) refer to a single particle. If actually more than one particle interact with a dislocation segment, summations have to be carried out before τ_{obst} is inserted into Eqs. (2.18) and (2.20). This will be done, for example, in Eq. (4.60).

Force profiles are presented in Figures 4.3c and 4.5c for a single particle having a positive lattice mismatch—that is, $a_p < a_s$ [Eqs. (3.16) and (3.32)]—and for a long-range-ordered particle, respectively. In Figure 4.5d also the combination of both interaction mechanisms is shown. The long-range-ordered γ'-particles in nickel-base superalloys actually have much smaller values of $|\varepsilon|$.

The interaction force F between a straight screw dislocation and a misfitting particle vanishes because the integration of Eq. (4.2) is extended over the positive and over the negative peak shown in Figure 4.3b. Consequently their contributions cancel. F of the edge dislocation (Figure 4.3c) is positive for $y < 0$, negative for $y > 0$ and vanishes for the symmetric position $y = 0$. The positive and negative peaks of τ_{obst} shown in Figure 4.3a are encountered in succession. As the edge dislocation approaches the particle from $y = -\infty$, it is attracted towards the particle. Once the dislocation has passed $y = 0$, it must be pushed forward by the external stress. In Section 4.5.1 it will be detailed how σ_{obst}, τ_{obst} and F depend on ε.

In their models of strengthening (Section 4.2.2.2) Schwarz and Labusch [4.8, 4.9] considered the following reduced interaction force profiles $F^*(y^*)$:

$$F(y - h_y, -h_z) = F_0(h_z) F^*(y^*) \tag{4.4a}$$

where

$$F_0(h_z) > 0 \tag{4.4b}$$

$$y^* = (y - h_y)/w_y \tag{4.4c}$$

$$\begin{aligned} F^*(y^*) &= A_n y^*[1 - (y^*)^n]^2 \qquad &\text{for } |y^*| \leqslant 1 \\ F(y^*) &= 0 &\text{for } |y^*| > 1 \end{aligned} \tag{4.4d}$$

$$A_n = \pm(2n + 1)^{(2n+1)/n}/(2n)^2 \tag{4.4e}$$

n has been varied between 2 and 12. When n is not an even integer, the term $(y^*)^n$ must be replaced by $(|y^*|)^n$. w_y is the range of the interaction force in the y-direction. If A_n is positive, the interaction is repulsive. The maximum value of $F^*(y^*)$ equals unity. The full width at half-maximum of $F^*(y^*)$ is close to 0.6. This width is almost independent of n.

The interaction energy $U(y - h_y, -h_z)$ between the obstacle and the straight dislocation is obtained by integrating Eq. (4.4a):

$$U(y - h_y, -h_z) = -\int_{-\infty}^{y} F(y - h_y, -h_z)\,dy \tag{4.5a}$$

The reduced energy U^* is defined in analogy to F^*:

$$U^*(y^*) = U(y - h_y, -h_z)/[w_y F_0(h_z)] \tag{4.5b}$$

F^* and U^* are plotted versus y^* in Figure 4.7. Evidently the interaction with this obstacle is elastic: $U^*(-\infty) = U^*(+\infty) = 0$. Because the plus sign of A_n has been chosen in Figure 4.7a, the interaction is repulsive. In Figure 4.7b, it is attractive because A_n is negative. The effect of n is demonstrated in Figure 4.7c. There n equals 12, whereas it is 2 in the two other figures. The higher the value of n, the steeper the function $F^*(y^*)$ next to $y^* = \pm 1.0$.

The sign of the interaction force needs some more attention. As the dislocation approaches the repulsive obstacle, whose profiles are shown in Figure 4.7a, from $y^* = -\infty$, it experiences a negative force. The external stress τ_{ext} has to push the dislocation forward until $y^* = 0$ is reached. From there on, one expects the obstacle to accelerate the dislocation. Actually, however, the motion of the dislocation is normally so highly damped that it does not acquire any kinetic energy in the range $y^* > 0$. When it approaches the next repulsive obstacle, the external stress τ_{ext} has to push the dislocation forward again; that is, the action of τ_{ext} is required everywhere where the slope of $U^*(y^*)$ is positive. If the interaction between the obstacle and the dislocation is attractive (Figure 4.7b), it is drawn from $y^* < 0$ towards $y^* = 0$, but it arrives at the minimum of $U^*(y^*)$ without any kinetic energy. τ_{ext} has to drive it uphill in the range $y^* > 0$. Evidently, attractive and repulsive obstacles produce the same effects.

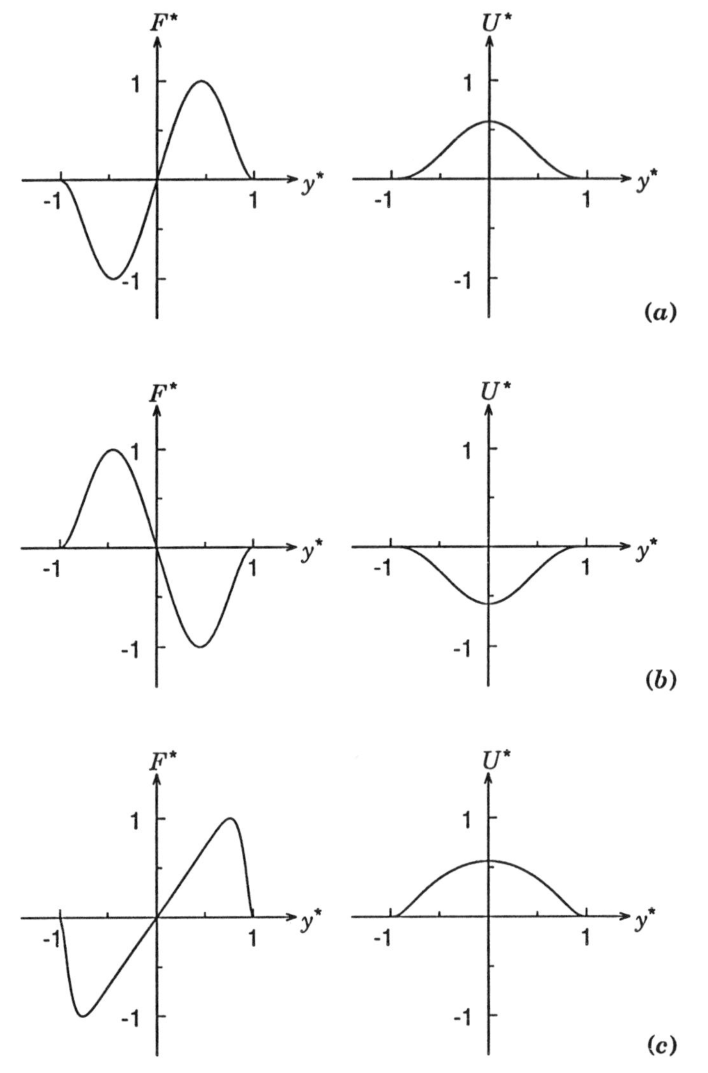

Fig. 4.7. Elastically interacting obstacle. F^* of Eq. (4.4d) and U^* of Eq. (4.5b) versus y^*. (*a*) A_n of Eq. (4.4e) is positive, $n = 2$. (*b*) $A_n < 0$, $n = 2$. (*c*) $A_n > 0$, $n = 12$.

In both cases the ascending flank of $U^*(y^*)$ is the decisive part. This will also become evident in Section 5.4.2.1, where lattice mismatch strengthening will be treated: The CRSS is governed by the absolute value of the constrained lattice mismatch ε. The rare case that the dislocation does acquire kinetic energy will be discussed in Sections 4.2.2.2.2 and 5.4.3.

$F^*(y^*)$ and $U^*(y^*)$ of an energy storing particle are shown in Figure 4.8. There is only a positive slope of $U^*(y^*)$. In the range $0 \leqslant y^* \leqslant 1$, $F^*(y^*)$ is the

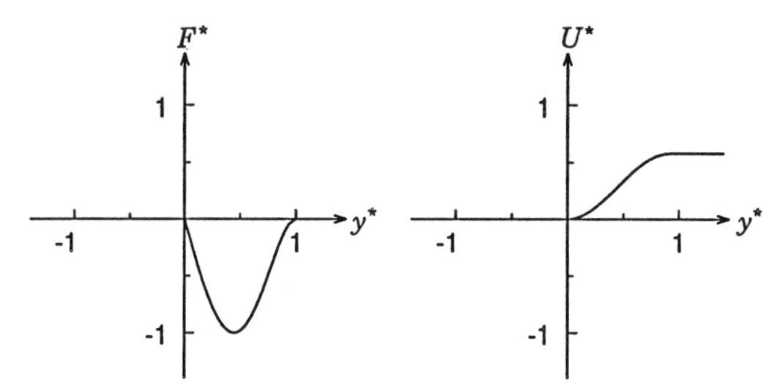

Fig. 4.8. Energy storing obstacle. F^* of Eq. (4.6) and U^* of Eq. (4.5b) versus y^*. $A_n < 0$, $n = 2$.

same as that in Eq. (4.4d):

$$F^*(y^*) = A_n y^* [1 - (y^*)^n]^2 \qquad \text{for } 0 \leqslant y^* \leqslant 1$$
$$F^*(y^*) = 0 \qquad \text{for } y < 0 \text{ and } y > 1 \tag{4.6}$$

A_n is negative, and it is defined in Eq. (4.4e). $U(y - h_y, -h_z)$ and $U^*(y^*)$ follow from Eqs. (4.5). There is one important exception to the rule that A_n of energy storing obstacles is negative: A_n of the trailing dislocation of a pair shearing a long-range-ordered particle is positive. This dislocation is attracted by the antiphase boundary created by the leading dislocation. The trailing dislocation releases the energy stored in the boundary.

4.1.2 Point Obstacles

Actually the shear stress $\tau_{obst}(x - h_x, y - h_y, -h_z)$ of a shearable particle has finite ranges w_x and w_y in the x- and y-direction, respectively. The integration [Eq. (4.2)] involved in the "straight-line approximation" eliminated w_x and led to the force profile $F(y - h_y, -h_z)$. Now also w_y of F is reduced to zero: The particle is supposed to interact with the straight dislocation only if the latter one's y-coordinate coincides with h_y. At this point the dislocation experiences the force $-F_0$; everywhere else the interaction force vanishes. This explains the term "point obstacle." F_0 is given by

$$F_0 = \text{Max}\{|F(y - h_y, -h_z)|\} \tag{4.7a}$$

The dependence of F_0 on h_z is not explicitly indicated. The force exerted on the dislocation is assumed to be $-F_0$, because negative forces impede the dislocation glide towards $+y$ (Section 4.1). Hence $\tau_{obst}(x - h_z, y - h_y, -h_z)$ of a point obstacle can be written with Dirac's δ-function as

$$\tau_{\text{obst}}(x - h_x, y - h_y, -h_z) = \delta(x - h_x)\,\delta(y - h_y)(-F_0)/b$$

Morris and Syn [4.10] introduced a more general definition of "point obstacles." They did not base it on the "straight-line approximation" as has been done in Eq. (4.7a), but allowed for the flexibility of the dislocation.

If a shearable particle interacts with the dislocation via various mechanisms i, $1 \leqslant i \leqslant n$, the maximum force F_0 of this combination is often very close to the maximum force of the strongest mechanism:

$$F_0 = \text{Max}\left\{\left|\sum_{i=1}^{n} F_i(y - h_y, -h_z)\right|\right\} \approx \text{Max}\{F_{0i}\}_{1 \leqslant i \leqslant n} \tag{4.7b}$$

The reason for this is that the forces $|F_i|$ are likely to reach their maxima F_{0i} at different coordinates y. This is demonstrated in Figures 4.5c and 4.5d: The force $|F_\gamma|$, which is due to the long-range order of the particle, peaks when the dislocation has reached the center of the intersection of the glide plane with the particle, whereas the force $|F_\varepsilon|$, which is caused by the lattice mismatch ε, has its maximum when the dislocation just touches the particle. F_0 of the sum of the two forces is equal to $F_{\varepsilon 0} = \text{Max}\{|F_\varepsilon|\}$. The range w_y of the combined forces may, however, exceed the individual ranges w_{yi}. In Figure 4.5, ε is about 10 times as large as ε of long-range-ordered γ'-precipitates in technical nickel-base superalloys. For them, F_0 is very close to $F_{\gamma 0}$.

Often it is convenient to replace F_0 by the breaking angle ϕ_c of the obstacle; the subscript c of ϕ_c stands for "critical":

$$F_0 = 2S\cos(\phi_c/2) \tag{4.8a}$$

where S is the dislocation line tension. Equation (4.8a) is interpreted with reference to Figure 4.9a. A dislocation bows out between three collinear, equispaced particles. Each of the two dislocation arcs exerts a force on the central particle. The y-component of either force is $[S\cos(\phi/2)]$. The stronger the particle, the higher the value of F_0 and the smaller the angle $\phi_c/2$ in the moment when the dislocation breaks free from the particle. In that moment Eq. (4.8a) holds; solving it for $\cos(\phi_c/2)$ yields

$$\cos(\phi_c/2) = \frac{F_0}{2S} \tag{4.8b}$$

If F_0 is large, the dislocation bows out strongly and the variation of S with the angle θ_d (Section 4.2.1.4.3) has to be allowed for. In principle, F_0 is subjected to no limitation. Equations (4.8), however, are only applicable as long as F_0 does not exceed $2S$. For large values of F_0, the Orowan process is

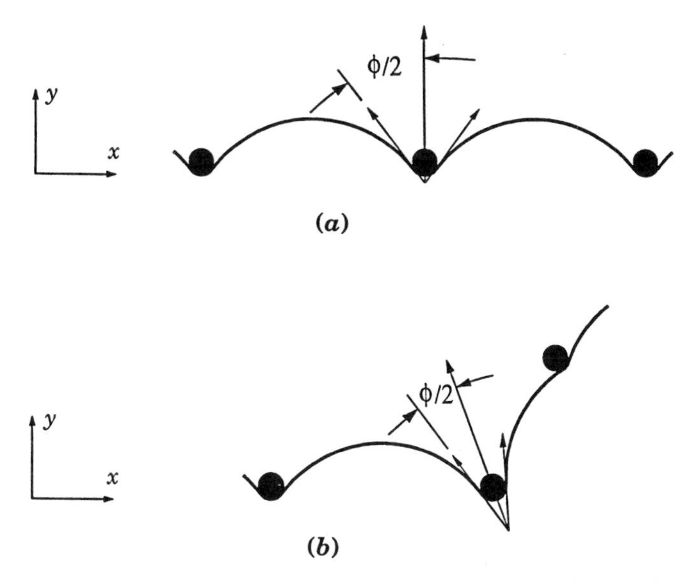

Fig. 4.9. Definition of the bowing angle $\phi/2$. (*a*) Symmetric dislocation configuration. (*b*) Asymmetric dislocation configuration.

expected to operate (Chapter 6). The exact value of F_0 at which the transition from particle shearing to circumventing them occurs depends on $S(\theta_d)$ and on the dislocation line energies entering the equations derived in Section 6.1.

If the dislocation bow-outs to the left and to the right of the central particle are not the same, the meaning of Eqs. (4.8) has to be generalized. In Figure 4.9b two asymmetric bow-outs are sketched: The three particles do not lie on a straight line, and their spacings differ from each other. In such cases ϕ and ϕ_c are defined as the angle enclosed by the two dislocation arms. The sum of the two forces exerted by them on the central particle is not parallel to the y-direction. Therefore one postulates that the particle can withstand the force F_0 in any direction. Thus Eqs. (4.8) are maintained.

When dislocation bending was discussed in Chapter 2, bending was assumed to be only slight. This primarily concerned the dislocation line tension and its applications—for example, Eqs. (2.11) and (2.20). Figure 4.9a offers a convenient means to set a quantitative, though somewhat arbitrary, limit for "slight bending" [4.11, 4.12]:

$$\cos(\phi/2) \leqslant 0.5 \tag{4.9a}$$

or

$$\phi/2 \geqslant 60^\circ \tag{4.9b}$$

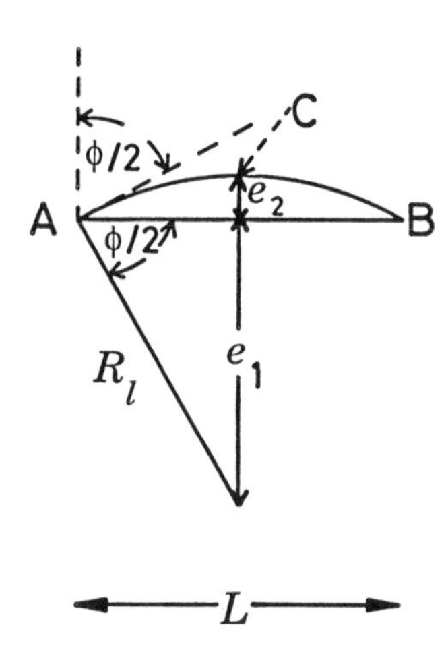

Fig. 4.10. Slightly bent dislocation. $\phi/2 = 60^\circ$

or for $\phi = \phi_c$:

$$F_0 \leqslant S \tag{4.9c}$$

Shearable obstacles whose maximum interaction force F_0 is less than S are called "weak"; they lead to only slight bending of dislocations. Figure 4.10 illustrates this situation for $\phi/2 = 60^\circ$—that is, for $F_0 = S$. The various lengths defined there are as follows:

Straight length AB:	L
Radius of curvature R_l:	L
Length of the arc ACB:	$\frac{\pi}{3}L = 1.047L$
Total length of the two straight lines AC plus CB:	$2[(L/2)^2 + e_2^2]^{1/2} = 1.035L$
Elevation e_2:	$(1 - \sin 60^\circ)L = 0.134L$

Relations (4.9) also afford a limit to the range of the applicability of the "straight-line approximation" introduced in Section 4.1.1. The configurations of the leading edge dislocations of the pairs shown in Figure 4.6 are now analyzed with reference to the above limits. This transmission electron micrograph (TEM) has been taken under full load after the dislocations assumed their critical configurations. The average ratio (e_2/L) (Figure 4.10) is about 0.3 [4.2]. Evidently the dislocation bow-outs exceed the limits set by relations (4.9). $F_0/(2S)$ of cobalt-rich particles in copper (Sections 3.3.5 and 5.4.2.2.2) is calculated to be less than 0.6 [4.13] provided that the dissociation of the dislocations into Shockley partials is allowed for AND (logic AND) the particle radius is below 3 nm; that is, dislocation bowing is rather slight in this material. This view is supported by direct TEM observations of the dislocation configurations in fully loaded thin foils of copper–cobalt [4.14]. Messerschmidt et al. [4.15] have published very detailed investigations of dislocation bowing in magnesium oxide.

4.2 DERIVATION OF THE CRITICAL RESOLVED SHEAR STRESS

The CRSS is measured in standard tensile or compression tests. Single-crystal specimens are deformed at a constant shear strain rate $\dot{\gamma}$. Its order of magnitude is 10^{-4}/s. Macroscopically the CRSS is defined as the resolved shear stress that produces a certain plastic strain—for example, 2×10^{-4}. Alternatively the CRSS may be calculated from the intersection of the linearly extrapolated elastic and first plastic part of the load versus elongation curve [4.12, 4.16, 4.17]. This latter method is sketched in Figure 4.11. Both methods lead to quite similar results.

The macroscopic shear strain rate $\dot{\gamma}$ has to be expressed in terms of the density ρ_d and the glide velocity v_n (Section 2.4) of mobile dislocations. ρ_d is their total length per unit volume. If there are N_d mobile dislocations in the crystal and each of them sweeps out its entire glide plane of area A_g, the total

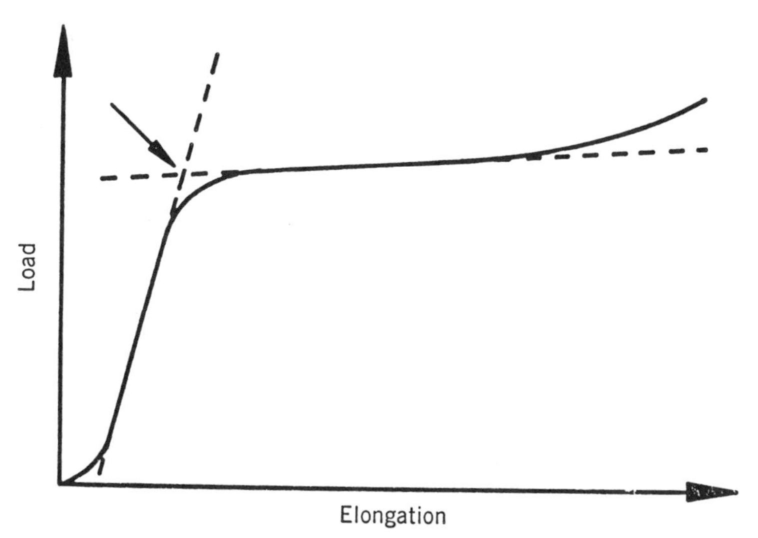

Fig. 4.11. Load versus elongation curve. The CRSS is calculated from the load marked by the arrow.

shear strain γ produced by them is

$$\gamma = N_d b A_g / V \tag{4.10}$$

where V is the volume of the specimen. Equation (4.10) is the definition of shear strain [4.18, 4.19]. In the time dt each dislocation covers only the area $(l_d v_n\, dt)$, where l_d is their mean length. The resulting shear strain $d\gamma = (\dot{\gamma}\, dt)$ is obtained by weighting γ of Eq. (4.10) by the ratio of the areas swept out—that is, by $[(l_d v_n\, dt)/A_g]$:

$$\dot{\gamma}\, dt = d\gamma = (N_d b A_g / V)[(l_d v_n\, dt)/A_g]$$

The term $(N_d l_d / V)$ equals ρ_d:

$$\dot{\gamma} = \rho_d b v_n \tag{4.11}$$

It must be stressed that ρ_d is the density of *mobile* dislocations. The CRSS is the shear stress that raises the product $\rho_d v_n$ to the level given by Eq. (4.11); that is, ρ_d as well as v_n have to be sufficiently high. ρ_d may be low for at least two reasons:

1. Only very few dislocations have been created during crystal growth. This is often the case in covalent crystals—for example, in germanium and silicon [4.20, 4.21].
2. The grown-in dislocations may be locked. The best-known example is the system α-iron with some carbon: Carbon clouds immobilize the dislocations [4.22–4.26]. Suzuki [4.27, 4.28] suggested that there should be similar effects in face-centered cubic (fcc) metals: Solute atoms may condense onto the stacking fault ribbons of dissociated dislocations.

Both of these mechanisms lead to yield points.

In particle-strengthened metals and alloys, ρ_d is normally sufficiently high. It is the function $v_n(\tau_{\text{ext}}, T_D)$ that is decisive. τ_{ext} is the external applied resolved shear stress and T_D is the temperature at which the specimen is deformed. The subscript D of T_D stands for "deformation." Because the energies involved in overcoming the particle–dislocation interactions are high, thermal activation can normally be disregarded when particle hardening is concerned. This has already been discussed in Section 2.5. Thus one may write $v_n(\tau_{\text{ext}}, T_D) \approx v_n(\tau_{\text{ext}}, T_D = 0\,\text{K}) = v_n(\tau_{\text{ext}})$. τ_{ext} must be high enough to drive dislocations across macroscopic areas, in which they have to overcome many particles. The CRSS is not governed by unlocking mechanisms. This is evident in Figure 4.6, where the leading dislocations of all pairs are scalloped even though they have already moved over large distances. Because there are particles everywhere in the glide plane, $|\tau_{\text{obst}}|$ reaches high levels at many spots.

Once τ_{ext} exceeds the minimum level required to overcome the particle–dislocation interactions in the entire glide plane, a further increase of τ_{ext} raises v_n sharply. Above the minimum stress level, v_n is governed by damping processes. As far as the glide over large distances is concerned, the function $v_n(\tau_{ext})$ has a threshold: Below the threshold v_n is zero, above the threshold $\partial v_n / \partial \tau_{ext}$ is positive. This latter point will be analyzed in greater detail in Section 4.2.2.2.2; it concerns dynamic dislocation effects. Disregarding them, one defines the CRSS τ_p as the threshold stress mentioned above. The view that the CRSS of particle-hardened materials equals the stress necessary to drive dislocations across large areas and over many particles is borne out by the success of the models based on this idea.

The CRSS τ_p of metals and alloys hardened by shearable particles is derived from Eqs. (2.18) or (2.20). In principle they can be solved as follows. At time $t = 0$ a nearly straight dislocation is placed in the stress field $\tau_{obst}(x, y)$ produced by the particles. For $t > 0$, each segment of the dislocation is required to move across the glide plane at a velocity not less than v_{n0}. This yields the functions $y(x, t)$ and $\tau_{ext}(t)$. τ_{ext} is constant throughout the entire specimen. τ_p equals the maximum of $\tau_{ext}(t)$. If thermal activation and dynamic effects are disregarded, τ_p is independent of v_{n0} and it suffices to derive τ_p in the limit that v_{n0} approaches zero. The neglect of thermal activation implies the exclusion of dislocation climb and cross-slip. There have been two different approaches to solve the two equations: analytical methods and computer simulations. They will be discussed in Sections 4.2.1 and 4.2.2, respectively. Their advantages and disadvantages are as follows:

Analytical Solutions

Advantages: The results are easy to handle formulae, which directly show the variation of the CRSS with, for example, the mean radius of the particles and their volume fraction.

Disadvantages: Given the complexity of the problems, rather grave simplifications have to be made.

Computer Simulations

Advantages: The effects of many parameters on the CRSS can be tested—for example, those of the profile of the particle–dislocation interaction force.

Disadvantages: The results are tables of data, which are often difficult to represent by simple formulae; this is even more difficult to do if different dependencies are to be expressed simultaneously.

The force F and the dislocation line tension depend on the character of the dislocation. This has been demonstrated in Figures 4.3 and 2.2, respectively. A dislocation loop emitted by a source can only expand if all of its parts can move. Therefore the macroscopic CRSS τ_p is the stress necessary to make that type of dislocation move which requires the highest stress [4.29]. In their analysis of strengthening of the superalloy NIMONIC PE16 by ordered γ'-particles, Nembach and Neite [4.7] inserted the geometric mean of the line tensions S of edge and screw dislocations. These authors had shown that this mean was very close to $(2/\pi)\int_0^{\pi/2} S(\theta_d)\,d\theta_d$.

The various analytical solutions of Eqs. (2.18) or (2.20) will be discussed in Section 4.2.1. The following point, which has already been mentioned in Section 4.1, is stressed again. If an obstacle interacts with dislocations via more than one mechanism, τ_{obst} and F are assumed to comprise *all* contributions—for example, that due to a particle's long-range order and that caused by its modulus mismatch. Depending on the approximations to be applied, τ_{obst} and F may represent the interaction either with a single particle or with all particles that simultaneously interact with the dislocation segment $d\mathbf{s}$.

In most of the derivations of the CRSS, only one dislocation will be considered. With few exceptions the effects of the mutual interactions between dislocations gliding in the same plane are disregarded. Such exceptions will be discussed in Sections 4.2.2.3, 5.3, and 5.5.4.1. Micrographs of many dislocations gliding in the same plane are shown in Figures 4.6 and 5.24.

4.2.1 Analytical Solutions of Eqs. (2.18) and (2.20)

4.2.1.1 Mott and Nabarro's Model. Mott and Nabarro [4.30] considered hardening by shearable particles that have a lattice mismatch. The authors disregarded thermal activation and dynamic effects and wrote for the shear stress τ at the distance R from the center of the particle of radius ρ, $R > \rho$ (Section 5.4.1):

$$\tau = C\varepsilon\rho^3/R^3 \tag{4.12}$$

The constant C involves the shear modulus of the matrix; ε is the constrained lattice mismatch [Eq. (3.32)]. Isotropy has been assumed. Mott and Nabarro obtained an estimate of the CRSS τ_p by identifying it with the mean stress $\bar{\tau}$, which in turn was derived from Eq. (4.12) by inserting the mean particle spacing $(1/n_v^{1/3})$ for R. n_v is the number of particles per unit volume, r is their average radius, and f is their volume fraction:

$$\tau_p = \bar{\tau} = C|\varepsilon|\omega_3 r^3/(1/n_v)$$

and with Eq. (3.2) one obtains

$$\tau_p = \frac{3}{4\pi}C|\varepsilon|f \tag{4.13}$$

The statistical factor ω_3 has been defined in Eq. (3.5b). Evidently τ_p turns out to be independent of r. This result is at variance with the experimental data presented in Figures 1.6 and 5.19 for the copper–cobalt system. There the CRSS is seen to vary with r. This system is the prime example for lattice mismatch strengthening; it will be discussed in Section 5.4.2.1.8 and 5.4.2.2.2. Mott and Nabarro [4.31–4.36] soon realized that the lengths over which τ is averaged have to be given careful consideration. These lengths depend on the finite flexibility of the dislocations—that is, on their line tension (Section 2.2).

4.2.1.2 Upper and Lower Limits for the CRSS. Upper and lower limits for τ_p are now presented. They are based on three assumptions:

1. Dynamic dislocation effects and thermal activation are disregarded.
2. The shearable particles can be treated as point obstacles; all of them yield the same maximum interaction force F_0 (Section 4.1.2).
3. The particles are uniformly spaced; let their spacing along the dislocation be L.

Instead of postulating uniformity in strength (assumption 2) and spacing (assumption 3), one might as well say that F_0 and L are the respective averages. L depends on the strength of the bow-out of the dislocation and thus on the external applied stress τ_{ext}. This has been explained in Section 1.3. τ_{ext} is slowly raised from zero until the dislocation breaks free from a particle. This is the critical moment: τ_{ext} and L equal the CRSS τ_p and the critical length L_c, respectively. Equation (2.9c) relates τ_p to L_c and F_0:

$$\tau_p b L_c = F_0 \tag{4.14}$$

This is the basic equation of strengthening by shearable particles. It applies also to single solute atoms—that is, to solid solution hardening—but only at 0 K. At finite temperatures the effects of thermal activation on solid solution strengthening must not be disregarded.

Because of the uniformity in particle strength and spacing, breaking away from one particle is equivalent to overcoming all of them. Each particle has to carry the force exerted by τ_{ext} on the dislocation arc lying to the right of the particle. The calculation of L_c is the crucial point of all theories of hardening. There are, however, two limits to L_c:

$$L_{min} \leqslant L_c \leqslant L_{max} \tag{4.15}$$

The lengths L_{min} and L_{max} have been defined in Section 3.1. L_{min} is the square lattice spacing and L_{max} the particle spacing along a straight line. L_c equals L_{min} if the dislocation is infinitely flexible—that is, if its line tension S vanishes. L_c assumes the maximum value L_{max} if the dislocation is infinitely stiff—that is, if S is infinite. This yields upper and lower bounds for τ_p, respectively:

$$\textit{Upper bound:} \qquad \tau_p = \frac{F_0}{bL_{\min}} \tag{4.16a}$$

$$\textit{Lower bound:} \qquad \tau_p = \frac{F_0}{bL_{\max}} \tag{4.17}$$

Equation (4.16a) lends itself to an alternative interpretation: If the obstacles form a square array and the dislocation attacks it along the edges of length $L_{\min}$, Eq. (4.14) leads directly to Eq. (4.16a). The important point is that the obstacles form a collinear array and that their spacing equals $L_{\min}$.

Evidently, τ_p increases with the flexibility of the dislocation. At $L_c \approx L_{\min}$ the transition from particle shearing to circumventing them may occur. In the present chapter only the first mechanism is dealt with. Substituting [$2S\cos(\phi_c/2)$] for F_0 [Eq. (4.8a)], Eq. (4.16a) becomes

$$\tau_p = \frac{2S\cos(\phi_c/2)}{bL_{\min}} \tag{4.16b}$$

For convenience the reduced CRSS τ_p^* is defined:

$$\tau_p^* = \tau_p bL_{\min}/(2S) \tag{4.18}$$

Hence for the reduced upper bound one obtains

$$\tau_p^* = \cos(\phi_c/2) \tag{4.16c}$$

Actually the applicability of Eqs. (4.16) is governed not only by S, but also by F_0. If F_0 is very high, the dislocation may be strongly bent even though S is finite. If $[F_0/(2S)] = \cos(\theta_c/2)$ is close to unity, L_c nearly equals $L_{\min}$ and Eqs. (4.16) yield good approximations for τ_p. It depends on the system under consideration whether the peak of the aging curve $\tau_p(r, f = \text{const.})$ can be described by Eqs. (4.16) (particle shearing, $L_c = L_{\min}$) or whether it is governed by the onset of the operation of the Orowan process. This will be detailed at the end of Section 4.2.1.3.2. In Section 4.2.1.4.1 it will be shown that if a random array of obstacles is considered instead of a uniform one, the right-hand sides of Eqs. (4.16) have to be multiplied by a statistical factor which is around 0.9. Actually the distinction between "uniform" and "random" means that in the former case (uniform) one disregards the distribution of interobstacle spacings and considers only their average, whereas in the latter case (random) this distribution is allowed for. In Figure 4.12a, τ_p^* is plotted versus $\phi_c/2$. In agreement with the supposition that Eq. (4.16c) yields an upper limit for τ_p^*, the respective curve lies above all others and also above the results obtained by computer simulations. The latter will be described in Section

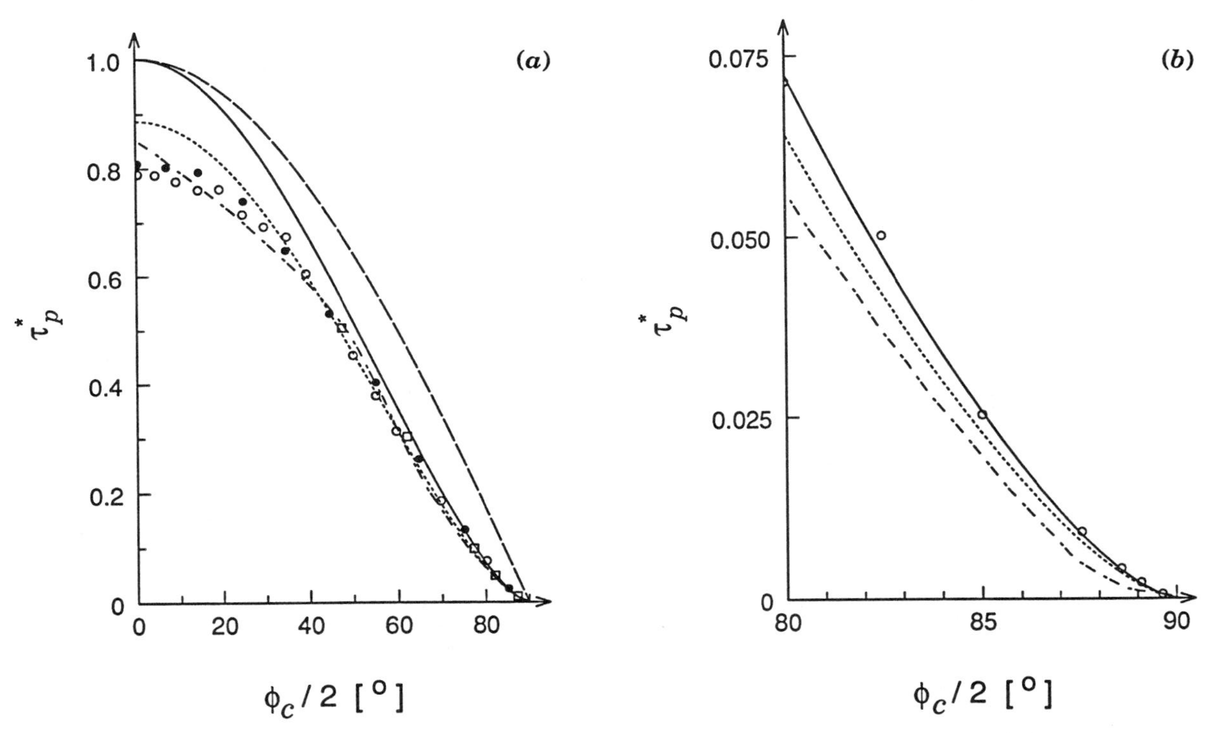

Fig. 4.12. Reduced CRSS τ_p^* defined in Eq. (4.18) versus the critical angle $\phi_c/2$. Computer simulations by Foreman and Makin [4.37, 4.38]: ○1000 point obstacles; ● 10,000 point obstacles. Computer simulations by Hanson and Morris [4.39]: □10,000 point obstacles. The curves represent the results of the various models: — — — Eq. (4.16c), upper limit; ——Eq. (4.26c), Friedel; --- Eq. (4.43), Hanson and Morris; -·- Kocks.

4.2.2.1. Three of the suppositions under which Eq. (4.16c) has been deduced have been listed at the beginning of this section: no dynamic effects, no thermal activation, and uniformity of the strength and spacing of the point obstacles. The fourth supposition is that $[F_0/(2S)]$ is large.

4.2.1.3 Friedel's Model. Friedel [4.40, 4.41] and somewhat later Fleischer and Hibbard [4.42, 4.43] published an analytical solution for Eq. (2.20). Originally the authors did not treat particle hardening but instead analyzed other mechanisms. Their ideas, however, can also be applied to hardening by shearable particles. To stress this generality, for the time being the term "obstacle" will be used instead of "particle." Friedel's original work primarily concerned the calculation of the obstacle spacing along the dislocation in the critical moment—that is, the derivation of Eq. (4.25). Once this length is known, Eq. (4.14) yields the CRSS.

4.2.1.3.1 Suppositions. Friedel made the following restrictive suppositions and approximations; most of them were not stated explicitly:

1. The obstacles to the glide of the dislocation are shearable point obstacles; that is, the ranges w_x and w_y are disregarded.
2. All obstacles produce the same maximum interaction force F_0.
3. The obstacles are uniformly spaced in the glide plane.
4. The propagation of the dislocation is steady; the meaning of "steady" is explained below.
5. Bowing of the dislocation is only slight [relations (4.9)]; that is, the obstacles are weak.
6. Thermal activation is disregarded.
7. The motion of the dislocation is overdamped.
8. The interaction between different dislocations is disregarded.

These suppositions and their implications require some comments. Points 2 and 3 mean that the respective averages are to be inserted and that the spread in strengths F_0 and spacings λ (Section 3.3) are disregarded. Except for the points 4 and 5, all others have been implied in the derivation of Eqs. (4.16). Point 8 had not been stated there explicitly. Point 5 permits one to disregard the variation of the dislocation line tension S along the bent dislocation—that is, to consider S as constant. The meaning of point 4 is explained with reference to Figure 4.13. Under the applied stress τ_{ext}, the dislocation bows out between the point obstacles A, B, and C; they are assumed to lie on a straight line. The dislocation forms two arcs: AB and BC. Their radius of curvature R_l is given by Eq. (2.11b). Because S is constant, R_l depends only on τ_{ext}. By virtue of the supposed uniformity in the spacing of the obstacles (point 3), the lengths AB and BC are identical: Both equal L. Once τ_{ext} is strong enough to release the

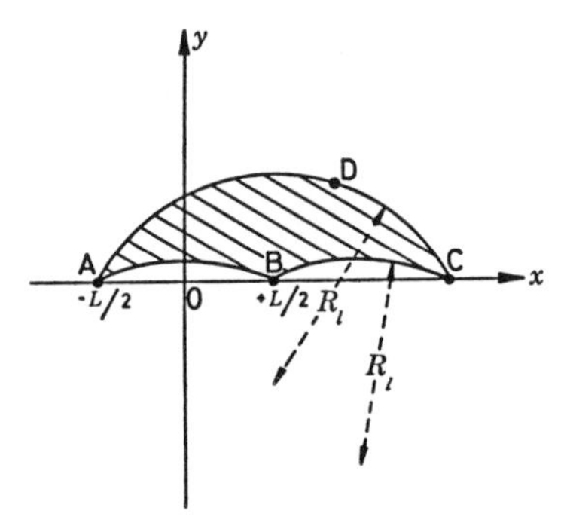

Fig. 4.13. A dislocation bows out between the three equidistant, collinear, shearable point obstacles A, B, and C. The area A_F is shown hatched. R_l is the radius of the circular arcs.

dislocation from obstacle B, it sweeps out the area A_F. A_F is shown hatched in Figure 4.13. The subscript F of A_F stands for "Friedel." The radius of curvature of the larger, new arc AC equals R_l again. The meaning of "steady propagation" is that the dislocation encounters exactly one new obstacle at the outer periphery of the area A_F: obstacle D. Evidently one obstacle has been overcome, namely B, and a new one has been met, namely D. If the dislocation did not meet obstacle D, the glide would become unstable: Obstacles A and C alone would be unable to withstand the force exerted by τ_{ext} on the arc AC and would give away. Such a succession of breakthroughs is often referred to as *unzipping*. Because the obstacles are supposed to be uniformly spaced, there is no chance of a later restabilization. The steady-state condition is described by

$$A_F n_s = 1 \tag{4.19}$$

where n_S is the number of obstacles per unit area.

By virtue of supposition 3 (uniform spacing of the obstacles), the obstacles are equispaced along the critical configuration of the dislocation. Thus the configuration shown in Figure 4.13 can be assumed to be representative of the entire glide plane. As in the derivation of Eq. (4.16a), the obstacles are supposed to be collinear and equispaced. The important difference lies in their critical spacing L_c. In the case of Eq. (4.16a), L_c was set equal to its minimum value L_{min}, whereas Friedel calculated the actual critical length. Supposition 6 (no thermal activation) means that the following solution of Eq. (2.20) actually applies to 0 K.

4.2.1.3.2 Solution of Eq. (2.20). The postulated strong damping, supposition 7, suppresses any dynamic dislocation effects. Therefore the terms τ_{inert} and τ_{drag} in Eq. (2.18) and those involving time derivatives in Eq. (2.20) can be eliminated from the respective equations. Because point obstacles are being dealt with, τ_{obst} vanishes everywhere except at the centers of the obstacles. There they exert $-F_0$ on the dislocation. In the range $|x| < L/2$—that is, for the arc AB (Figure 4.13)—Eq. (2.20) reduces to

$$b\tau_{ext} = -S(\partial^2 y/\partial x^2) \tag{4.20}$$

Because the dislocation is supposed to bow out only slightly (supposition 5), the line tension S is constant. Due to the uniformity of the particle spacing, breaking through one obstacle is equivalent to breaking through all of them; that is, the situation shown in Figure 4.13 is representative of the entire glide plane. The boundary conditions are

$$y(x = -L/2) = y(x = +L/2) = 0$$

The solution of the differential Eq. (4.20) is

$$y(x) = \frac{b\tau_{ext}}{8S}(L^2 - 4x^2) \tag{4.21}$$

To implement Eq. (4.19), the areas under the dislocation arcs have to be calculated first. The area between the x-axis and an arc is referred to by A_{AB}, where the subscripts A and B indicate the base line of the arc. Because the lengths AB and BC are equal, the same holds for the areas A_{AB} and A_{BC}. For A_{AB} one obtains

$$A_{AB}(\tau_{ext}) = \int_{-L/2}^{L/2} y\,dx \tag{4.22}$$

τ_{ext} is quoted as argument of A_{AB} in order to emphasize that A_{AB} is governed by τ_{ext}. The results for the areas A_{AB} and A_{AC} are

$$A_{AB} = \frac{b\tau_{ext}}{12S}L^3 \tag{4.23a}$$

and

$$A_{AC} = \frac{b\tau_{ext}}{12S}(2L)^3 = 8A_{AB} \tag{4.23b}$$

Equation (4.23b) follows from Eq. (4.23a) by substituting $2L$ for L. The area A_F, which the dislocation sweeps out after breaking free from obstacle B, is given by

$$A_F = A_{AC} - A_{AB} - A_{BC}$$

In the moment of the breakthrough, τ_{ext} and L equal the CRSS τ_{pF} and the critical length L_{cF}, respectively; the subscript F stands for "Friedel":

$$A_F = 6A_{\text{AB}} = \frac{b\tau_{pF}}{2S} L_{cF}^3 \tag{4.24}$$

This is combined with Eqs. (4.19) and (3.1a):

$$A_F n_s = 1$$

$$A_F = 1/n_s = L_{\min}^2$$

and

$$L_{\min}^2 = A_F = \frac{b\tau_{pF}}{2S} L_{cF}^3$$

$$L_{cF} = \left(\frac{2SL_{\min}^2}{b\tau_{pF}}\right)^{1/3} \tag{4.25}$$

Inserting this into Eq. (4.14) yields

$$\tau_{pF} b \left(\frac{2SL_{\min}^2}{b\tau_{pF}}\right)^{1/3} = F_0$$

$$\tau_{pF} = \frac{2S}{bL_{\min}} \left(\frac{F_0}{2S}\right)^{3/2} \tag{4.26a}$$

F_0 may also be expressed in terms of S and $\cos(\phi_c/2)$ [Eq. (4.8a)]:

$$\tau_{pF} = \frac{2S}{bL_{\min}} [\cos(\phi_c/2)]^{3/2} \tag{4.26b}$$

Or if τ_{pF} is replaced by the reduced CRSS τ_{pF}^* defined in Eq. (4.18), one obtains

$$\tau_{pF}^* = [\cos(\phi_c/2)]^{3/2} \tag{4.26c}$$

Inserting for $L_{\min}$ from Eq. (3.3c) yields the following instead of Eq. (4.26a):

$$\tau_{pF} = \frac{1}{b} \frac{F_0^{3/2}}{(2S\pi\omega_q)^{1/2}} \frac{f^{1/2}}{r} \tag{4.26d}$$

where r is the average radius of the particles, f their volume fraction, and ω_q the statistical coefficient defined in Eq. (3.5d).

Evidently τ_{pF} increases with F_0 and $\cos(\phi_c/2)$ more rapidly than linearly. The reason is that strong obstacles require a high stress to overcome them and the higher the stress is, the more strongly bent is the dislocation and the more obstacles are interacting with it simultaneously (Figure 1.5). This has already been mentioned in Section 1.3. In Figure 4.12, τ_{pF}^* is plotted versus $\phi_c/2$. Because slight bowing of the dislocation has been supposed (supposition 5 in Section 4.2.1.3.1), $\phi_c/2$ should exceed 60° [relation (4.9b)]. In spite of this limitation, τ_{pF}^* of Eq. (4.26c) covers the whole range $0° \leqslant \phi_c/2 \leqslant 90°$ in Figure 4.12a. For $\phi_c/2 > 80°$, Eq. (4.26c) is in excellent agreement with the results of the computer simulations, which will be discussed in Section 4.2.2.1. The latter were performed under the same assumptions under which Eq. (4.26c) had been deduced and which have been compiled in Section 4.2.1.3.1, *except* that the following two assumptions were not necessary in the computer simulations:

3. Uniform spacing of the point obstacles (i.e., fluctuations in the obstacle spacings are disregarded; hence the configuration shown in Figure 4.13 is representative of the entire glide plane)
4. Steady propagation of the dislocation

It is remarkable that the simulations and Eq. (4.26c) lead to the same results for large $\phi_c/2$. For the smallest values of $\phi_c/2$, i.e. for very strong obstacles, τ_{pF}^* of Eq. (4.26c) is about 20% too high. It will be shown in Section 4.2.1.4.1 that this is due to assumption 3. Except at $\phi_c/2 = 0°$, τ_{pF}^* of Eq. (4.26c) is below τ_p^* of Eq. (4.16c), which gives an upper limit for τ_p^*. For $\phi_c/2 = 0°$, both equations yield the same result for τ_p^*. If $\phi_c/2$ is of medium magnitude, the correct choice of S is important. In the derivations of Eqs. (4.26) as well as in the computer simulations, S was assumed to be constant. This may lead to deviations. The effects of S on τ_p will be discussed in Section 4.2.1.4.3.

In general, F_0 and S depend on the character of the dislocation. S of edge dislocations is smaller than that of screw dislocations (Figure 2.2). The actual CRSS is τ_p of that type of dislocation that requires the highest stress. This point has already been discussed in Section 4.2.

Fusenig and Nembach [4.13] measured the CRSS of copper single crystals that had been strengthened by coherent misfitting cobalt-rich particles. Some of their data are reproduced in Figure 4.14. There τ_p is plotted versus $\sqrt{f}$, where f is the particles' volume fraction. The mean particle radii r of all specimens were the same within $\pm 1.6\%$. Evidently τ_p is proportional to $\sqrt{f}$. This is in agreement with Eq. (4.26d). In this comparison the slight variation of $\sqrt{S}$ with f via the dislocation's outer cutoff radius R_0 is disregarded. This will be discussed at the end of Section 4.2.1.4.3. The dependence of τ_p on r cannot be analyzed at the present stage because the averaging procedures involved in deriving the function $\tau_p(r, f = \text{const.})$ are rather complex. They will be presented in Section 5.4.2. It can be concluded that Eq. (4.26d) describes the variation of τ_p with f well.

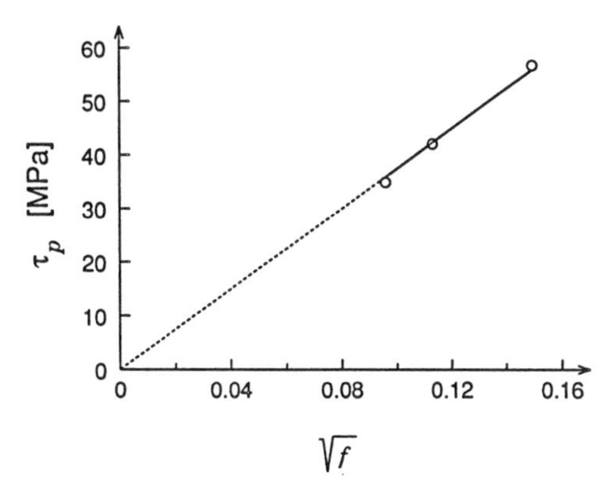

Fig. 4.14. CRSS τ_p of copper single crystals hardened by cobalt-rich particles, which have a lattice mismatch, versus $\sqrt{f}$, $T_D = 283\,\text{K}$. f is the particles' volume fraction, and their mean radii were between 2.83 nm and 2.92 nm. τ_p has been derived from the total CRSS τ_t with the aid of Eq. (5.58) with k equal to 1.25. (After Ref. 4.13.)

The critical length L_{cF} is found by inserting τ_{pF} of Eq. (4.26a) into Eq. (4.25):

$$L_{cF} = \left[\frac{2SL_{\min}^2}{\dfrac{2S}{L_{\min}} \left(\dfrac{F_0}{2S} \right)^{3/2}} \right]^{1/3}$$

$$L_{cF} = L_{\min} \left(\frac{2S}{F_0} \right)^{1/2} \tag{4.27a}$$

With Eq. (4.8b), one obtains

$$L_{cF} = L_{\min} \frac{1}{[\cos(\phi_c/2)]^{1/2}} \tag{4.27b}$$

or with Eq. (4.26c) one gets

$$L_{cF} = L_{\min}/(\tau_{pF}^*)^{1/3} \tag{4.27c}$$

L_{cF} has to stay within the limits given in relation (4.15): $L_{\min} \leqslant L_{cF} \leqslant L_{\max}$. According to Eqs. (4.27b) and (4.27c), in which F_0 has been replaced by $[2S\cos(\phi_c/2)]$, L_{cF} will not be smaller than $L_{\min}$. But Eq. (4.27a) does not guarantee this. If F_0 is sufficiently high, the Orowan process (Chapter 6) may operate. Moreover, L_{cF} of Eqs. (4.27) may exceed $L_{\max}$ for very small values of F_0 and $\cos(\phi_c/2)$. Therefore the limits for L_{cF} are stated explicitly:

$$L_{\min} \leqslant L_{cF} \leqslant L_{\max} \tag{4.27d}$$

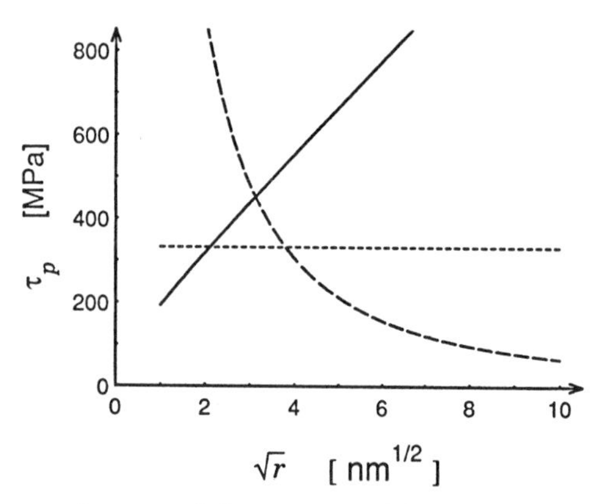

Fig. 4.15. CRSS τ_p versus the root of the mean particle radius r. $f = 0.1$, $b = 0.25\,\text{nm}$, and $\gamma = 0.25\,\text{J/m}^2$. ——Friedel's equation [Eq. (4.26a)], $K_S = 3.4\,\text{GPa}$; --- upper bound, Eq. (4.16a); — — — Orowan's equation [Eq. (6.14d)], $Y = 0.9$ and $K_{Eg} = 6.7\,\text{GPa}$.

If F_0 is large, L_{cF} is close to its minimum value $L_{\min}$ and τ_{pF} assumes its upper limit given by Eq. (4.16a). This limitation of τ_{pF} is written as

$$\tau_{pF} \leqslant \tau_p \text{ of Eq. (4.16a)}$$

During an Ostwald ripening treatment, F_0 normally increases; hence by virtue of Eq. (4.27a) the ratio $L_{cF}/L_{\min}$ decreases and τ_{pF} approaches its upper limit.

Because Eqs. (4.26) and (4.27) rest on the supposition that dislocation bowing is only slight (supposition 5 in Section 4.2.1.3.1), relation (4.9a) requires that $\cos(\phi_c/2)$ does not exceed 0.5. Inserting this limit into Eq. (4.27b) yields

$$L_{cF} \geqslant \sqrt{2} L_{\min} \tag{4.27e}$$

The ranges in which Eqs. (4.16a) (upper bound), (4.26a) (Friedel's equation), and (6.14d) (Orowan's equation) are applicable are illustrated in Figure 4.15. There the CRSS τ_p is plotted versus the root of the average particle radius r. The material is assumed to be subjected to Ostwald ripening (Section 3.3.1.2): The precipitated volume fraction f stays constant, but r increases. $L_{\min}$ is proportional to r [Eq. (3.3c)]. The following parameters have been inserted:

General:

$f = 0.1$
$1\,\text{nm} \leqslant r \leqslant 100\,\text{nm}$
$b = 0.25\,\text{nm}$
The statistical factors ω_r and ω_q are those listed in Table 3.2 for g_{WLS}.

Friedel's equation [Eq. (4.26a)]:	The particles are sheared, L_c is larger than $L_{\min}$. $F_0 = 2\omega_r r\gamma$ K_S (Section 2.2) $R_0 = L_{\min}$ = outer cutoff radius in the dislocation line tension S (Section 2.2) $R_i = b$ = inner cutoff radius in S (Section 2.2) Because F_0 enters Eq. (4.26a) as $F_0^{3/2}$, the effect of r on F_0 outweighs that on $L_{\min}$: τ_p increases with r.
Upper bound [Eq. (4.16a)]:	The particles are sheared, L_c equals $L_{\min}$. $F_0 = 2\omega_r r\gamma$ The effect of r on F_0 just cancels that on $L_{\min}$: τ_p is independent of r.
Orowan's equation [Eq. (6.14d)]:	The Orowan [4.44] process operates, the particles are bypassed. K_{Eg} [Eq. (6.12b)] τ_p is roughly proportional to $1/L_{\min}$: τ_p decreases as r increases.

The CRSSs derived from these three equations are shown in Figure 4.15. The actual CRSS at given r is represented by the lowest curve. The three equations describe the actual CRSS in the following approximate ranges of r:

Friedel's equation [Eq. (4.26a)]:	$1\,\text{nm} \leqslant r \leqslant 4\,\text{nm}$
Upper bound [Eq. (4.16a)]:	$4\,\text{nm} < r \leqslant 15\,\text{nm}$
Orowan's equation [Eq. (6.14d)]:	$15\,\text{nm} < r$

Equation (4.26a) had been based on the supposition that dislocations bow out only slightly. Therefore in the present case, relations (4.9) require that r does not exceed 2 nm. Because here only an exemplary comparison is given, the violation of the conditions (4.9) is of no consequence. If K_{Eg} were much lower, the curves representing Friedel's equation [Eq. (4.26a)] and Orowan's equation [Eq. (6.14d)] would intersect below the horizontal line, which indicates the upper bound for the CRSS [Eq. (4.16a)]. In this case, Eq. (4.16a) would not be applicable in any range of r. Evidently the maximum CRSS is either given by Eq. (4.16a) or by the intersection of Friedel's and Orowan's curves, whichever is less. Three states of aging have been defined in Section 1.4:

Underaged:	$r < r_{\max}$
Peak-aged:	$r = r_{\max}$
Overaged:	$r > r_{\max}$

r_{max} is that mean particle radius that yields the maximum CRSS. As in Figure 4.15, r_{max} may stand for a range of r. The three states of aging are very distinct in Figure 4.15 and in the experimental data presented in Figure 1.6. There is no conformity in the literature about the definition of "peak-aged." This expression may refer either to the maximum CRSS in general or to that part of the $\tau_p(r, f = \text{const.})$-curve where Eq. (4.16a) holds. As the curve representing the underaged state approaches either of the two other curves shown in Figure 4.15, the dislocations are strongly bent and Friedel's equations [Eqs. (4.26)] have to be amended. It depends on the specific parameters of the particle-hardened material under consideration whether the actual maximum CRSS is given by either (1) the intersection of the curves representing Eq. (4.26a) (particle shearing: $L_c > L_{min}$, underaged state) and Eq. (6.14d) (circumventing of particles, overaged state) [4.45–4.47] or (2) Eq. (4.16a) (particle shearing: $L_c = L_{min}$). The lower one of the two stresses following from 1 and 2 is the actual maximum CRSS. In Section 5.5.4.2 the nickel-base superalloy NIMONIC PE16 will be discussed. The maximum CRSS of this material is in agreement with Eq. (5.70), which is based on Eq. (4.16a).

If the coherent particles have an appreciable lattice mismatch, they may become semicoherent (Section 3.2.1.1) and finally incoherent during aging. The ensuing effects have been demonstrated by Orlova et al. [4.48] for an austenitic steel strengthened by carbides; Brown and Ham [4.49] gave a theoretical analysis of these phenomena.

At least in theory, even particle shearing may lead to negative derivatives $\partial\tau_p/\partial r$. This happens, for example, if F_0 in Friedel's equation [Eq. (4.26d)] is proportional to r^x with x less than 2/3. Examples are Eqs. (5.6) and (5.25). Therefore in the following the term "overaged" will refer to all states of aging in which f is constant and $\partial\tau_p/\partial r$ and $\partial\tau_p/\partial t$ are negative. t is the duration of the aging treatment. In the underaged state, $\partial\tau_p/\partial r$ and $\partial\tau_p/\partial t$ are positive. The quoted limit 2/3 for x holds only if the slight variation of $\sqrt{S}$ with r via the dislocation's outer cutoff radius R_0 is disregarded. The dependence of R_0 on r will be discussed at the end of Section 4.2.1.4.3.

If $\phi_c/2$ is of medium magnitude, Eqs. (4.26) may show deficiencies because the variation of S with the strength of the bow-out of the dislocation has not been allowed for. This point will be considered in Section 4.2.1.4.3.

4.2.1.4 Advanced Analytical Solutions of Eq. (2.20). There have been attempts to dispense with at least some of the restrictive suppositions on which Friedel's model has been based (Section 4.2.1.3.1). These refinements concerned the arrangement of the particles in the glide plane, their finite size, and to a minor extent the supposition that the dislocation should be only slightly bent. "Slightly bending" implies that the dislocation line tension can be considered as constant. Moreover, Schwarz [4.9] treated dynamic dislocation effects analytically (Section 4.2.1.4.4).

4.2.1.4.1 Random Distribution of Point Obstacles in the Glide Plane. Kocks [4.29, 4.50–4.52] and Hanson and Morris [4.39] eliminated Friedel's suppositions 3 and 4 listed in Section 4.2.1.3.1. These authors considered randomly instead of uniformly spaced (supposition 3) point obstacles and did not assume that the dislocation motion is steady (supposition 4). The implications of "randomly" and "uniformly" need some clarification. Friedel considered only the mean interobstacle spacing and disregarded a possible distribution of spacings. This is equivalent to having uniformly spaced obstacles. Therefore all of them can be represented by A, B, and C sketched in Figure 4.13. In the present section the distribution of the interobstacle spacings of a random planar array of point obstacles is allowed for. The statistical approach of Kocks and that of Hanson and Morris are entirely different.

Kocks' [4.50, 4.51] treatment of impenetrable (i.e., nonshearable) point obstacles is presented first because it is more easily understood than that of penetrable (i.e., shearable) obstacles. Let the spacing of two obstacles be L. Then the minimum radius of curvature of a dislocation bowing out *between* them equals $L/2$. Inserting this into Eq. (2.11a), one obtains the stress τ needed to make the dislocation pass between the impenetrable obstacles:

$$\tau = \frac{2S}{bL} \tag{4.28}$$

S is the line tension of the dislocation. Equation (4.28) yields a rather crude estimate of the Orowan stress; more sophisticated relations will be presented in Section 6.1.

Kocks connected neighboring obstacles in the glide plane by lines, which he referred to as "links." They are dashed in Figure 4.16. These links divide the plane into cells. A link for which τ derived from Eq. (4.28) is below the actually applied stress τ_{ext} will be called transparent. Kocks defined the following parameters:

$P(\tau_{ext})$	Probability that a given link is transparent under the applied stress τ_{ext}
N_l	Number of links, per unit volume, which are occupied by dislocation line; in Figure 4.16 these are the links 1, 2, and 3
A_k	Area swept out by the dislocation once it breaks through a link; after a breakthrough the dislocation may traverse several cells
dA	Total area swept out by dislocations once τ_{ext} is raised by $d\tau_{ext}$
a_c	Mean area of a cell
$i+1$	Number of sides of a cell
γ	Shear strain
V	Volume of the specimen

Raising τ_{ext} by $d\tau_{ext}$ produces the additional shear strain $d\gamma$:

$$d\gamma = b\,dA/V \tag{4.29a}$$

This follows directly from Eq. (4.10). Substituting $[A_k N_l dP]$ for dA/V, one gets

$$d\gamma = bA_k N_l dP \tag{4.29b}$$

There is an alternative derivation of $d\gamma$. Once the dislocation breaks for instance through link 2 in Figure 4.16, it will touch links 5 and 6. The number, per unit volume,

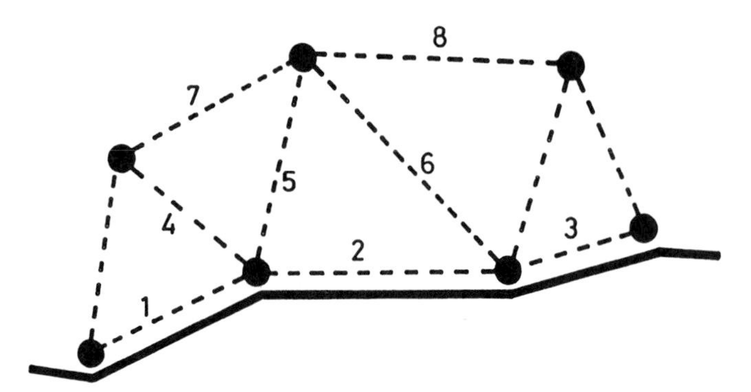

Fig. 4.16. Schematic sketch of point obstacles in the glide plane. The "links" connecting neighboring obstacles are dashed. The drawn-out line indicates a dislocation. (After Ref. 4.51.)

of newly contacted links is dN_l. They may have been transparent already before the raise in stress. Thus there are two contributions to the number of links through which the dislocation breaks after the raise in stress:

$$d\gamma = ba_c(N_l dP + P dN_l) \tag{4.30}$$

dN_l is related to i and dA:

$$dN_l = i\frac{dA}{a_c}\frac{1}{V}$$

Substituting for dA/V from Eq. (4.29a), this becomes

$$dN_l = i\frac{1}{a_c b}d\gamma \tag{4.31}$$

Equations (4.29b), (4.30), and (4.31) can be solved for A_k:

$$A_k = \frac{a_c}{1 - iP} \tag{4.32}$$

Evidently there is a critical value P_c of P:

$$P_c = 1/i \tag{4.33}$$

For P approaching P_c, the area A_k, which a dislocation sweeps out once it has broken through a link, diverges; that is, at P_c the dislocation traverses its entire glide plane. P_c is related to the CRSS τ_p:

$$P(\tau_p) = P_c \tag{4.34}$$

This answers the question whether there is a critical stress beyond which the dislocation can glide over unlimited areas. Kocks [4.50] has pointed out that this result is by no means trivial for a random array of obstacles.

It remains to derive the function $P(\tau_{ext})$. First Kocks showed that $P(\tau_{ext})$ has the appearance of a step function. P is small as long as τ_{ext} is below a critical value; as τ_{ext} approaches the critical value, P increases sharply to unity. Kocks [4.50] demonstrated this by arranging 550 points at random in a plane and marking the links by lines. Then he eliminated those links which were transparent at the given stress τ_{ext1} and drew heavy border lines between the link-free and the "linked" areas. Reduced stresses τ^*_{ext} are used, and they are defined in analogy to τ^*_p [Eq. (4.18)]:

$$\tau^*_{ext} = \tau_{ext} b L_{min}/(2S) \tag{4.35}$$

At $\tau^*_{ext1} = 0.74$, there was a continuous impenetrable region that enclosed link-free islands. A dislocation would not have been able to glide over long distances. At $\tau^*_{ext1} = 0.90$, more than half of the array was link-free. But the different large link-free areas were still separated from each other by "linked" areas. At $\tau^*_{ext1} = 1.04$, nearly the entire array was free of links. The heavy border lines enclosed only some very small "linked" areas. A dislocation loop emitted by a source could sweep out nearly the entire glide plane. This suggests that the reduced CRSS is between 0.90 and 1.0.

A statistical analysis led to the following conclusions:

$$i \approx 3 \tag{4.36}$$

and

$$P(\tau^*_{ext}) = \exp\left[-\frac{\pi}{4}\frac{1}{(\tau^*_{ext})^2}\right] \tag{4.37}$$

Combining Eqs. (4.33)–(4.37) yields for the reduced CRSS τ^*_{pK}, where the subscript K stands for "Kocks":

$$\tau^*_{pK} = 0.85 \tag{4.38}$$

This result agrees satisfactorily with those obtained in the computer simulations for $\phi_c/2$ close to zero (Figure 4.12a). Kocks [4.51] stressed that care has to be taken to insert the proper line tension S: that of a half-loop. Actually the expressions derived in Section 6.1 should be used. The importance of Eq. (4.38) lies in the statistical factor 0.85, which allows for the random arrangement of the point obstacles in the glide plane. In Eqs. (6.14b)–(6.14d), (6.16), and (6.17), which describe the Orowan [4.44] process, this factor is represented by Y.

Now one can proceed to shearable point obstacles [4.51]. Because only reduced stresses τ^*_{ext} [Eq. (4.35)] are considered, the variation of S with the bowing angle of the dislocation is not quite apparent. The probability that the dislocation can break through an obstacle if the stress τ_{ext} is applied will be denoted by $Q(\tau_{ext})$. N_o is the number, per unit volume, of obstacles occupied by dislocation line. Then Eq. (4.30) assumes the form

$$d\gamma = ba_c(N_l dP + N_o dQ + P dN_l + Q dN_o) \tag{4.39}$$

The terms involving N_o and Q are new: Raising τ_{ext} by $d\tau_{ext}$ allows the dislocation to break through some obstacles. Equations (4.29b) and (4.31) are replaced by a new equation, respectively, by a pair of equations:

$$d\gamma = bA_k N(dP + dQ) \tag{4.40}$$

$$dN_l = i\frac{1}{a_c b}d\gamma \tag{4.31}$$

$$dN_o = j\frac{1}{a_c b}d\gamma \tag{4.41}$$

The meaning of j is analogous to that of i. Because a dislocation is a line, the number of links is equal to the number of point obstacles on it:

$$N = N_l = N_o$$

The increase dN_l of the number of links encountered after a bow-out or after a breakthrough process and the increase in the number of points at which the angle ϕ changed after either process are not necessarily identical. Therefore Eq. (4.41) has to be considered. Equation (4.32) becomes

$$A_k = \frac{a_c}{1 - iP - jQ} \tag{4.42}$$

Kocks derived the function $Q(\tau_{ext})$ by a rather intricate statistical procedure. The results for the reduced CRSS τ_p^* are shown in Figure 4.12. For small breaking angles $\phi_c/2$—that is, for strong obstacles (Figure 4.12a)—Kocks' curve is the best representation of the data obtained in computer simulations. For large $\phi_c/2$—that is, for weak obstacles (Figure 4.12b)—Friedel's model agrees best with the data. If the obstacles are strong, bowing of the dislocation is also strong and the spatial distribution of the obstacles in the glide plane is important. For $\phi_c/2 = 0^0$, the results for τ_p^* are as follows [4.52]:

Friedel [Eq. (4.26c)]:	1.0
Kocks [Eq. (4.38)]:	0.85
Foreman and Makin [4.37, 4.38]:	0.81

The reduction of τ_p^* from 1.0 to 0.85 or 0.81 is due to the randomness of the obstacle arrays. In Section 6.1, improved formulae will be presented for the Orowan process. There too the factor $Y \approx 0.85$ has to be inserted. As stated above, Friedel's equations [Eqs. (4.26)] are in good agreement with the results of the computer simulations if $\phi_c/2$ is large—that is, if the obstacles are weak. In this range of $\phi_c/2$, the assumption that the dislocation line tension S is constant is justified. Therefore Friedel's equations [Eqs. (4.26)] are correct as they stand, if $\phi_c/2$ is large. For $\phi_c/2$ of medium magnitude, there remains a problem with S. If τ_p^* is quoted instead of τ_p, this problem is less apparent. It will be discussed further in Section 4.2.1.4.3. Because Eq. (4.16a) does not involve S, it suffices to multiply the right-hand side of Eq. (4.16a) by about 0.85 to allow for the randomness of the array. This will be done in Eq. (4.56).

Kocks [4.51] has pointed out that $\partial\tau_p^*/\partial\phi_c$ stays finite as $\phi_c/2$ approaches 0°. This point will be taken up in Section 4.2.2.1. Dorn et al. [4.53] have published a statistical treatment of unzipping (Section 4.2.1.3.1). Unless the obstacles were very strong, this group's results were close to those obtained by Kocks.

In their statistical analysis, Hanson and Morris [4.39] followed the procedures established in the computer simulations of the dislocation motion through a planar random array of shearable point obstacles of identical strength [4.37, 4.38, 4.54–4.56]. These procedures are described first. The glide plane is the plane $z = 0$; it extends from $-x_0$ to $+x_0$ and from $-y_0$ to $+y_0$. The coordinates of the point obstacles are chosen by a pseudo-random-number generator. The boundary conditions for solving Eq. (4.20) are $\partial y/\partial x = 0$ for $x = \pm x_0$. In contrast to Friedel's treatment, Eq. (4.20) must now be solved in the entire glide plane. Only if the obstacles are uniformly spaced, it suffices to consider the representative obstacle configuration shown in Figure 4.13. Because S and the applied stress τ_{ext} are postulated to be constant, the present solution $y(x)$ of Eq. (4.20) is a succession of circular arcs of identical curvature. Their radius R_l follows from Eq. (2.11b). Actually the term that involves S in Eq. (4.20) has to be based on the more accurate Eq. (2.16a) instead of on Eq. (2.16b). An example of a resultant dislocation configuration is sketched in Figure 4.17.

At the start of the simulation, τ_{ext} is zero and the dislocation is straight: $y(x) = -y_0$. After the application of a small stress τ_{ext1}, the dislocation assumes the configuration C1. It is constructed as follows. A circle of radius R_l is first centered at $(-x_0, -y_0 - R_l)$ and then slowly moved upwards along the line $x = -x_0$ until it touches an obstacle. This will be referred to as O_1. The circular arc connecting O_1 with the boundary at $-x_0$ is called A_1. Because the center of the circle lies on the boundary, A_1 fulfills the boundary condition $\partial y/\partial x = 0$ at $x = -x_0$. Now another circle with R_l is rotated counterclockwise around O_1 until it contacts another obstacle, called O_2. The arc A_2 connects O_1 with O_2. The angle $\phi_{1,2}$ included by A_1 and A_2 is calculated and compared with the critical angle ϕ_c, which characterizes the uniform strength of the obstacles. If $\phi_{1,2}$ is larger than ϕ_c, O_1 can withstand the forces exerted on it by A_1 and A_2 and the circle rolling process is continued around O_2, and so on. If the angle $\phi_{i,i+1}$ between the arcs A_i and A_{i+1} is smaller than ϕ_c, the dislocation breaks free from O_i. In this case, O_i is eliminated from the array and the circle rolling process resumed around O_{i-1} until a new obstacle O_i' is found. If finally the right boundary at $x = x_0$ is reached, the stable dislocation configuration C1 has been constructed. This will always be possible if τ_{ext1} is small and thus R_l is large. After a stable configuration has been found, τ_{ext} is raised by a small increment and the described procedure is repeated. The highest stress τ_{ext} that still leads to a stable configuration is identified with the

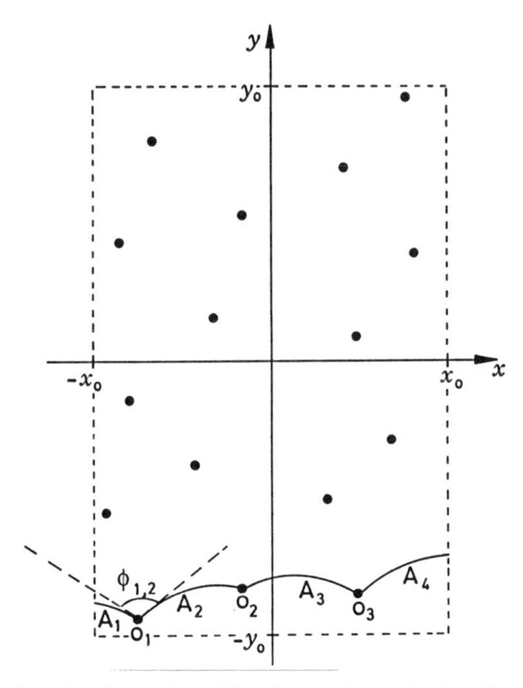

Fig. 4.17. Point obstacles O_i in the glide plane, schematically. The dislocation forms circular arcs A_i between them. It assumes this configuration C1 under the applied stress $\tau_{ext\,1}$.

CRSS τ_p. Provisions have to be made so that the dislocation does not intersect itself. The strongest stable dislocation configuration will be referred to as CMAX.

Hanson and Morris analyzed the circle rolling process and the search for CMAX within the framework of classical probability theory. They treated this process as a branching process [4.57], derived an estimate of the reduced CRSS τ^*_{pHM} defined in Eq. (4.18), and calculated the mean obstacle spacing L_{cHM} along CMAX. The subscripts HM of τ^*_{pHM} and of L_{cHM} stand for "Hanson and Morris":

$$\tau^*_{pHM} = 0.887[\cos(\phi_c/2)]^{3/2} = 0.887\tau^*_{pF} \tag{4.43}$$

$$L_{cHM} = L_{\min}\frac{0.764}{[\cos(\phi_c/2)]^{1/2}} \tag{4.44a}$$

Or with Eq. (4.43) one obtains

$$L_{cHM} = 0.734L_{\min}/(\tau^*_{pHM})^{1/3} \tag{4.44b}$$

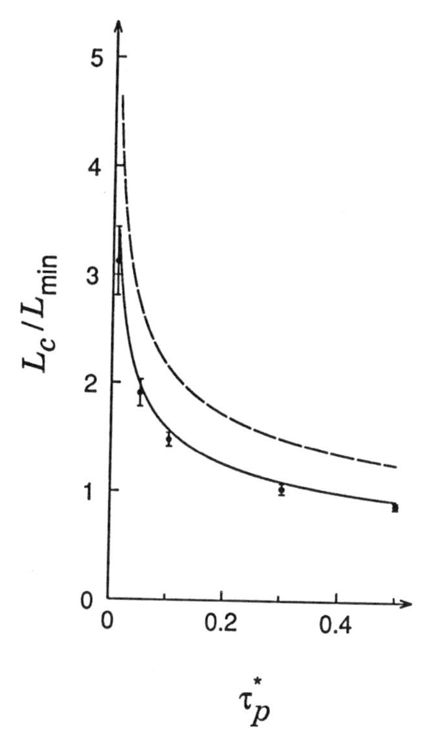

Fig. 4.18. $L_c/L_{\min}$ versus the reduced CRSS τ_p^*. L_c is the mean obstacle spacing along the strongest dislocation configuration CMAX. Data points and drawn-out curve: $L_{cHM}/L_{\min}$ versus τ_{pHM}^* of Eqs. (4.44b) and (4.43). Dashed curve: $L_{cF}/L_{\min}$ versus τ_{pF}^* of Eqs. (4.27c) and (4.26c). The data points are the results of Hanson and Morris' [4.39] computer simulations, and the data bars indicate the scatter of 10 arrays with 10,000 obstacles each.

All three equations are meant only for weak obstacles—that is, for large $\phi_c/2$ or for small τ_{pHM}^*. Therefore the assumption that S is constant is justified. τ_{pHM}^* is 11% smaller than Friedel's result τ_{pF}^* [Eq. (4.26c)]. Labusch [4.58] has criticized Hanson and Morris' treatment of the degeneracy problem.

The critical length L_{cHM} must be between $L_{\min}$ and $L_{\max}$ [relation (4.15)]. Friedel's length L_{cF} of Eqs. (4.27) had been subjected to the same limitation. $L_c/L_{\min}$ is plotted versus τ_p^* in Figure 4.18. Equation (4.44b) represents the results of Hanson and Morris' computer simulations very well, whereas Friedel's equation [Eq. (4.27c)] yields values that are too high.

Except for the numerical factors, Eqs. (4.43), (4.44a), and (4.44b) are identical with Eqs. (4.26c), (4.27b), and (4.27c), respectively, which had been derived on the basis of Friedel's model. In Figure 4.12, τ_p^* is plotted versus $\phi_c/2$. For $\phi_c/2$ above 80°, Friedel's equation [Eq. (4.26c)] represents the results of the computer simulations somewhat better than does Hanson and Morris' equation [Eq. (4.43)]. But at lower breaking angles, Eq. (4.43) is definitely a better representation of the data than is Eq. (4.26c). Though Eq. (4.43) had been derived under the supposition that $\phi_c/2$ is large, the curve representing Eq. (4.43) has been extended down to $\phi_c/2 = 0°$ in Figure 4.12a; the analogue holds for Eq. (4.26c).

With the only exception of the curve representing Eq. (4.16c), all data and curves shown in Figure 4.12 are based on the assumption that S is constant, even though the dislocation may bow out strongly. As stated above, this may lead to deficiencies of the results, if $\phi_c/2$ is of medium magnitude. None of the three models discussed so far represents the data best over the entire range of $\phi_c/2$. For different ranges, different models have to be chosen:

$80° < \phi_c/2$:	Friedel's model, Eq. (4.26c)
$30° \leqslant \phi_c/2 \leqslant 80°$:	Kocks' or Hanson and Morris' model, Eq. (4.43)
$\phi_c/2 < 30°$:	Kocks' model

Friedel considered uniformly spaced point obstacles, whereas the other authors studied random arrays. The results communicated above lead to the conclusion that for $\phi_c/2$ below about 80°, the CRSS of a random array of point obstacles is smaller than that of a uniform one. The TEM work described in Section 3.3.1.5 indicated that precipitates that have been subjected to Ostwald ripening are actually not randomly distributed in space. There are neither very closely nor very widely spaced nearest neighbors. Only as long as the precipitation process is governed by homogeneous nucleation and hardly any Ostwald ripening has taken place, the spatial distribution of precipitates may be truly random.

Both τ^*_{pHM} of Eq. (4.43) and L_{cHM} of Eq. (4.44a) are smaller than the respective values calculated from Friedel's equations [Eqs. (4.26c) and (4.27b)]. This seems to be hard to reconcile with Eq. (4.14), which relates the product $\tau_p L_c$ to F_0. This apparent discrepancy is due to the fact that in the case of a uniform array of point obstacles, all of them exert the maximum force F_0 in the critical moment, whereas in the case of a random array not all obstacles are stressed up to their maximum capacity F_0. Therefore the more densely spaced obstacles of the latter case yield a lower CRSS.

4.2.1.4.2 Finite Size of the Obstacles. Ham [4.59] eliminated Friedel's supposition that the obstacles should be point obstacles and allowed to some extent for the variation of the line tension S with the character of the dislocation. This concerns the suppositions 1 and 5 listed in Section 4.2.1.3.1. Figure 4.13 is replaced by Figure 4.19. The area A_H swept out by the dislocation once it breaks through the central obstacle, is shown hatched in Figure 4.19. The subscript H of A_H, τ_{pH}, and L_{cH} refers to "Ham." The position of the dislocation inside of the central obstacle depends on the interaction mechanism and on the line tension S. In the critical moment, L equals L_{cH}. r_r and q are the mean radius and the mean area, respectively, of the intersection

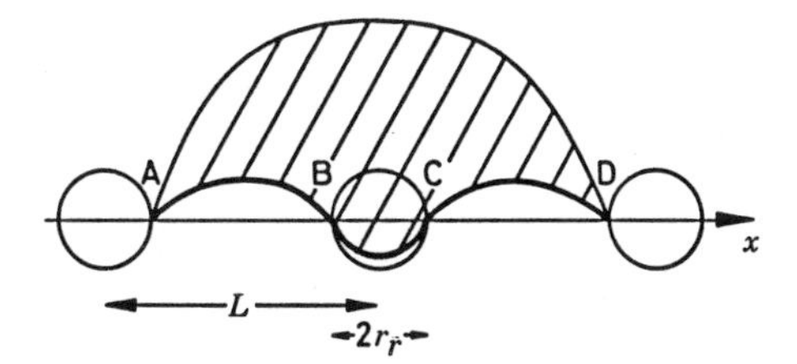

Fig. 4.19. A dislocation (thick portion) bows out between three equidistant, collinear, shearable obstacles of finite, but identical, size. The area A_H is shown hatched. (After Ref. 4.59.)

between the glide plane and particles. r_r and q have been defined in Section 3.1. Ham did not distinguish between r_r and $\sqrt{q/\pi}$(Section 3.1).

As in Section 4.2.1.3.2, the area between the x-axis and the dislocation arc over the length AB is designated A_{AB}. Thus one gets

$$A_H = A_{\mathrm{AD}} + A_{\mathrm{BC}} - A_{\mathrm{AB}} - A_{\mathrm{CD}} \tag{4.45}$$

A_{BC} lies below the x-axis. Because the obstacles are supposed to be uniformly spaced (supposition 3 in Section 4.2.1.3.1), A_{AB} is the same as A_{CD}. Equation (4.45) refers to the critical moment when the dislocation breaks free from the central obstacle. Therefore τ_{ext} and L equal the critical parameters τ_{pH} and L_{cH}, respectively.

First a simplified version of Ham's derivations is given. The following additional assumptions and approximations are made: A_{BC} is disregarded, S is assumed to be constant, and the ratio $[2r_r/L_{cH}]$ is approximately by the volume fraction f, which is assumed to be small. Recalling Eq. (4.23a), one gets the following for the critical configuration:

$$A_H = \frac{b\tau_{pH}}{12S}[(2L_{cH} - 2r_r)^3 - 2(L_{cH} - 2r_r)^3] \tag{4.46a}$$

Or after substituting f for $2r_r/L_{cH}$ one obtains

$$A_H = \frac{b\tau_{pH}}{12S}[8L_{cH}^3(1 - f/2)^3 - 2L_{cH}^3(1 - f)^3]$$

and after linearization one has

$$A_H = \frac{b\tau_{pH}}{12S}[8L_{cH}^3(1 - 3f/2) - 2L_{cH}^3(1 - 3f)]$$

$$A_H = \frac{b\tau_{pH}}{2S}L_{cH}^3(1 - f) \tag{4.46b}$$

From here on, the derivation of Eq. (4.26a) is followed; A_H replaces A_F. Because τ_{pH} acts on the entire length L_{cH} of the dislocation and not only on the length $[L_{cH} - 2r_r]$, L_{cH} has to be inserted into Eq. (4.14). The result is

$$\tau_{pH} = \frac{1}{bL_{\min}} \frac{F_0^{3/2}}{(2S)^{1/2}} (1 - f)^{1/2} \tag{4.47a}$$

$$\tau_{pH} = \frac{2S}{bL_{\min}} \left(\frac{F_0}{2S}\right)^{3/2} (1 - f)^{1/2} \tag{4.47b}$$

Or after another linearization one obtains

$$\tau_{pH} = \frac{2S}{bL_{\min}} \left(\frac{F_0}{2S}\right)^{3/2} (1 - f/2) \tag{4.47c}$$

Evidently, here only the obstacles' finite extension in the x-direction (Figure 4.19) comes into play. It leads to a reduction of the CRSS. Because the dislocation cannot bow out over the entire length L_{cH} but only over the length $L_{cH} - 2r_r$, the dislocation appears to be more stiff. In Section 4.2.1.2 it has been deduced that the CRSS decreases as the stiffness of the dislocation increases.

Ham [4.59] correctly allowed for the variation of S with θ_d, which is the angle between the dislocation line and its Burgers vector. In order to calculate the area A_{BC}, the dislocation–particle interaction had to be specified. Ham made the following assumptions: (1) There is small bow-out of the dislocation, and therefore the function $S(\theta_d)$ can be linearized. In all other models presented so far, S has been assumed to be constant. (2) There is elastic isotropy with $\nu = 1/3$. (3) The obstacles are long-range-ordered precipitates, and their specific antiphase boundary energy is γ. Ham's result for A_H of a screw dislocation is

$$A_H = \frac{2}{E_l(\theta_d = 0°)} \frac{(\gamma r_r)^3}{(\tau_{\text{ext}} b)^2} (1 - 2r_r/L_{cH}) \tag{4.48}$$

$E_l(\theta_d = 0°)$ is the dislocation line energy of a screw dislocation. Equation (4.48) replaces Eq. (4.46b); the further deductions remain the same as above. In materials strengthened by shearable long-range-ordered precipitates, dislocations actually glide in pairs (Sections 4.1 and 5.5). Hence Eq. (4.48), which refers to a unit screw dislocation, is not quite to the point. The current state of modeling of order strengthening will be detailed in Section 5.5.

In general the obstacle–dislocation interaction force has finite ranges w_x and w_y in the x- and y-direction, respectively. Their effects on the CRSS are quite different. Those of w_y are demonstrated here for particles that repel the dislocation elastically. They may, for example, have a lattice mismatch. Those particles that lie within w_y ahead of the dislocation impede its glide. Those of them, which have already been passed by the dislocation but lie no more than

w_y behind it, drive it forward; that is, they enhance τ_{ext}. The strength of the ensuing effects are governed by the dimensionless parameter η_0 introduced by Labusch [4.8, 4.36, 4.60–4.63]:

$$\eta_0 = \frac{w_y}{L_{min}}\left(\frac{2S}{F_0}\right)^{1/2}$$

Or with the critical angle $\phi_c/2$ given in Eq. (4.8b) one obtains

$$\eta_0 = \frac{w_y}{L_{min}} \frac{1}{[\cos(\phi_c/2)]^{1/2}}$$

Evidently η_0 depends not only on geometric parameters like w_y and L_{min}, but also on the strength of the obstacles. The ratio w_y/L_{min} characterizes the angle that the dislocation makes by zigzagging through the array of obstacles. The ratio $F_0/2S$ is a measure of the angle through which the dislocation bows out in the critical moment.

The statistics that have to be applied to derive the CRSS depend on η_0. The two limiting cases are: η_0 approaching zero and η_0 approaching infinity. The respective statistics are connected with the names Friedel and Mott [4.8, 4.63].

Friedel regime: $\eta_0 \to 0$
The obstacle–dislocation interaction force is strong and short-ranged. The obstacles are point obstacles.

Mott regime: $\eta_0 \to \infty$
The obstacle–dislocation interaction force is weak and long-ranged. The obstacles are diffuse.

η_0 of individual solute atoms in a solid solution is large. Labusch [4.60, 4.61] derived an analytical expression for the CRSS of solid solutions. It is not presented here, but it will be quoted as Eq. (4.65) in connection with Schwarz and Labusch's [4.8] computer simulations.

η_0 of shearable particles tends to be small, and often it is far below unity. Examples will be given in Chapter 5. It can easily be rationalized that η_0 decreases during an Ostwald ripening treatment (Section 3.2.1.2), which raises the average radius r of the particles but leaves their volume fraction virtually constant. Let their maximum force F_0 and their range w_y be proportional to r. Then it follows from the above definition that η_0 is proportional to $1/\sqrt{r}$; that is, η_0 decreases during the ripening treatment. η_0 of solid solutions is normally large; it is lowered when the solute atoms form precipitates.

4.2.1.4.3 Dislocation Line Tension. This title implies a simplification of the actual problem: What is the shape of a dislocation bowing out under the resolved stress τ? The dislocation is supposed to stay in its glide plane. τ may comprise τ_{ext}, τ_{obst}, and perhaps the stress due to the interaction with other dislocations. The term "line tension" refers to deWit and Koehler's [4.64] concept. A general approach to an answer to the above question has been taken by Brown [4.65], who introduced the concept of the self-stress τ_{self}. In the state of static equilibrium, τ_{self} balances τ at any point of the dislocation line. τ_{self} is similar to τ_{back} defined in Section 2.3. There are several contributions to τ_{self}, and they have been mentioned in Section 2.2. If an initially straight dislocation is bent, its energy changes because its length, its character, and the self-interaction between its different parts change. The importance of the core contribution has been stressed by Schmid and Kirchner [4.66].

For later reference the various models to describe the resistance of dislocations against bending are numbered. θ_d is the angle between the dislocation segment $d\mathbf{s}$ and its Burgers vector:

1. Brown's concept of the self-stress τ_{self}. The elastic anisotropy may be allowed for.
2. deWit and Koehler's concepts of the dislocation line tension $S(\theta_d)$:

$$S_2 = S(\theta_d) \tag{4.49a}$$

   $S(\theta_d)$ has been defined in Section 2.2. The elastic anisotropy of the material can be taken into account. The core contribution can be allowed for by the choice of the inner cutoff radius R_i. Two expressions for the outer radius R_0 have been given in Eqs. (2.8). The self-interaction is disregarded.
3. Constant line tension:

$$S_3 = S(\theta_{d0}) \tag{4.49b}$$

   θ_{d0} is the angle θ_d before the dislocation bows out. Again the elastic anisotropy may be taken into account. If a dislocation bows out only slightly, S_3 is a good approximation.
4. The variation of S with θ_d is entirely disregarded:

$$S_4 = \mu_s b^2/2 \tag{4.49c}$$

   where μ_s is the shear modulus of the matrix.

Except for a numerical factor, the results for the CRSS are the same for S_3 and S_4. In Friedel's [4.40, 4.41] (Section 4.2.1.3) and in Hanson and Morris' [4.39] (Section 4.2.1.4.1) models, S has been assumed to be constant; that is, S_3 or S_4 has been inserted. According to Eq. (2.11b), the radius R_l of the dislocation's local curvature is proportional to S. If and only if S and τ are spatially constant, the bent dislocation forms a circular arc. Because no analytical expressions, but only numerical data, are available for τ_{self}, this concept cannot be incorporated into analytical solutions of Eqs. (2.18) and (2.20). In principle this is possible for the line tension S_2. Its variation with θ_d can be allowed for in such solutions. One may choose either Eq. (2.7) or Bacon and Scattergood's [4.67] Fourier series, which have been meant for elastically isotropic and anisotropic materials, respectively. The mathematical problems encountered in the further deductions of the CRSS are, however, prohibitive. In the derivations of Eqs. (4.48) and (5.69), $S_2(\theta_d)$ has been used, but linearization were necessary.

Some authors used appropriate averages over the function $S_2(\theta_d)$. Nembach and Neite [4.7] chose the effective line tension $S_{eff\ 1}$ for the superalloy NIMONIC PE16:

$$S_{eff\ 1} = [S(\theta_{d0} = 0°) \cdot S(\theta_{d0} = 90°)]^{1/2} \tag{4.50a}$$

This geometric mean is very close to the following average:

$$S_{eff\ 1} \approx \frac{2}{\pi} \int_0^{\pi/2} S(\theta_d)\, d\theta_d \tag{4.50b}$$

This has already been mentioned in Section 4.2. The elastic anisotropy has been allowed for in the calculation of S. Kocks et al. [4.68] inserted weights in the integral in Eq. (4.50b) and related the range over which the integration has to be extended to the critical angle $\phi_c/2$. For elastic isotropy the authors' results were

$$S_{eff\ 2}(\theta_{d0} = 0°) = S(\theta_{d0} = 0°) \sin^2(\phi_c/2) + E_l(\theta_{d0} = 90°) \cos^2(\phi_c/2) \tag{4.51a}$$

$$S_{eff\ 2}(\theta_{d0} = 90°) = S(\theta_{d0} = 90°) \sin^2(\phi_c/2) + E_l(\theta_{d0} = 0°) \cos^2(\phi_c/2) \tag{4.51b}$$

E_l is the dislocation line energy defined in Section 2.2. If the obstacles are weak, $\phi_c/2$ is close to 90°; in such cases $S_{eff\ 2}(\theta_{d0})$ approaches $S(\theta_{d0})$. In the other extreme case, where $\phi_c/2$ nearly vanishes, $S_{eff\ 2}(\theta_{d0} = 0°)$ nearly equals $E_l(\theta_{d0} = 90°)$ and $S_{eff\ 2}(\theta_{d0} = 90°)$ is close to $E_l(\theta_{d0} = 0°)$. These latter results will be discussed further in Section 6.1 in connection with the Orowan process.

Melander and Persson [4.69] started out with Eqs. (2.7) and wrote for the effective line tension parameter $K_{S\ eff}$ [Eqs. (2.7)]:

$$K_{S\ eff} = \frac{\mu}{4\pi(1-\nu)}[1 + \nu - 3\nu\, \varphi(\phi_c/2, \theta_{d0})] \tag{4.52}$$

μ and ν are the shear modulus and Poisson's ratio of the matrix, respectively. φ depends on $\phi_c/2$ and on θ_{d0}. Elastic isotropy has been assumed in the derivation of Eq. (4.52).

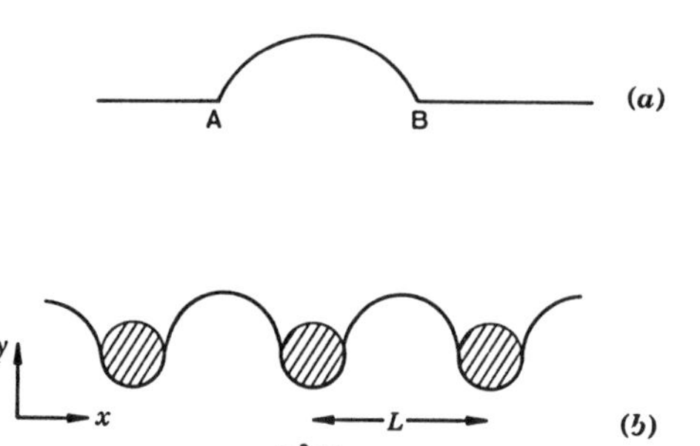

Fig. 4.20. Schematic diagram depicting a bent dislocation. (*a*) The dislocation length AB of an otherwise straight dislocation bows out. (*b*) The dislocation bows out between collinear, equidistant, impenetrable obstacles.

Here only some aspects of the results of more advanced treatments of bent dislocations are discussed. Most authors aimed at deriving the critical stress τ_c of a Frank–Read source [4.70, 4.71] or that of the Orowan process [4.44], but not the CRSS of materials containing shearable particles. τ_c of these two processes is the stress needed to make a dislocation segment pass between two points. The dislocation configuration found for stresses just below τ_c is the critical one. Two of the relevant situations are illustrated in Figure 4.20. In the following the homogeneous external stress τ_{ext} is assumed to be the only stress acting on the dislocation.

Foreman [4.72] applied Brown's self-stress concept to the situation sketched in Figure 4.20a. The configurations were numerically established by allowing the dislocation to relax to the state of equilibrium. Elastic isotropy has been assumed. Some of the results obtained for dislocations of edge, mixed, and screw character are presented in Figure 4.21. Three points warrant mentioning:

1. At given stress, the edge dislocation bows out more strongly than does the screw dislocation.
2. The area A_c that a dislocation sweeps out before it reaches the critical configuration is larger for the edge than for the screw dislocation.
3. The dislocation of mixed character does not bow out normal to its original direction.

Point 1 is in qualitative agreement with Eq. (2.11b), which is based on the line tension model: $S(\theta_d = 90°)$ is smaller than $S(\theta_d = 0°)$. All three results can also be understood if one recalls that for the dislocation line energies E_l the opposite holds: $E_l(\theta_d = 90°)$ is larger than $E_l(\theta_d = 0°)$. Consequently the dislocation avoids being of edge character and tries to align itself with its Burgers vector. Brown's self-stress and deWit and Koehler's line tension concept lead to similar critical configurations.

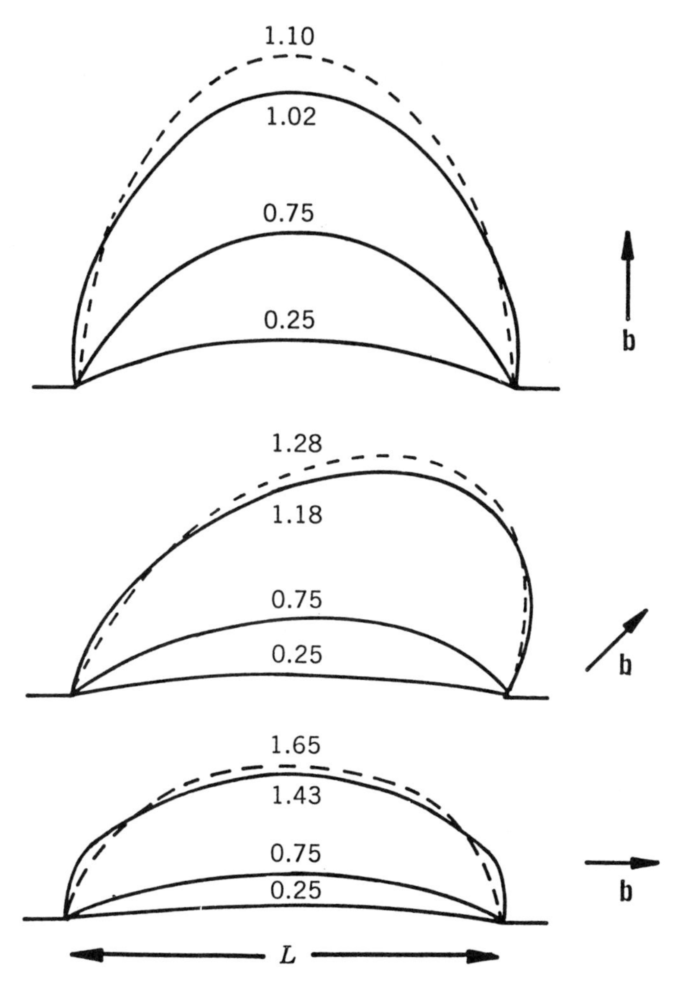

Fig. 4.21. An otherwise straight dislocation bows out between two points. The equilibrium configurations have been established on the basis of Brown's self-stress concept, and elastic isotropy is assumed; the applied stress is indicated in units of $\mu_s b/L$, $\nu = 1/3$, $L = 1000R_i$, and $R_i \sim b$. The most strongly bent, drawn-out configuration is the critical one. The broken lines indicate the critical configurations derived from de Wit and Koehler's line tension $S_2 = S(\theta_d)$ of Eq. (4.49a). The angles θ_{d0} of the originally straight dislocation are 90°, 45°, and 0°, from top to bottom. (After Ref. 4.72.)

Bacon et al. [4.73] and Scattergood and Bacon [4.74] investigated the Orowan process on the basis of the self-stress concept. These groups considered elastically isotropic and anisotropic crystals, respectively. The geometry is shown in Figure 4.20b. There the length L and the radius r are defined. The critical configurations have been found to be sensitive to the elastic anisotropy

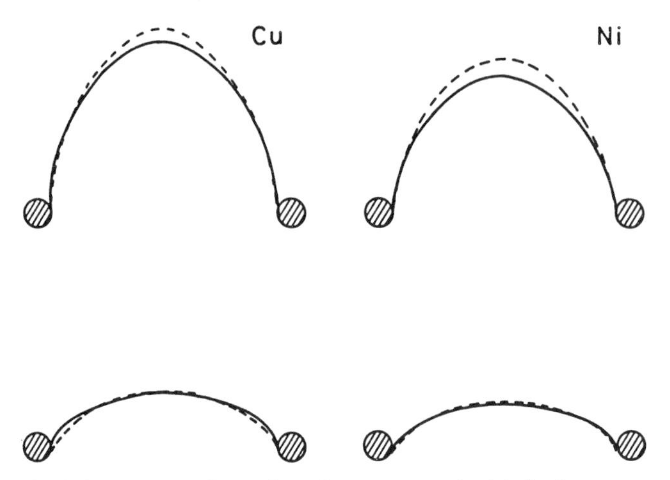

Fig. 4.22. Critical Orowan configurations in copper and nickel; the geometry is that shown in Figure 4.20b. The elastic anisotropy has been allowed for. $L = 1100R_i$, $2r = 100R_i$, and R_i = inner cutoff radius. *Upper row:* Edge dislocations. *Lower row:* Screw dislocations. ——Self-stress model, --- deWit–Koehler line tension model. (After Ref. 4.74.)

and to the parameter D:

$$\frac{1}{D} = \frac{1}{2r} + \frac{1}{L - 2r} \tag{4.53}$$

The CRSS τ_c as well as the critical area A_c increase with D. The elastic interaction between the two dislocation arms that touch the particles from the right and from the left side has profound effects on τ_c and on A_c. The smaller the value of r, the larger this attractive interaction. The critical configurations of copper and nickel are shown in Figure 4.22. The difference in shape reflects that in anisotropy. The anisotropy factors $[2C_{44}/(C_{11} - C_{12})]$ are 3.2 (copper) and 2.4 (nickel) [4.75–4.77]. The C_{ik} are the elastic stiffnesses. Scattergood and Bacon found that the critical configurations derived from the self-stress model and those based on the line tension $S_2 = S(\theta_d)$ defined in Eq. (4.49a) are nearly the same provided that $(L - 2r)/5$ is inserted for the dislocation's outer cutoff radius R_0. This good agreement is evident in Figure 4.22. This result will be discussed further in Section 6.1 when the Orowan process is analyzed in greater detail.

The effects of various parameters on the shape of dislocations in nickel are demonstrated in Figure 4.23. These configurations are solutions of Eq. (2.18) for an isolated dislocation segment of length L. Because the state of static equilibrium is concerned, only τ_{ext}, τ_{back}, and τ_{drag} are kept. The choice of the drag coefficient B merely affects the time needed to approach the equilibrium configuration. 2×10^{-5} Pa · s has

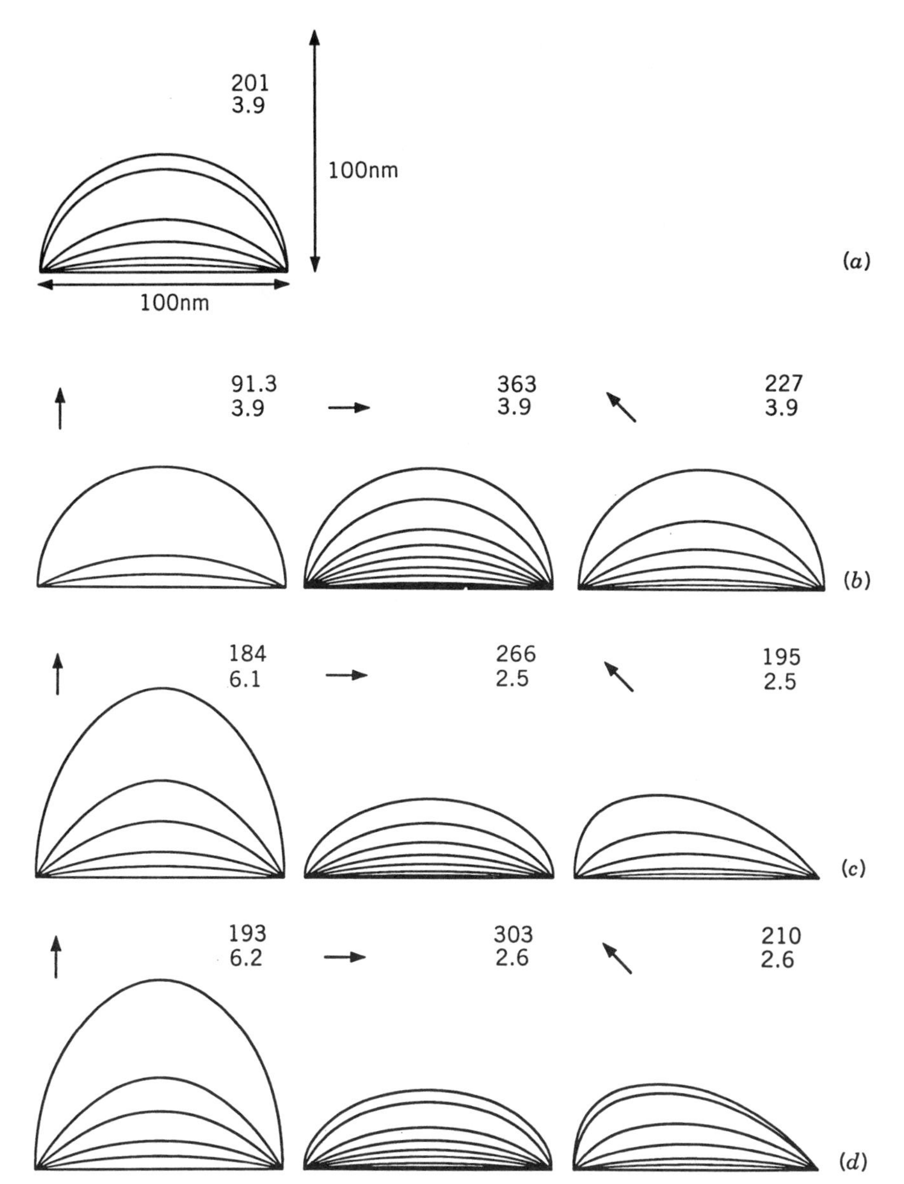

Fig. 4.23. A dislocation of length L bows out between two points; the self-interaction is disregarded. The material is nickel. $L = 100\,\mathrm{nm}$, $b = 0.255\,\mathrm{nm}$, $\mu = 79\,\mathrm{GPa}$, and $\nu = 1/3$. τ_{ext} equals 25 MPa, 50 MPa, 100 MPa, 150 MPa, 200 MPa, ..., τ_c. τ_c (*top*) and A_c are quoted in the upper right in MPa and in $10^3(\mathrm{nm})^2$, respectively. $\theta_{d0} = 90°, 0°, 45°$ from left to right; the short arrow indicates the direction of the Burgers vector. (*a*) S_4 of Eq. (4.49c). (*b*) Elastic isotropy: $S_3 = S(\theta_{d0})$ of Eq. (4.49b); $K_S(\theta_{d0})$ and R_0/R_i are given by Eqs. (2.7) and (2.8b), respectively. (*c*) As (*b*), but $S_2 = S(\theta_d)$ of Eq. (4.49a). (*d*) As (*c*), but elastic anisotropy, $K_S(\theta_d)$ is given in Figure 2.2. (Courtesy of D. Rönnpagel, Braunschweig.)

been inserted for B. The following conditions have been imposed; t is the time and v_n the local velocity of the dislocation:

$$t < 0: \quad y(x) = 0, \quad \tau_{ext} = 0, \quad v_n = 0$$

$$t \geqslant 0: \quad y(x = \pm L/2) = 0, \quad \tau_{ext} = \text{const.} = 25\,\text{MPa}, \quad 50\,\text{MPa}, \quad 100\,\text{MPa}, \quad 150\,\text{MPa}, \quad 200\,\text{MPa}, 250\,\text{MPa}, \ldots, \tau_c$$

The fact that the dislocation is fixed at the points ($x = \pm L/2$, $y = 0$) is expressed by the equation $y(x = \pm L/2,\ t \geqslant 0) = 0$. Figure 4.23 shows the solutions $y(x, t \to \infty)$. All configurations are close to segments of ellipses. As to be expected on the basis of Eq.

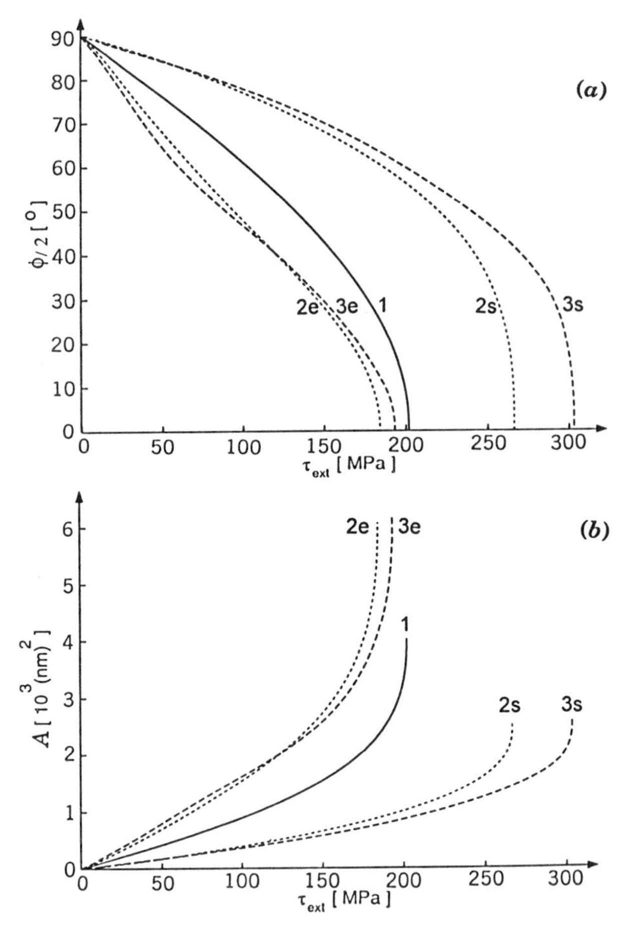

Fig. 4.24. Angle $\phi/2$ (*a*) and area A (*b*) versus the applied stress τ_{ext}. The material is nickel, and all parameters are the same as in Figure 4.23. 1, S as in Figure 4.23a; 2, S as in Figure 4.23c; 3, S as in Figure 4.23d. e, edge dislocation; s, screw dislocation. (Courtesy of D. Rönnpagel, Braunschweig.)

(2.11b) the radius R_l of the local curvature is smallest for the edge segments of the dislocations. This follows from the fact that $S(\theta_d = 90°)$ is smaller than $S(\theta_d = 0°)$. One might as well say that dislocations try to align themselves with their Burgers vector. In Sections 4.2.1.4.1 and 4.2.2.1, bent dislocations are assumed to form circular arcs. Figure 4.23 clearly demonstrates the deficiencies of this approximation.

On the basis of Eqs. (4.28) and (4.23a), one expects the ratios $\tau_c(\theta_{d0} = 0°)/\tau_c(\theta_{d0} = 90°)$ and $A(\theta_{d0} = 0°, \tau_{ext})/A(\theta_{d0} = 90°, \tau_{ext})$ of nickel to be 4.2 and 0.24, respectively. These values have been read from Figure 2.2; the elastic anisotropy is allowed for. $A(\theta_{d0}, \tau_{ext})$ is the area swept out by the dislocation if τ_{ext} is applied. From Figure 4.23c it follows that the actual ratio of the CRSSs and the ratio $A_c(\theta_{d0} = 0°)/A_c(\theta_{d0} = 90°)$ are 1.4 and 0.42, respectively. These values apply to S_2 of Eq. (4.49a), elastic isotropy, and $R_0 = 4R_l/e$ with $e = 2.718$. If the elastic anisotropy is allowed for, the respective ratios are 1.6 and 0.42. All these ratios are much closer to unity than those estimated on the basis of Eqs. (4.28) and (4.23a). Evidently the difference in the behavior of screw and edge dislocations becomes less pronounced as the bow-out increases. In Figure 4.24, the angle $\phi/2$ and the area A are plotted versus τ_{ext}. $\phi/2$ is defined in analogy to $\phi/2$ in Figure 4.9a: $\phi/2$ is the angle that the dislocation and the y-axis includes at $x = \pm L/2$. Allowing (Figure 4.23d) for the anisotropy of nickel or disregarding (Figure 4.23c) it leads to very similar configurations. The same holds for A_c and τ_c. The relative differences are larger for screw than for edge dislocations.

The choice of the outer cutoff radius R_0 in the dislocation line tension needs some further attention. In Section 2.2, two alternatives have been presented for the ratio R_0/R_i: $R_0/R_i = L/b$ [Eq. (2.8a)] and $R_0/R_i = [1.47R_l/b]$ [Eq. (2.8b)], with R_i = inner cutoff radius, b = length of the Burgers vector, L = length of the dislocation before it bows out, and R_l = radius of the local curvature of the bent dislocation. Equation (2.8b), which is probably the best choice, has been suggested by Rönnpagel [4.78]. In the case of particle strengthening, different approximations have been used for L appearing in Eq. (2.8a): (1) $L = L_{max}$ = particle spacing along a straight line (Section 3.1), (2) $L = L_{min}$ = square lattice spacing (Section 3.1), (3) $L = L_{cF}$ = Friedel length [Eq. (4.27a)], (4) $L = L_{cL}$ = Labusch length [Eq. (4.67)], and (5) $L = zb$, where z is a constant (e.g., 100 or 1000). Moreover, alternatives 1–4 may be reduced by $2r_r$; in this case, L equals the respective *free* length between the particles. r_r has been defined in Section 3.1. Because L_{max}, L_{min}, L_{cF}, and L_{cL} are functions of the particles' size and volume fraction, Ostwald ripening (Section 3.2.1.2) has an additional, indirect effect on the CRSS—namely, through the dislocation line tension S, which depends on R_0. An example for the importance of this point will be given in Section 5.5.4.1; there the choice of R_0 determines which one of several alternative models describes the same set of experimental CRSS data best.

Above, the length b of the Burgers vector has been inserted for the inner cutoff radius R_i. Several times b can be justified as well. Because R_i is independent of the particle dispersion, the exact value of R_i is of only minor importance. Moreover, it will be shown in Chapter 5 that in many cases the CRSS depends only on the square root of the dislocation line tension.

4.2.1.4.4 Dynamic Dislocation Effects. Schwarz [4.9] calculated the stress τ_{ext} that is needed to drive a dislocation at the velocity v_n. Because he used the normalized parameters that he and Labusch [4.8] had introduced in their computer simulations of the motion of a dislocation through a random array of obstacles, Schwarz' results will be discussed in the respective context in Section 4.2.2.2.2. Here only the analytical solution for extremely high velocities v_n is given. If τ_{drag} outweighs τ_{obst}, τ_{inert}, and τ_{back} in Eq. (2.18), Eq. (2.20) reduces to

$$b\tau_{ext} = Bv_n \tag{4.54}$$

where B is the drag coefficient.

4.2.2 Computer Simulations

Computer simulations of the motion of dislocations through planar random arrays of obstacles have led to a deeper understanding of the relevant microscopic processes. Such simulations yield detailed information pertaining, for example, to the following problems and questions:

- How does the spatial distribution of obstacles affect the CRSS?
- What are the effects of the finite ranges of the obstacle–dislocation interaction forces on the CRSS?
- How are the obstacles distributed along the hardest dislocation configuration CMAX (Section 4.2.1.4.1)?

4.2.2.1 Point Obstacles, No Dynamic Dislocation Effects. Foreman and Makin [4.37, 4.38] and Morris and coworkers [4.39, 4.55, 4.56] simulated the motion of a dislocation through planar random arrays of point obstacles. The suppositions were the same ones that Kocks [4.29, 4.50, 4.51] and Hanson and Morris [4.39] had made in their statistical treatments. With only two exceptions, all of Friedel's suppositions listed in Section 4.2.1.3.1 have been maintained. Supposition 3 (uniformly spaced obstacles) and supposition 4 (steady-state propagation of the dislocation) have been eliminated. Instead of Friedel's supposition of "only slight bowing of the dislocation" (supposition 5), constancy of the dislocation line tension has been postulated; that is, S_4 of Eq. (4.49c) has been inserted. The critical angle $\phi_c/2$ may cover the entire range 0–90°. The simulation procedures have already been described in Section 4.2.1.4.1. At the start, the straight dislocation is placed at the bottom of the rectangular random array of point obstacles sketched in Figure 4.17. After the small reduced stress $\tau^*_{ext\,1}$ [Eq. (4.35)] has been applied, the stable dislocation configuration C1 is calculated. Then $\tau^*_{ext\,1}$ is slightly raised to $\tau^*_{ext\,2}$ and C2 is constructed. Finally, after the application of $\tau^*_{ext\,max}$, the dislocation penetrates the top of the array. The dislocation configuration just prior to the application of $\tau^*_{ext\,max}$ has been called CMAX in Section 4.2.1.4.1. τ^*_{ext} needed to reach

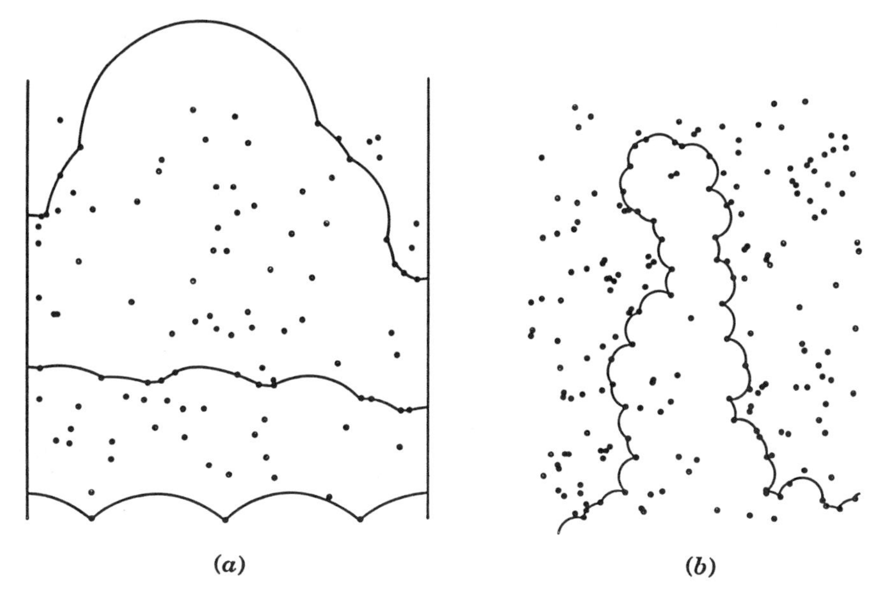

Fig. 4.25. Simulated dislocation configurations. After Foreman and Makin [4.37]. (*a*) $\phi_c/2 = 65°$. The second configuration from the bottom is CMAX. (*b*) $\phi_c/2 = 5°$.

CMAX is the reduced CRSS τ_p^*. To obtain reliable data for τ_p^*, the number n_0 of obstacles in the array has to exceed 1000 [4.37, 4.38]. The authors often chose $n_0 = 10{,}000$. One may also take averages over several computer runs.

Figure 4.25 shows some stable dislocation configurations. Because the obstacles are weak in Figure 4.25a, the dislocation is only slightly bent. The mean obstacle spacing along the dislocation is appreciably larger than L_{min}. This agrees with Eqs. (4.27b) and (4.44a). "Unzipping" occurs frequently: Once an obstacle gives way, its neighbors cannot sustain the force exerted on them by the dislocation and also break. This is a kind of domino effect. Unzipping has already been mentioned in Section 4.2.1.3.1 when Friedel's supposition of the "steady state" has been discussed. Because random arrays of obstacles are considered, unzipping may stop after a relatively short distance when a harder obstacle configuration is met. Figure 4.25b illustrates a situation frequently encountered with strong obstacles, here $\phi_c/2$ equals 5°. They require high stresses, which lead to small radii R_l of curvature. In many places, $2R_l$ is smaller than the spacing of neighboring obstacles and the dislocation can pass between them. It penetrates the array along paths of easy movement. Difficult groups of obstacles are encircled and dislocation loops are generated. Some of them collapse after severance from the main dislocation.

Foreman and Makin [4.37, 4.38] found that the hardest configuration CMAX is encountered well before the top of the array is reached. This proves that it is sufficiently large. In Figure 4.25a, CMAX is the second configuration from the bottom. In Figure 4.12a, Foreman and Makin's results for the reduced CRSS τ_p^* are shown together with those obtained by Hanson and Morris [4.39]. The agreement between the two sets of data is remarkably good. Arrays with $n_0 = 1000$ and with $n_0 = 10{,}000$ yield quite similar results. The stress increment was varied with n_0: It was 2% for $n_0 = 1000$ and 1% for $n_0 = 10{,}000$.

Foreman and Makin described the results that they obtained for arrays with $n_0 = 10{,}000$, by an empirical relation:

$$\tau_{p\,\mathrm{sim}}^* = \tau_{pF}^* \left(0.80 + \frac{\phi_c}{5\pi}\right) \tag{4.55}$$

where "sim" stands for "simulation." τ_{pF}^* refers to Friedel's equation [Eq. (4.26c)]. So far $\phi_c/2$ has been expressed in degrees; however, in Eq. (4.55) it is expressed in radians. The authors stressed that Eq. (4.55) is accurate over the entire range of $\phi_c/2$. Only for $\phi_c/2$ approaching zero, there is an insignificant deviation. If the obstacles are very weak (i.e., $\phi_c \approx \pi$), $\tau_{p\,\mathrm{sim}}^*$ is very close to τ_{pF}^*. This has already been mentioned at the end of Section 4.2.1.4.1. It is also evident in Figure 4.12b. Because S has been assumed to be constant in the simulations, all results presented in this section may be deficient if the obstacles are of medium strength. This has been discussed in Section 4.2.1.4.1. Brown and Ham [4.49] criticized on general grounds that $\partial\tau_{p\,\mathrm{sim}}^*/\partial\phi_c$ of Eq. (4.55) does not vanish for ϕ_c approaching zero. These authors suggested that one should use Friedel's equation [Eq. (4.26c)] if $\phi_c/2$ is above 50°, and Eq. (4.56) elsewhere:

$$\tau_p^* = [\cos(\phi_c/2)]^{3/2} \qquad \text{for } \phi_c/2 > 50^\circ \tag{4.26c}$$

and

$$\tau_p^* = 0.8\cos(\phi_c/2) \qquad \text{for } \phi_c/2 \leqslant 50^\circ \tag{4.56}$$

Except for the factor 0.8, Eq. (4.56) is identical with Eq. (4.16c), which yields an upper limit for τ_p^*. Because Eq. (4.55) covers the whole range of ϕ_c smoothly, it is preferred to Brown and Ham's combination of Eqs. (4.26c) and (4.56). It is emphasized that the curves and data shown in Figure 4.12 are the results of models and of computer simulations. Hence the good agreement does not prove that the models accurately describe actual particle-hardened materials. This point will be discussed further in Chapter 5.

The distribution of obstacle spacings along CMAX can be directly evaluated, and the same holds for the distribution of forces actually exerted by the dislocation on the obstacles of CMAX. Hanson and Morris [4.39] have published such distribution functions and the respective averages. The mean obstacle spacing along CMAX is plotted versus τ_p^* in Figure 4.18. The agreement with Eq. (4.44b) is satisfactory.

So far all point obstacles of the array have been supposed to be of the same strength; this was Friedel's supposition 2 listed in Section 4.2.1.3.1. Foreman and Makin [4.38] dispensed with it and studied random mixtures of point obstacles having various strengths:

1. Mixtures of obstacles of two different strengths characterized by the breaking angles $\phi_{c1}/2$ and $\phi_{c2}/2$ and by the respective numbers (per unit area) n_{s1} and n_{s2}. Often such mixtures of two classes of obstacles are called *bimodal.*
2. Spectra of obstacle strengths. The spectra were centered at ϕ_{co}, and their full widths were ϕ_{cw}. Within these ranges, all breaking angles occurred with the same probability. $\phi_c/2$ stayed within its natural limits 0° and 90°.

Binary Mixtures. Three definitions are needed:

$f'_j = n_{sj}/(n_{s1} + n_{s2})$.
τ^*_{pj} = reduced CRSS of the class j obstacles with $\phi_{cj}/2$ and n_{sj}; the CRSS of the array would be equal to τ_{pj} if n_{sj} were kept, but n_{si}, $i \neq j$, were reduced to zero.
τ^*_p = reduced CRSS of the entire, bimodal array.

τ^*_{pj} and τ^*_p are calculated with the aid of Eq. (4.18), in which $L_{\min}$ appears. $L_{\min}$ in turn is defined in Eq. (3.1a). There n_{sj} and $(n_{s1} + n_{s2})$ have to be inserted for τ^*_{pj} and τ^*_p, respectively.

Foreman and Makin performed computer simulations for the following combinations of obstacle strengths:

High/Medium:	$(\phi_{c1}/2 = 5°)/(\phi_{c2}/2 = 45°)$
High/Low:	$(\phi_{c1}/2 = 5°)/(\phi_{c2}/2 = 65°)$
Medium/Low:	$(\phi_{c1}/2 = 45°)/(\phi_{c2}/2 = 65°)$

Class 2 obstacles are the weaker ones. In Figure 4.26, τ^*_p is plotted versus f'_2. The curves represent Eq. (4.57) with various choices of the exponent k:

$$(\tau^*_p)^k = (\tau^*_{p1}\sqrt{f'_1})^k + (\tau^*_{p2}\sqrt{f'_2})^k \tag{4.57}$$

$k = 1.0$, Figure 4.26a	This is an acceptable representation of the data only in the important case that few strong obstacles are mixed in with many weak ones; central curve, lower right. Even in this case, $k = 1.36$ yields a better representation of the data (Figure 4.26d).
$k = 2.0$, Figure 4.26b	This is a good representation of the data in two cases: (1) upper curve, left part: Few obstacles of medium strength are mixed in with many strong ones, and (2) lower curve: Weak and medium strong obstacles are mixed, and there are more of the latter than of the former.

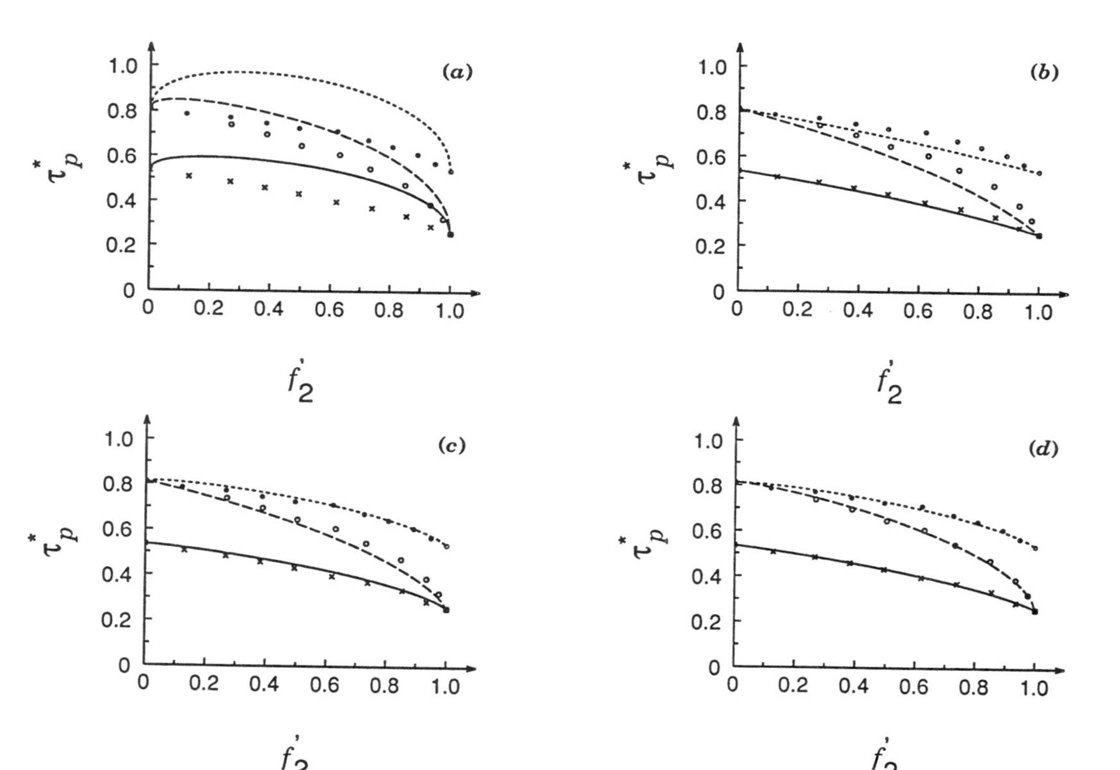

Fig. 4.26. Results of the computer simulations of the motion of a dislocation through a random bimodal array of point obstacles. The reduced CRSS τ_p^* [Eq. (4.18)] is plotted versus the fraction f'_2 of class 2 obstacles. The breaking angle combinations are, from top to bottom: ● · · · · · · · 5°/45°, ○----- 5°/65°, × —— 45°/65°. The curves represent Eq. (4.57): (*a*) $k = 1.0$; (*b*) $k = 2.0$; (*c*) $k = 1.61$; (*d*) ● 5°/45°, $k = 1.67$; ○5°/65°, $k = 1.36$; × 45°/65°, $k = 1.79$. (Data taken from Ref. 4.38.)

$k = 1.61$, Figure 4.26c	k has been fitted to all three combinations of obstacle strengths simultaneously. This yields acceptable representations of the data, if the strengths of the obstacles differ only slightly from each other.
$k = 1.67$ for $5°/45°$ $k = 1.36$ for $5°/65°$ $k = 1.79$ for $45°/65°$ Figure 4.26d	k has been fitted to each of the three combinations of obstacle strengths separately. The curves represent the data well. But k depends on the strength of the obstacles.

Reppich et al. [4.79] and Ardell [4.80] also fitted Eq. (4.57) to Foreman and Makin's data and obtained results for k which were close to those used in Figure 4.26d. These results will be discussed further in a more general context in Chapter 7.

Spectra of Obstacle Strengths. The reduced CRSS τ_p^* of an obstacle array with a spectrum of strengths is quite accurately given by $\tau_p^*(\phi_{co}/2)$, which is the reduced CRSS of an array of particles of uniform strength $\phi_{co}/2$. $\tau_p^*(\phi_{co}/2)$ can be read from Figure 4.12. Ardell [4.81] has pointed out that an improved representation of the results is obtained by generalizing Eq. (4.57) and choosing $k = 2.0$. If k exceeds 1.0, Eq. (4.57) weights strong contributions more heavily than weak ones. The most important result of these simulations is that τ_p^* equals 0.58 for $\phi_{co}/2 = 45°$ and $\phi_{cw} = 90°$ – that is, if all breaking angles between 0° and 90° occur with the same probability [4.38]. ϕ_{cw} is the full width of the spectrum.

Foreman also simulated the effects of the spatial distribution of the obstacles. He generated their coordinates by a pseudo-random-number generator, but imposed the condition that the minimum spacing of obstacles had to exceed a fraction α of L_{min}. If α vanishes, a truly random array is obtained. High values of α result in a two-dimensional liquid-like regularity of the obstacle array. Foreman's results have been communicated by Brown and Ham [4.49]. For $\phi_c/2 < 30°$, τ_p^* increases with α—that is, with the regularity of the array. This is in agreement with the results of Kocks' and Hanson and Morris' statistical analyses described in Section 4.2.1.4.1. The computer simulations yielded for very strong obstacles with $\phi_c/2 = 0°$: for $\alpha = 0.0$; $\tau_p^* = 0.82$, and for $\alpha = 0.90$; $\tau_p^* = 0.94$. If Eq. (4.16c), which gives an upper limit for τ_p^*, is deduced for a regular hexagonal array instead of for a regular square one, a factor 0.93 appears on the right-hand side of this equation. This follows from Eq. (3.1b). This factor is in excellent agreement with Foreman's result for $\phi_c/2 = 0°$ and $\alpha = 0.90$. Strunin and Popov [4.82] confirmed in their more mathematical treatment that τ_p^* increases with the regularity of the obstacle array. This point will be discussed further in Section 4.2.2.2.1. The TEM work described in Section 3.3.1.5 indicated that precipitates that have been subjected to an Ostwald ripening treatment are actually not randomly distributed in space. There are neither very closely nor very widely spaced nearest neighbors.

Only as long as the precipitation process is governed by homogeneous nucleation and hardly any Ostwald ripening has taken place, the spatial distribution of precipitates may be truly random.

4.2.2.2 Finite Obstacle Size, Dynamic Dislocation Effects. In addition to those of Friedel's suppositions that had already been dispensed with by Foreman and Makin [4.37, 4.38] and by Hanson and Morris [4.39] in their computer simulations, Schwarz and Labusch [4.8, 4.83] eliminated two more suppositions: They studied the motion of dislocations through planar random arrays of obstacles of finite size and they also allowed for dynamic dislocation effects. For the sake of clarity, Friedel's suppositions are listed again and it is indicated which ones are still kept and which ones are relaxed or entirely dispensed with. Their numbers are the same ones as in Section 4.2.1.3.1. Because obstacles of finite size are considered, their volume fraction f has to be subjected to a limitation. This is stated in the additional supposition 9. The straight-line approximation (Section 4.1.1) is performed. After the integration [Eq. (4.2)] involved in it, the range w_x in the x-direction does not appear in the subsequent derivations.

SUPPOSITIONS

1. Friedel:	Shearable point obstacles; that is, the range w_y is disregarded.
Schwarz/Labusch:	Shearable one-dimensional obstacles; that is, w_y is allowed for.
2. Friedel:	All obstacles produce the same maximum force F_0.
Schwarz/Labusch:	All obstacles produce the same one-dimensional force–distance profile.
3. Friedel:	The obstacles are uniformly spaced.
Schwarz/Labusch:	The obstacles are distributed at random.
4. Friedel:	The propagation of the dislocation is steady.
Schwarz/Labusch:	This supposition is dispensed with.
5. Friedel *and* Schwarz/Labusch:	Bowing of the dislocation is slight; that is, the obstacles are weak and their critical angle $\phi_c/2$ is large. The dislocation line tension is assumed to be constant.
6. Friedel *and* Schwarz/Labusch:	Thermal activation is disregarded.
7. Friedel:	The motion of the dislocation is overdamped; consequently there are no dynamic effects.

Schwarz/Labusch:	The actual strength of the damping processes is allowed for; there are dynamic dislocation effects.
8. Friedel *and* Schwarz/Labusch:	The interaction between different dislocations is disregarded.
9. Friedel:	There is no limit to the volume fraction f.
Schwarz/Labusch:	The volume fraction f of the particles should not be too high: $f \leqslant$ about 0.2.

Schwarz and Labusch did not explicitly state supposition 9, but it was implied by the choice of obstacle arrays that they simulated. Their results cannot be expected to be applicable if f exceeds about 0.2. If the obstacles have no extension—that is, if they are point obstacles—such a limitation is not necessary. Schwarz and Labusch [4.8, 4.83] considered obstacle–dislocation interaction force profiles of the types described by Eqs. (4.4) and (4.6). These equations refer to elastically interacting obstacles and to energy-storing obstacles, respectively. In order to demonstrate the value of these theoretical profiles, they are compared with real ones in Figure 4.27. The straight-line approximation described in Section 4.1.1 has been applied to the real obstacle–dislocation interactions.

Figure 4.27a: A particle that has a lattice mismatch interacts elastically with an edge dislocation, only $F(y/b > 0)$ is shown. The interaction is attractive. Equations (4.4) have been least-squares fitted to the actual profile. The adjustable parameters are as follows:
$n = 1.89$
$A_n F_0 = -5.74 \times 10^{-9}\,\mathrm{N}$
$w_y = 10.4\,\mathrm{nm} = 2.04\rho$
ρ is the radius of the particle.

Figure 4.27b: A long-range-ordered particle is sheared by an edge dislocation. Energy is stored in the particle. Equation (4.6) has been fitted to the actual profile. The adjustable parameters are as follows:
$n = 0.907$
$A_n F_0 = -20.2 \times 10^{-9}\,\mathrm{N}$
$w_y = 14.3\,\mathrm{nm} = 2.81\rho$

The latter profile is quite well described by Eq. (4.6), whereas the sharp peak shown in Figure 4.27a is not reproduced by Eqs. (4.4). In general, Eqs. (4.4) and (4.6) can be expected to be acceptable representations of the actual one-dimensional force–distance profiles. Often—for example, in Section 5.2.1—their full widths at half-maximum will be inserted for w_y. Though this may not be quite exact, the functional variation of w_y with r is, however,

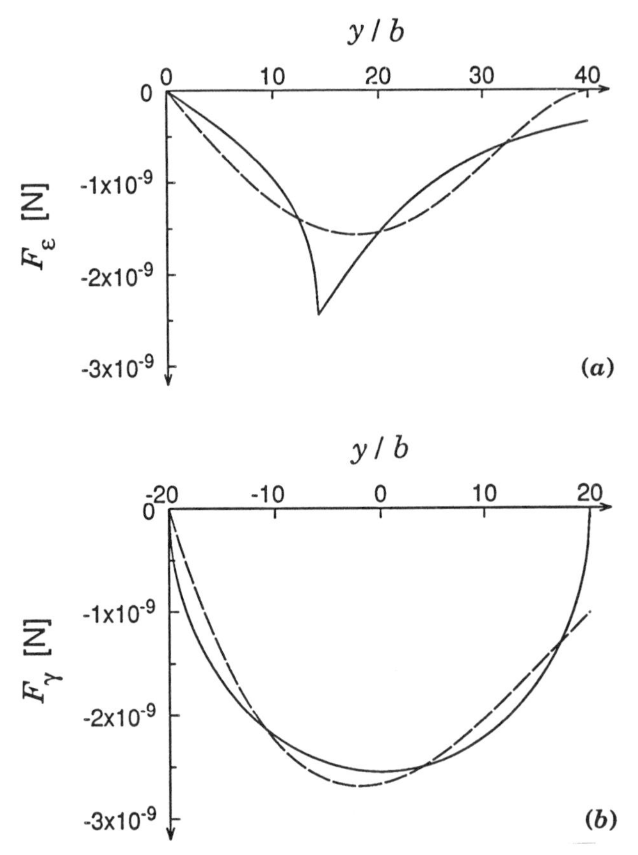

Fig. 4.27. Particle–edge dislocation interaction force profiles; the dislocation is undissociated. Force F versus distance y/b. Drawn out curves: Actual profile after performing the straight-line approximation (Section 4.1.1). Dashed curves: Least-squares fit of Eqs. (4.4), respectively (4.6), to the actual profiles. Burgers vector length b is 0.25 nm, particle radius ρ is $20b$. (a) The particle has the constrained lattice mismatch $\varepsilon = 0.015$, $\mu_s = 31$ GPa. Coordinates of its center: $(0, 0, \rho/\sqrt{2})$. (b) The particle is long-range-ordered: Specific antiphase boundary energy γ is 0.25 J/m^2. Coordinates of its center: (0, 0, 0).

believed to be well represented by this choice. Sometimes w_y is considered to be an adjustable parameter; this will be done in Section 5.5.4.1.

Equation (2.20), which describes the motion of a slightly bent dislocation through a planar array of obstacles, involves seven parameters: the inertial mass m_{eff}, the drag coefficient B, the dislocation line tension S, the maximum interaction force F_0, the two ranges w_x and w_y, and the square lattice spacing $L_{\min}$. Schwarz and Labusch reduced this number to two by introducing normalized parameters and coordinates. As before, the glide plane is the plane $z = 0$. Therefore only x- and y-coordinates are given. Those of the center of particle i are (h_{xi}, h_{yi}). The new parameters and coordinates are as follows; f_0 is only an abbreviation:

$$f_0 = \frac{F_0}{2S} = \cos(\phi_c/2) \tag{4.58a}$$

Space:

$$\xi = x\frac{\sqrt{f_0}}{L_{\min}} \tag{4.58b}$$

$$\eta = y\frac{1}{L_{\min}\sqrt{f_0}} \tag{4.58c}$$

$$\xi_i = h_{xi}\frac{\sqrt{f_0}}{L_{\min}} \tag{4.58d}$$

$$\eta_i = h_{yi}\frac{1}{L_{\min}\sqrt{f_0}} \tag{4.58e}$$

$$\xi_0 = w_x\frac{\sqrt{f_0}}{L_{\min}} \tag{4.58f}$$

$$\eta_0 = w_y\frac{1}{L_{\min}\sqrt{f_0}} \tag{4.58g}$$

Time:

$$\theta = t\frac{1}{L_{\min}}\left(\frac{Sf_0}{m_{\text{eff}}}\right)^{1/2} \tag{4.58h}$$

Velocity:

$$\omega = v_n\left(\frac{m_{\text{eff}}}{S}\right)^{1/2}\frac{1}{f_0} \tag{4.58i}$$

Resolved shear stress:

$$\tau^{\otimes} = \tau\frac{bL_{\min}}{2Sf_0^{3/2}} = \tau/\tau_{pF} \tag{4.58j}$$

Drag coefficient:

$$\gamma = B\frac{L_{\min}}{(4Sm_{\text{eff}}f_0)^{1/2}} \tag{4.58k}$$

The normalized range η_0 has already been introduced in Section 4.2.1.4.2. τ_{pF} is the CRSS calculated by Friedel; it is given in Eq. (4.26a). The normalized effective mass per unit length of the dislocation and its normalized line tension both equal 1/2. The definitions of the *normalized* CRSS $\tau_p^{\otimes}$ [Eq. (5.58j)] and of the *reduced* CRSS τ_p^* [Eq. (4.18)] are different.

Equation (2.20) assumes the form

$$\tau_{\text{ext}}^{\otimes} + \tau_{\text{obst}}^{\otimes} = \frac{1}{2}\frac{\partial^2 \eta}{\partial \theta^2} + \gamma \frac{\partial \eta}{\partial \theta} - \frac{1}{2}\frac{\partial^2 \eta}{\partial \xi^2} \tag{4.59}$$

Let $\tau_{\text{obst i}}^{\otimes}$ be the normalized shear stress exerted by obstacle i on a dislocation segment. $\tau_{\text{obst}}^{\otimes}$ in Eq. (4.59) is the sum over the contributions of all the obstacles interacting simultaneously with the dislocation segment. τ_{obst} in Sections 2.5 and 4.1 had been defined correspondingly:

$$\tau_{\text{obst}}^{\otimes}(\xi, \eta) = \sum_i \tau_{\text{obst}\, i}^{\otimes}\left(\frac{\xi - \xi_i}{\xi_0}, \frac{\eta - \eta_i}{\eta_0}\right) \tag{4.60}$$

Schwarz and Labusch carried out the "straight-line approximation" described in Section 4.1.1:

$$\tau_{\text{obst}\, i}^{\otimes}\left(\frac{\xi - \xi_i}{\xi_0}, \frac{\eta - \eta_i}{\eta_0}\right) = \delta(\xi - \xi_i) F^*\left(\frac{\eta - \eta_i}{\eta_0}\right) \tag{4.61}$$

δ is Dirac's δ-function. F^* is given in Eqs. (4.4) and (4.6) for elastically interacting and for energy storing obstacles, respectively. Equation (4.61) corresponds to Eq. (4.3). The range w_x has been eliminated by performing the straight-line approximation. The importance of w_y increases with the obstacles' volume fraction.

Evidently Eq. (4.59) involves only two parameters that are characteristic of the system: the normalized range η_0 and the normalized drag coefficient γ. This greatly simplifies the solution of Eq. (4.59) and makes it possible to present the results in the form of general equations and graphs, which can be applied to quite different types of particle-hardened systems. Actually there is a third parameter besides γ and η_0: the exponent n appearing in Eqs. (4.4) and (4.6). But it will be shown below that it leads to only minor effects.

Schwarz and Labusch solved the differential equation (4.59) numerically. They simulated the motion of a dislocation through a planar random array of 900 obstacles in a computer. According to Foreman and Makin [4.37, 4.38], this number is somewhat low. The boundary conditions in the ξ-direction were made periodic. The glide plane was also made periodic in the η-direction. Thus the dislocation can be thought of moving on the surface of a drum that is periodic along its axis.

The drag coefficient B and thus also γ comprise several components: B_{ph}, B_e, B_m, and B_s. They represent the contributions of phonons, conduction electrons, magnons, and solute atoms, respectively [4.13, 4.84, 4.85]. B_m vanishes in nonmagnetic materials. In most cases, B of particle-strengthened materials is high enough to preclude any dynamic dislocation effects. At or above the Debye temperature, often B_{ph} alone suffices for this. If the particles have been formed in a precipitation process, at least some of the solute atoms will stay solved in the matrix and raise B_s. The rare case of a precipitation-strengthened material that exhibits dynamic dislocation effects will be discussed in Section 5.4.3. In dilute solid solutions and in superconductors they have been frequently observed [4.86–4.92].

Before the results of the computer simulations are presented, dynamic dislocation effects are shortly visualized; they will be discussed in Section 4.2.2.2.2 in greater detail. If B and consequently γ are low, the dislocation acquires kinetic energy while gliding in between the obstacles. This energy helps to overcome them.

Strong Dynamic Effects. The drag coefficient B is very small and the dislocation velocity v_n is very high. The kinetic energy ($\approx \frac{1}{2} m_{\text{eff}} v_n^2 w_x$), which the dislocation length w_x accumulates during its glide in between the obstacles, suffices to let the dislocation length w_x break through the obstacles directly on impact. Therefore the dislocation overcomes each obstacle independently of the positions of the other ones and the dislocation stays nearly straight.

Weak Dynamic Effects. B and v_n are of medium magnitude. In contrast to the case of strong dynamic effects, the kinetic energy of the dislocation is not high enough to break through the obstacles directly on impact. It bows out between them. Because of its kinetic energy it overshoots its equilibrium position. Three angles $\phi/2$ (Figure 4.9) are considered now: the critical one, $\phi_c/2$, which characterizes the strength of the obstacles [Eqs. (4.8)]; the equilibrium one, $\phi_e/2$, which the dislocation would assume if it moved quasistatically under the same external stress; and $\phi_k/2$, which is reached at the moment of the farthest overshoot. The subscripts c, e, and k of $\phi/2$ stand for "critical," "equilibrium," and "kinetic" energy, respectively. If the relation

$$\phi_k/2 < \phi_c/2 < \phi_e/2$$

is fulfilled, the dislocation can overcome the obstacles only dynamically, but not quasistatically.

No Dynamic Effects. If B is large, the dislocation motion is overdamped and there are no dynamic effects. In particle-strengthened materials this is the standard case.

If there are dynamic effects, one has to distinguish two different normalized CRSSs:

The upper CRSS: $\tau_{pu}^{\otimes}$
The lower CRSS: $\tau_{pl}^{\otimes}$

The respective non-normalized CRSSs are $\tau_{pu} = \tau_{pu}^{\otimes} \cdot \tau_{pF}$ and $\tau_{pl} = \tau_{pl}^{\otimes} \cdot \tau_{pF}$ [Eq. (4.58j)], where τ_{pF} is Friedel's result for the CRSS (Eqs. (4.26)]. The computer simulations yield $\tau_{pu}^{\otimes}$ if at their start both the normalized external stress $\tau_{\text{ext}}^{\otimes}$ and the normalized velocity ω vanish. $\tau_{\text{ext}}^{\otimes}$ is raised in small steps; that is, $\tau_{\text{ext}}^{\otimes}$ is applied quasistatically. For each stress the stable dislocation configuration is calculated. In this respect, Schwarz and Labusch's procedure is similar to those reported in Sections 4.2.1.4.1 and 4.2.2.1. $\tau_{pu}^{\otimes}$ is the highest normalized stress that still leads to a stable dislocation configuration. $\tau_{pl}^{\otimes}$ is obtained if the simulations are started at stresses $\tau_{\text{ext}}^{\otimes}$ well above $\tau_{pu}^{\otimes}$ and at finite velocities ω. Then $\tau_{\text{ext}}^{\otimes}$ is slowly decreased. $\tau_{pl}^{\otimes}$ is the lowest value of $\tau_{\text{ext}}^{\otimes}$ that still allows the dislocation to move indefinitely. There is some similarity between $\tau_{pu}^{\otimes}$ and $\tau_{pl}^{\otimes}$ on the one side and static and kinetic friction on the other one. Because under standard experimental conditions τ_{pu} is obtained, τ_{pu} will be identified with the CRSS. In the following chapters, often τ_{pL} will be written instead of τ_{pu}. The subscript L stands for "Labusch."

Schwarz and Labusch found that the dislocation motion is overdamped if the normalized drag coefficient γ exceeds about 3.0. For $\gamma > 3.0$, there are no dynamic effects and $\tau_{pl}^{\otimes}$ equals $\tau_{pu}^{\otimes}$. In this case, both CRSSs are functions of the normalized range η_0 only. This is the only parameter characteristic for the system that enters Eq. (4.59). For the time being, the minor effects of the exponent n appearing in Eqs. (4.4) and (4.6) are disregarded. They will be demonstrated in Figure 4.28. For γ below about 3.0, $\tau_{pl}^{\otimes}$ is smaller than $\tau_{pu}^{\otimes}$; that is, there are dynamic dislocation effects. The authors covered the whole possible range of η_0: $0 \leqslant \eta_0 \leqslant \infty$; that is, they treated the Friedel as well as the Mott case [4.8, 4.63] (Section 4.2.1.4.2). The results for the CRSSs will be presented for high and low values of γ in Sections 4.2.2.2.1 and 4.2.2.2.2, respectively.

4.2.2.2.1 Overdamped Dislocation Motion, No Dynamic Effects. In this section the normalized drag coefficient γ is supposed to be high enough to suppress any dynamic dislocation effects. It has been stated above that this is the case if γ is not less than about 3.0. γ of most particle strengthened metals and alloys exceeds this limit. For $\gamma > 3.0$, the two CRSSs $\tau_{pu}^{\otimes}$ and $\tau_{pl}^{\otimes}$ coincide; therefore only the former one is discussed. The two types of obstacles, those interacting elastically with dislocations and those that store energy, have to be treated separately.

Elastic Interaction Between Obstacles and Dislocations. The respective force–distance profile $F^*(y^*)$ is given in Eqs. (4.4). The interaction is supposed to be repulsive, and therefore A_n of Eq. (4.4e) is positive. In Figure 4.28, $\tau_{pu}^{\otimes}$ is plotted versus the normalized range η_0 defined in Eq. (4.58g). η_0 is limited to the range $0 \leqslant \eta_0 \leqslant 0.8$; more extended ranges of η_0 are discussed below. The exponent n appearing in Eqs. (4.4) has been varied between 2 and 12. Each data point represents the average over four obstacle arrays. The function $\tau_{pu}^{\otimes}(\eta_0)$ is linear:

$$\tau_{pu}^{\otimes} = C_1(1 + C_2\eta_0) \tag{4.62}$$

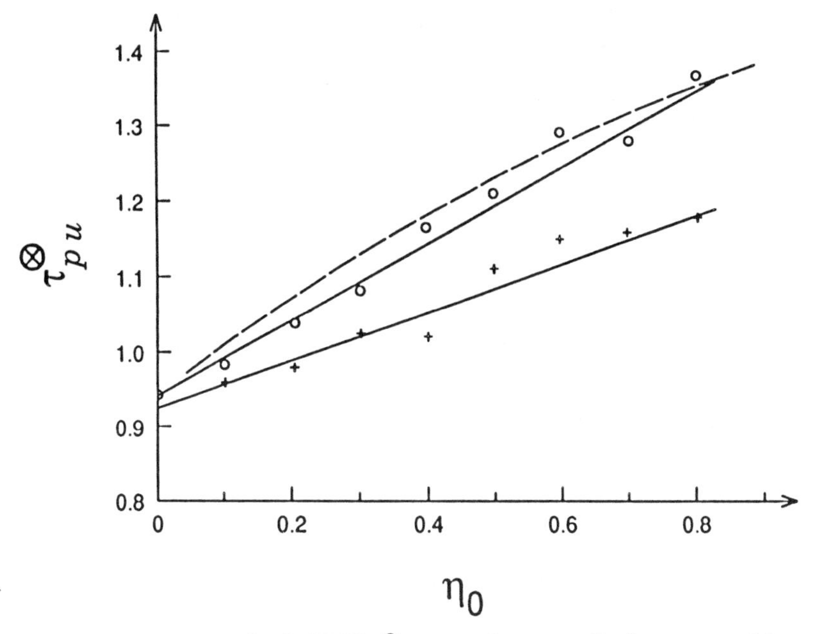

Fig. 4.28. Upper normalized CRSS $\tau_{pu}^{\otimes}$ versus the normalized range η_0. Normalized drag coefficient $\gamma = 3.0$. The obstacle–dislocation interaction is elastic. Exponent n of Eqs. (4.4): ○$n = 2$, $+n = 12$. The straight lines represent Eq. (4.62) fitted to the data. The dashed curve shows Eq. (4.63) fitted to the data with $n = 2$. (After Ref. 4.8.)

Only the coefficient C_2 depends on n. C_1 is independent of it, because as η_0 approaches zero, n becomes irrelevant. C_2 decreases as n increases. Least-squares fits of Eq. (4.62) to the data yielded:

$$
\begin{array}{ll}
n = 2 \text{ or } 12: & C_1 = 0.93 \pm 0.01 \\
n = 2: & C_2 = 0.55 \\
n = 12: & C_2 = 0.34
\end{array}
$$

The fit of Eqs. (4.4) and (4.6) to actual force–distance profiles (Figure 4.27) yielded exponents n below 2.0. Since even for the extreme range $2 \leqslant n \leqslant 12$, C_2 varies only between 0.55 and 0.34, the dependence of C_2 on n can be neglected. Hence $n = 2$ will be chosen as standard.

The statistical factor 0.887, which appears in Eq. (4.43), has a similar origin as C_1. Both coefficients have been brought about by the random distribution of the obstacles in the glide plane. The agreement is satisfactory.

A detailed analysis of the stable dislocation configurations found for $\tau_{ext}^{\otimes} \leqslant \tau_{pu}^{\otimes}$ and $\eta_0 < 0.5$ led to two conclusions:

1. If the obstacles repel the dislocation, it contacts most of them at their repulsive front edge. For n equal to 2 and A_n positive, the dislocation lies at y^* with $-1.0 < y^* <$ about -0.5 (Figure 4.7a).

2. The number of obstacles that simultaneously interact with the dislocation increases with η_0. This is expressed by Eq. (4.67) given below. It indicates that the increase of $\tau_{pu}^{\otimes}$ with η_0 is primarily due to a redistribution of the applied stress among a higher number of obstacles.

In non-normalized parameters, Eq. (4.62) read

$$\tau_{pu} = \tau_{pF}\left[C_1\left(1 + C_2\frac{w_y}{L_{\min}}\sqrt{\frac{2S}{F_0}}\right)\right]$$

τ_{pF} is Friedel's result given in Eqs. (4.26). The term in square brackets may be considered as a correction factor.

If the range of η_0 is extended from 0.8 up to 4.0, $\tau_{pu}^{\otimes}$ is no longer a linear function of η_0 but it is related to η_0 by

$$\tau_{pu}^{\otimes} = C_1(1 + C_2\eta_0)^{1/3} \tag{4.63}$$

The coefficients have been obtained by least squares fitting Eq. (4.63) to the data presented in Figure 4.29; there n equals 2.0: $C_1 = 0.94$, $C_2 = 2.5$. For $\eta_0 \leqslant 0.8$, Eq. (4.62) (the linear equation) represents the results of the simulations better than Eq. (4.63) does. This is illustrated in Figure 4.28. All values quoted above for C_1 and C_2 are listed in Table 4.1.

For η_0 approaching zero—that is, for point obstacles—Eqs. (4.62) and (4.63) yield

$$\tau_{pu}^{\otimes} = 0.93 \pm 0.01$$

and

$$\tau_{pL} = \tau_{pu} = (0.93 \pm 0.01)\ \tau_{pF}$$

Schwarz and Labusch's result is 7% lower than that of Friedel. This agrees qualitatively with Kock's [4.50–4.52], Hanson and Morris' [4.39], and Foreman and Makin's [4.37, 4.38] findings. Friedel obtained CRSS values that were too high because he considered uniformly spaced obstacles instead of randomly arranged ones. This point has already been discussed in Section 4.2.1.4.1.

Ohser [4.93] applied Schwarz and Labusch's simulation procedures to aluminum-rich aluminum–silver alloys that are strengthened by silver-rich precipitates. Their interaction with dislocations is believed to be due to differences in the stacking fault energy: It is lower in the precipitates than in the matrix [4.94–4.96]. This interaction mechanism will be discussed in Section 5.3. The following parameters have been inserted: $f_0 = F_0/(2S) = 0.17$, $\eta_0 = 1.35$, $\gamma \approx 2.0$. Ohser specifically studied the effects of the spatial particle distribution on the CRSS. The result was that the more uniform the particle

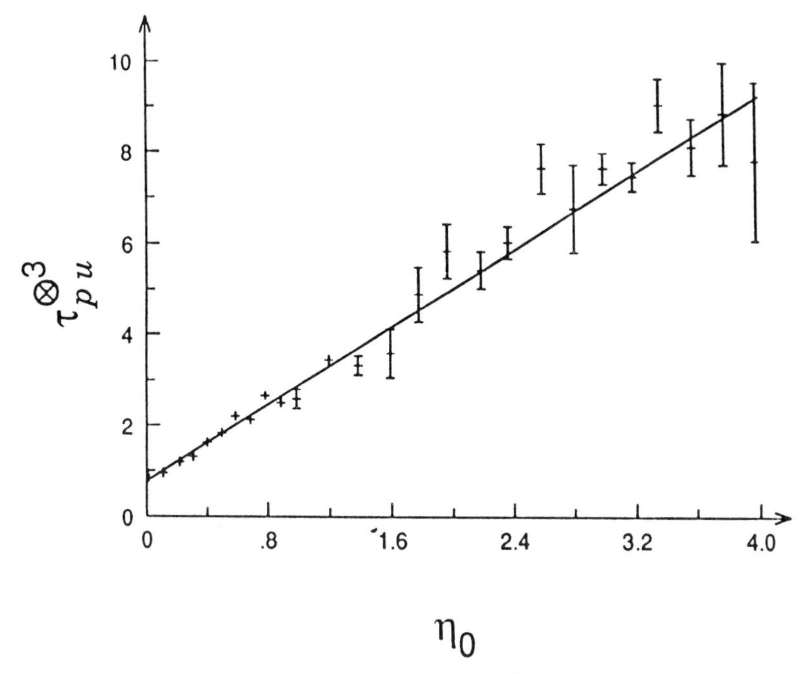

Fig. 4.29. Cube of the upper normalized CRSS $\tau_{pu}^{\otimes}$ versus the normalized range η_0. Normalized drag coefficient $\gamma = 3.0$, exponent $n = 2$. The obstacle–dislocation interaction is elastic. The vertical bars indicate the scatter of four independent simulations. The line represents Eq. (4.63) with $C_1 = 0.94$ and $C_2 = 2.5$. (After Ref. 4.8.)

arrangement, the lower the CRSS. This means that C_1 appearing in Eq. (4.62) decreases with the regularity. This result is at variance with those reported in Sections 4.2.1.4.1 and 4.2.2.1. Ohser's explanation was that the irregularity may lead to hard to overcome particle configurations. Moreover, the results presented in Sections 4.2.1.4.1 and 4.2.2.1 primarily concerned relatively strong obstacles, whereas the precipitates in aluminum–silver alloys are weak obstacles. This point needs further investigation.

Equation (4.63) can be used in the Mott regime where η_0 is large (Section 4.2.1.4.2). If the product $C_2\eta_0$ is much larger than unity, Eq. (4.63) reduces to

$$\tau_{pu}^{\otimes} = C_1(C_2\eta_0)^{1/3} \tag{4.64a}$$

or in non-normalized form:

$$\tau_{pu} = C_1 C_2^{1/3} \tau_{pF} \left[\frac{w_y}{L_{\min}} \left(\frac{2S}{F_0} \right)^{1/2} \right]^{1/3} \tag{4.64b}$$

Table 4.1. Equations and Parameters Relating the Upper Normalized CRSS $\tau_{pu}^{\otimes}$ to the Normalized Interaction Range η_0[a]

Type of Interaction	Range of η_0	Exponent n	Equation Number	Coefficients	
				C_1	C_2
Elastic	0.0–0.8	2	(4.62)	0.93 ± 0.01	0.55
	0.0–0.8	12	(4.62)	0.93 ± 0.01	0.34
	0.0–4.0	2	(4.63)	0.94	2.5
Energy storing	0.0–0.8	2	(4.62)	0.96	0.64
	0.0–1.0	2	(4.62)	0.94	0.72
	0.0–1.2	2	(4.62)	0.94	0.82

[a]The normalized damping constant γ equals 3.0.

This equation is applicable to solid solutions at 0 K. f has to be replaced by the solute concentration c_s. Then $L_{\min}^2$ equals a^2/c_s, where a^2 is the area per atom. Inserting for τ_{pF} from Eq. (4.26a) yields

$$\tau_{pu} = C_1 C_2^{1/3} \frac{c_s^{2/3} F_0^{4/3} w_y^{1/3}}{b(2S)^{1/3} a^{4/3}} \tag{4.65}$$

This formula had already been derived by Labusch [4.60, 4.61] analytically, but the numerical factor $C_1 C_2^{1/3} = 1.28$ could not be calculated analytically.

Energy-Storing Obstacles. Figure 4.30 gives the results for energy-storing obstacles. γ is again 3.0 and the exponent n is 2.0. As long as η_0 does not exceed 0.5, $\tau_{pu}^{\otimes}$ is nearly the same for energy storing and for repulsive, elastically interacting obstacles. Hence it is concluded that that part of the elastic interaction force profile in which the force F^* [Eqs. (4.4)] drives the dislocation forward comes into play only if η_0 is larger than 0.5. This agrees with the above observation that the dislocation tends to lie at the repulsive front edges of elastically interacting obstacles. For energy storing obstacles $\tau_{pu}^{\otimes}(\eta_0)$ can be satisfactorily represented by Eq. (4.62) (the linear equation). In Figure 4.30 it can be seen that the data actually exhibit a slight upward curvature. Therefore C_2 increases somewhat with the range of η_0 to which Eq. (4.62) is to be applied. All results for C_1 and C_2 of energy-storing obstacles are compiled in Table 4.1.

Schwarz and Labusch [4.8] published no results for the obstacle spacing L_{cL} along the strongest dislocation configuration CMAX; the subscript L of L_{cL} stands for "Labusch." Therefore L_{cL} is estimated on the basis of Eqs. (4.14) and (4.58j):

$$\tau_{pF} \tau_{pu}^{\otimes} b L_{cL} = F_0 \tag{4.66}$$

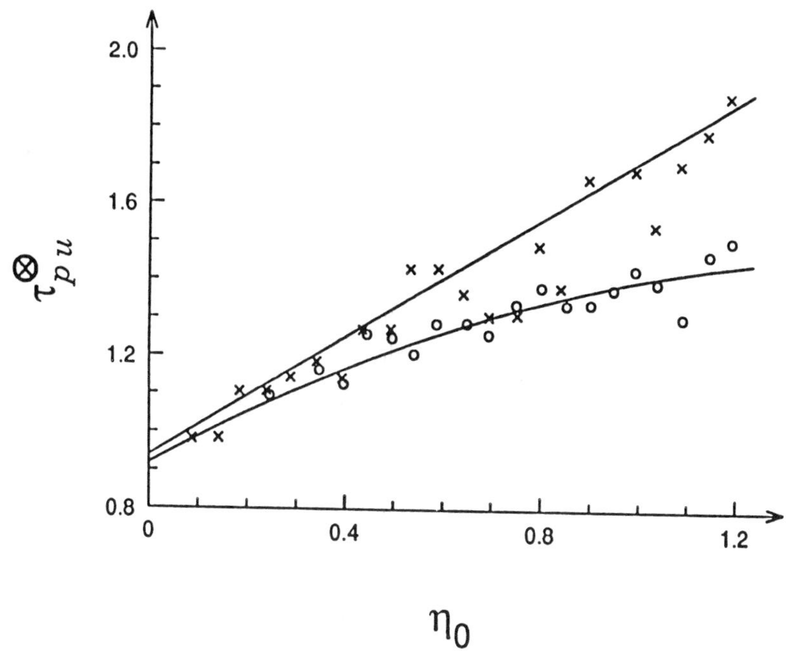

Fig. 4.30. Upper normalized CRSS $\tau_{pu}^{\otimes}$ versus the normalized range η_0. Normalized drag coefficient $\gamma = 3.0$, exponent $n = 2$. $\times$: Energy-storing obstacles; the straight line represents Eq. (4.62) with $C_1 = 0.94$ and $C_2 = 0.82$. ○: Elastically interacting obstacles; the bent line has been traced through the latter set of data. (After Ref. 4.8.)

Substituting for τ_{pF} and $\tau_{pu}^{\otimes}$ from Eqs. (4.25), (4.27a), and (4.62) yields [4.7]

$$L_{cL} = \frac{L_{cF}}{C_1(1 + C_2\eta_0)} \tag{4.67}$$

The coefficients C_1 and C_2 are those listed in Table 4.1. L_{cF} is Friedel's corresponding result given in Eqs. (4.27). Equation (4.67) applies to elastically interacting obstacles as well as to energy-storing ones; η_0 has to stay in the limits in which Eq. (4.62) is applicable. At the end of Section 4.2.1.4.1 it has been mentioned that the application of Eq. (4.14) is not quite correct if the obstacles are distributed at random in the glide plane. Since in this case not all obstacles experience the maximum force F_0, Eq. (4.67) overestimates L_{cL} somewhat. L_{cF} derived from Eq. (4.27b) is about 24% larger than the corresponding length L_{cHM}, which follows from Hanson and Morris' [4.39] equation [Eq. (4.44a)]. L_{cF} and L_{cHM} differ by a constant statistical factor. Equation (4.67) could be corrected by raising C_1 correspondingly.

The denominator of Eq. (4.67) increases with η_0. This expresses the above-mentioned result of Schwarz and Labusch's computer simulations—

namely, that the number of obstacles that simultaneously interact with the dislocation increases with η_0.

4.2.2.2.2 Underdamped Dislocation Motion, Dynamic Effects. If the normalized drag coefficient γ is less than about 3.0, dynamic dislocation effects occur. In this case the upper normalized CRSS $\tau_{pu}^{\otimes}$ is larger than the lower one $\tau_{pl}^{\otimes}$. Therefore both CRSSs have to be quoted as functions of η_0 and γ. Only elastic obstacle–dislocation interactions are considered.

First the results that Schwarz and Labusch [4.8] obtained for $\eta_0 = 0$—that is, for point obstacles—are discussed. $\tau_{pu}^{\otimes}(\gamma, \eta_0 = 0)$ and $\tau_{pl}^{\otimes}(\gamma, \eta_0 = 0)$ are shown in Figure 4.31. For $\gamma > 3$, $\tau_{pu}^{\otimes}$ and $\tau_{pl}^{\otimes}$ coincide: Both asymptotically tend to $C_1 = [0.93 \pm 0.01]$ of Eq. (4.62). For $\gamma < 3$, $\tau_{pu}^{\otimes}$ as well as $\tau_{pl}^{\otimes}$ decrease with decreasing γ. $\tau_{pu}^{\otimes}$ asymptotically reaches the value 0.37, whereas $\tau_{pl}^{\otimes}$ decreases monotonically. For $\gamma \ll 1$, $\tau_{pl}^{\otimes}(\gamma, \eta_0 = 0)$ approximately equals γ. There are strong dynamic effects (Section 4.2.2.2). Since for $\gamma \ll 1$ the dislocation breaks through the obstacles directly on impact, it stays nearly straight and the distribution of the obstacles in the glide plane has hardly any effect. Though each entry in Figure 4.31 has been derived from a different obstacle array, $\tau_{pl}^{\otimes}$ shows no scatter for $\gamma < 0.5$. $\tau_{pu}^{\otimes}$, however, scatters widely for $\gamma < 0.5$. The reason is that the weakest point of the first obstacle configuration met by the

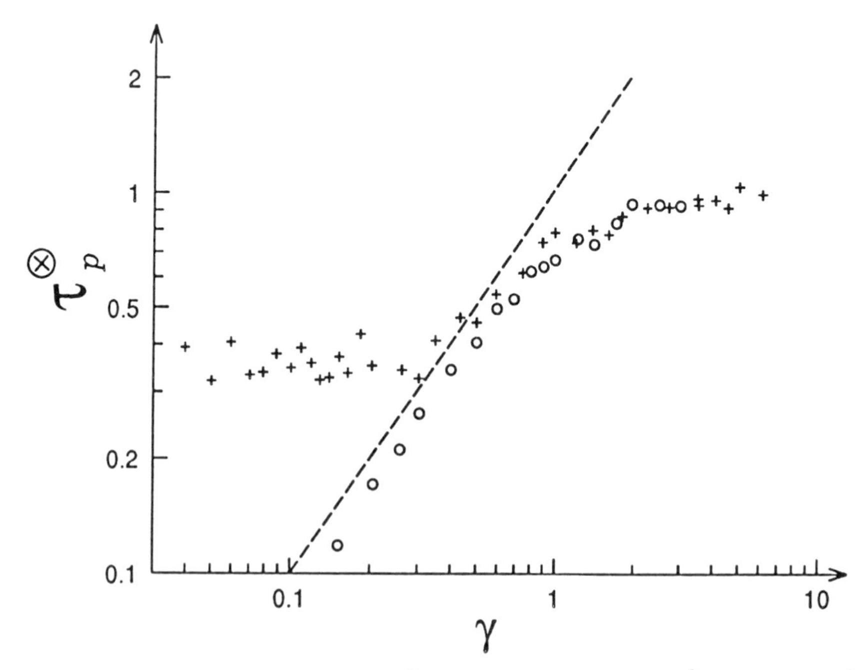

Fig. 4.31. Upper normalized CRSS $\tau_{pu}^{\otimes}$ (+) and lower CRSS $\tau_{pl}^{\otimes}$ (○) versus the normalized drag coefficient γ. Normalized range $\eta_0 = 0$, elastic interaction. Both axes are logarithmically divided. The dashed line corresponds to $\tau_{pl}^{\otimes} = \gamma$. (After Ref. 4.8.)

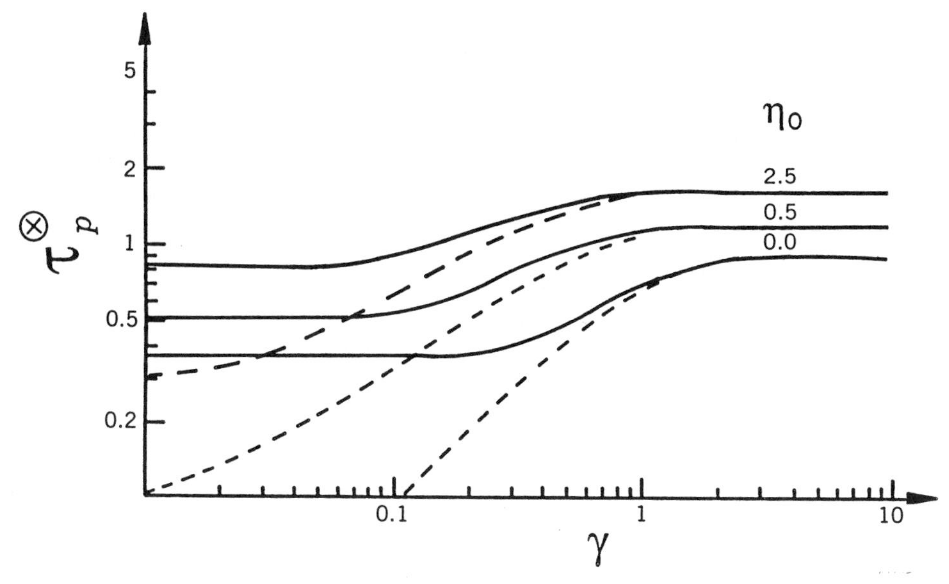

Fig. 4.32. Upper normalized CRSS $\tau_{pu}^{\otimes}$ (drawn out) and lower CRSS $\tau_{pl}^{\otimes}$ (dashed) versus the normalized drag coefficient γ. Exponent $n = 2$, elastic interaction. Both axes are logarithmically divided. The normalized range η_0 is given on the right-hand side. (After Ref. 4.8.)

dislocation determines $\tau_{pu}^{\otimes}$. Hence $\tau_{pu}^{\otimes}$ is likely to depend on the size of the array.

The variation of $\tau_{pu}^{\otimes}$ and $\tau_{pl}^{\otimes}$ with γ and η_0, $0 \leqslant \eta_0 \leqslant 2.5$, is shown in Figure 4.32. For the sake of clarity, only smoothed curves are presented. The scatter of the actual data was similar to that in Figure 4.31. There is a critical value γ_c at which $\tau_{pu}^{\otimes}$ and $\tau_{pl}^{\otimes}$ start to decrease from their high-γ-values, Schwarz and Labusch described the variation of γ_c with η_0 by

$$\frac{\gamma_c(\eta_0)}{\gamma_c(\eta_0 = 0)} = \frac{1}{(1 + 3\eta_0)^{1/2}} \tag{4.68}$$

$\gamma_c(\eta_0 = 0)$ is close to 3.0; this follows from Figure 4.31.

Three different ranges of γ can be distinguished in Figure 4.32; all ranges listed below are only approximate. The upper normalized CRSS $\tau_{pu}^{\otimes}$ and the lower CRSS $\tau_{pl}^{\otimes}$ increase with η_0 for $0 \leqslant \eta_0 \leqslant 2.5$.

1. Very low damping: $\gamma < 0.1$

   $\tau_{pu}^{\otimes}$ exceeds $\tau_{pl}^{\otimes}$; that is, $\tau_{pu}^{\otimes} > \tau_{pl}^{\otimes}$.
   $\tau_{pu}^{\otimes}$ is independent of γ; that is, $\partial\tau_{pu}^{\otimes}/\partial\gamma = 0$.
   $\tau_{pl}^{\otimes}$ increases with γ; that is, $\partial\tau_{pl}^{\otimes}/\partial\gamma > 0$.

   There are dynamic dislocation effects.

2. Damping of medium strength: $0.1 \leqslant \gamma \leqslant 3.0$

   $\tau_{pu}^{\otimes}$ exceeds $\tau_{pl}^{\otimes}$ for $\gamma \leqslant 1.0$, whereas for larger values of γ both normalized CRSSs are approximately the same; that is, $\tau_{pu}^{\otimes} > \tau_{pl}^{\otimes}$ for $\gamma \leqslant 1.0$ and $\tau_{pu}^{\otimes} \approx \tau_{pl}^{\otimes}$ for $\gamma > 1.0$.
   $\tau_{pu}^{\otimes}$ and $\tau_{pl}^{\otimes}$ increase with γ; that is, $\partial \tau_{pu}^{\otimes}/\partial \gamma > 0$ and $\partial \tau_{pl}^{\otimes}/\partial \gamma > 0$.

   For $\eta_0 \leqslant 2.5$, $\tau_{pu}^{\otimes}$ approximately doubles as γ increases from 0.1 to 3.0. There are dynamic dislocation effects.
3. Strong damping: $\gamma > 3.0$

   $\tau_{pu}^{\otimes}$ and $\tau_{pl}^{\otimes}$ coincide; that is, $\tau_{pu}^{\otimes} = \tau_{pl}^{\otimes}$.
   $\tau_{pu}^{\otimes}$ as well as $\tau_{pl}^{\otimes}$ are independent of γ; that is, $\partial \tau_{pu}^{\otimes}/\partial \gamma = 0$ and $\partial \tau_{pl}^{\otimes}/\partial \gamma = 0$.

   The dislocation motion is overdamped and there are no dynamic effects. Equations (4.62) and (4.63) relate $\tau_{pu}^{\otimes} = \tau_{pl}^{\otimes}$ to η_0. Most theories of strengthening are meant for this range of γ.

Now strong dynamic dislocation effects are considered. It has been stated in Section 4.2.2.2 that in this case (1) each obstacle is overcome independently of the others and (2) the dislocation stays approximately straight. Schwarz [4.9] treated this case analytically for elastically interacting obstacles. By virtue of point 2, the term $\partial^2 \eta/\partial \xi^2$ can be eliminated from Eq. (4.59). Averaging it over the normalized time interval θ_1 to θ_2 yields

$$\tau_{\text{ext}}^{\otimes} = \gamma\omega - \frac{1}{\theta_2 - \theta_1} \sum_i \int_{\theta_1}^{\theta_2} F^* \left(\frac{\eta(\theta) - \eta_i}{\eta_0} \right) d\theta \tag{4.69a}$$

where ω is the steady-state average velocity of the dislocation. F^* is defined in Eq. (4.4d). The summation is to be carried out over all obstacles lying in the area of unit width in the ξ-direction and of length $[\omega(\theta_2 - \theta_1)]$, $[\omega(\theta_2 - \theta_1)] \gg 1$, in the η-direction. Equation (4.69a) relates $\tau_{\text{ext}}^{\otimes}$ to ω. Because Eq. (4.69a) applies to the steady-state average velocity; $\partial^2 \eta/\partial \theta^2$ does not enter it.

In the limit of very high velocities, Eq. (4.69a) reduces to

$$\tau_{\text{ext}}^{\otimes} = \gamma\omega \tag{4.69b}$$

This is equivalent to Eq. (4.54).

Schwarz solved Eq. (4.69a) under two suppositions: $\gamma \ll 1.0$ and the force–distance profile is of antisymmetric triangular shape:

$$\tau_{\text{ext}}^{\otimes} = \gamma\omega + \eta_0\{\omega^2 \ln[(\omega + 1)/(\omega - 1)] - 2\omega\} \tag{4.70}$$

This function is shown in Figure 4.33 for $\gamma = 0.1$ and $\eta_0 = 0.5$. ω is not a unique function of $\tau_{\text{ext}}^{\otimes}$. $\tau_{pl}^{\otimes}$ and $\tau_{pu}^{\otimes}$ have been derived from computer simulations of the dislocation motion through random arrays of obstacles. Under conditions of constant stress $\tau_{\text{ext}}^{\otimes}$, the lower branch of the curve $\omega(\tau_{\text{ext}}^{\otimes})$ cannot be reached. Dislocations either move at velocities exceeding ω_l or do not move at all. Equation (4.69b), in which the effect of the obstacles is disregarded, overestimates ω.

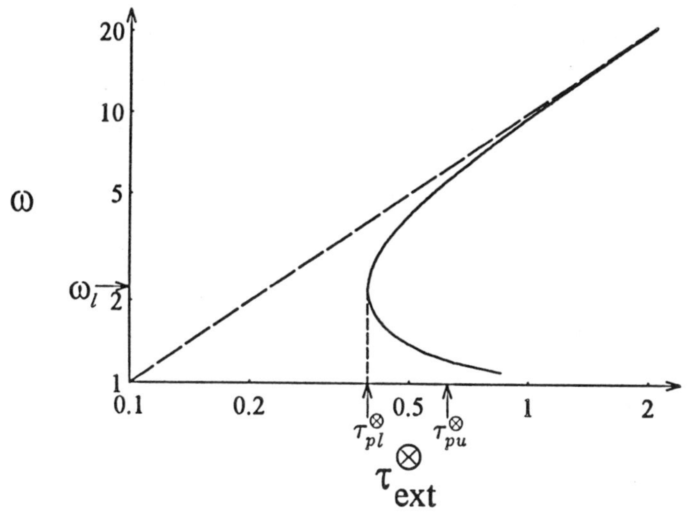

Fig. 4.33. Normalized velocity ω versus normalized stress $\tau^{\otimes}_{\text{ext}}$. The drawn-out curve represents Eq. (4.70), and the dashed line represents Eq. (4.69b). The profile of the obstacle–dislocation interaction force is of antisymmetric triangular shape. Both axes are logarithmically divided. Normalized drag coefficient $\gamma = 0.1$, normalized range $\eta_0 = 0.5$. The arrows mark the minimum velocity ω_l, the upper CRSS $\tau^{\otimes}_{pu}$, and the lower CRSS $\tau^{\otimes}_{pl}$. (After Ref. 4.9.)

4.2.2.3 Simulation of the Dislocation Motion Through Three-Dimensional Obstacle Arrangements. Rönnpagel and associates [4.78, 4.97–4.99] simulated the motion of dislocations through three-dimensional obstacle arrangements, which were generated as described in Section 3.3.1.5. These authors dispensed with all but two of Friedel's and Schwarz and Labusch's suppositions listed in Sections 4.2.1.3.1 and 4.2.2.2: Only supposition 6 (thermal activation can be disregarded) and supposition 9 (the particles' volume fraction f is rather low) have been kept in most cases. In one example the authors eliminated even supposition 6 [4.98]. Given the limitations of today's computers, Rönnpagel et al. normally chose to allow for various other effects and to neglect those of thermal activation. Its neglection in the case of particle strengthening has been justified at the end of Section 2.5.

Rönnpagel et al.'s advanced computer simulations involved four important points:

1. Three-Dimensional Obstacle Arrangements. The particles were arranged in space as described in Section 3.3.1.5. Their radii ρ were distributed according to the function g_{WLS} defined in Eq. (3.7). $\tau_{\text{obst}}(x, y)$, which was introduced in Section 4.1, is the sum over the stresses exerted by all those particles that simultaneously interact with the dislocation segment positioned at (x, y).

Because the glide plane is characterized by $z = 0$, the third coordinate is dropped. If the particles' interaction with dislocations is due to their long-range order, only those particles contribute to τ_{obst} that intersect the glide plane. In this case $\tau_{\text{obst}}(x, y)$ vanishes everywhere except inside of the particles. $\tau_{\text{obst}}(x, y)$ due to such a particle is shown in Figure 4.5a. If the particles have a lattice mismatch, even those of them may interact with dislocations that do not even contact their glide plane. $\tau_{\text{obst}}(x, y)$ produced by such a particle is presented in Figure 4.3. The ranges w_x and w_y of the particle–dislocation interaction force are fully allowed for in these simulations.

2. *Back Stress.* The stress τ_{back} [Eq. (2.11c)] varies along the bent dislocation. Elastic isotropy is assumed. The self-interaction between different parts of the bent dislocation is disregarded.

3. *Planar Faults.* A moving dislocation may create a planar fault along its entire length. This is allowed for by the introduction of τ_{fault} [4.97]:

$$b\tau_{\text{fault}} = \Gamma_{\text{fault}} \tag{4.71}$$

where Γ_{fault} is the energy per unit area of the fault. Equation (4.71) follows from Eq. (2.19). If a dislocation segment $d\mathbf{s}$ glides the distance dy, the energy $\Gamma_{\text{fault}}|d\mathbf{s}|dy$ is stored in the fault. Hence the force required to make the segment glide equals $\Gamma_{\text{fault}}|d\mathbf{s}|$. Inserting this result into Eq. (2.19) and substituting the subscript "fault" for "obst" yields Eq. (4.71). Examples for such faults are the stacking faults between the two Shockley partial dislocations in fcc metals and alloys and the antiphase boundaries created by nonperfect dislocations gliding in single-phase long-range-ordered materials. Though τ_{fault} has not been meant to describe the shear stresses due to particles, in some cases Eq. (4.71) will be used to calculate particle–dislocation interactions; examples are Eqs. (5.31) and (5.62).

4. *Mutual Interaction Between Different Dislocations.* If more than one dislocation glides in a plane, their mutual elastic interaction has to be allowed for. It is described by τ_{mut}. Let there be n_d dislocations. They are referred to by Dj, $1 \leqslant j \leqslant n_d$. $\tau_{\text{mut}}(x_i, y_i; x_j, y_j)$, $1 \leqslant i, j \leqslant n_d$, $i \neq j$, is the resolved shear stress exerted by Dj (sender) on a segment of Di (receiver). Because all dislocations are supposed to lie in the plane $z = 0$, all z-coordinates vanish and are not quoted. In the derivation of τ_{mut} the dislocation Dj (sender) is assumed to be straight. The Di segment has the coordinates (x_i, y_i); (x_j, y_j) are those of that point of Dj which is closest to the Di segment. $\tau_{\text{mut}}(x_i, y_i; x_j, y_j)$ is given here for two cases [4.97, 4.100]. $\mathbf{b}_m$ is the Burgers vector of Dm, $|\mathbf{b}_m| = |\mathbf{b}_1|$, $1 \leqslant m \leqslant n_d$. $\mathbf{s}_i$ and $\mathbf{s}_j$, $|\mathbf{s}_i| = |\mathbf{s}_j| = 1$, characterize the directions of the Di segment and of Dj, respectively.

(a) $\mathbf{b}_i \neq \mathbf{b}_j$, $\mathbf{s}_i = \mathbf{s}_j$. This applies, for example, to the two Shockley partial dislocations of a dissociated dislocation in an fcc material:

$$\tau_{\text{mut}}(x_i, y_i; x_j, y_j) = \frac{\mu_s}{2\pi b|\mathbf{r}_{ij}|}\{(\mathbf{b}_i \cdot \mathbf{s}_j)\cdot(\mathbf{b}_j \cdot \mathbf{s}_j) + \frac{1}{1-\nu}(\mathbf{b}_i \times \mathbf{s}_j)\cdot(\mathbf{b}_j \times \mathbf{s}_j)\}q \tag{4.72a}$$

(b) $\mathbf{b}_i = \mathbf{b}_j$, $\mathbf{s}_i \neq \mathbf{s}_j$. This describes, for example, the situation of a dislocation pair in a material which is strengthened by long-range-ordered particles:

$$\tau_{\text{mut}}(x_i, y_i; x_j, y_j) = \frac{\mu_s}{2\pi b|\mathbf{r}_{ij}|}\{(\mathbf{b}_j \cdot \mathbf{s}_j)^2 + \frac{1}{1-\nu}(\mathbf{b}_j \times \mathbf{s}_j)^2\}q \tag{4.72b}$$

Elastic isotropy is assumed; μ_s and ν are the shear modulus and Poisson's ratio of the matrix, respectively. The vector $\mathbf{r}_{ij}$ has the components $(x_i - x_j)$ and $(y_i - y_j)$. q governs the sign of τ_{mut}: $|q|$ equals unity. Now Eq. (2.18) assumes the following form for Di:

$$\tau_{\text{ext}} + \tau_{\text{obst}} + \sum_{\substack{j=1 \\ j \neq i}}^{n_d} \tau_{\text{mut}}(x_i, y_i; x_j, y_j) = \tau_{\text{inert}} + \tau_{\text{drag}} + \tau_{\text{back}} + \tau_{\text{fault}} \tag{4.73}$$

If one wants to include the effects of thermal activation, the stochastic stress τ_{therm} can be added on the left-hand side of Eq. (4.73). Actually this is a differential equation. τ_{inert}, τ_{drag}, and τ_{back} are given by Eqs. (2.13), (2.14), and (2.11c). The glide plane extends from $-x_0$ to $+x_0$ and from $-y_0$ to $+y_0$. The boundary conditions are

$$t < 0: \quad y(x) = -y_0, \quad \partial y/\partial t = v_n = 0, \quad \tau_{\text{ext}} = 0$$

$$t \geqslant 0: \quad \partial y/\partial x = 0 \text{ for } x = \pm x_0, \quad \tau_{\text{ext}} > 0$$

τ_{ext} is raised stepwise. After each step, Eq. (4.73) is solved numerically. This yields the function $y(x, t)$. Only after the dislocation's velocity v_n has become negligible, τ_{ext} is incremented again. The simulation is stopped once the dislocation has touched the line $y = +y_0$. The highest stress that still leads to a stable dislocation configuration is the CRSS. This procedure is the same one as those described in Sections 4.2.1.4.1 and 4.2.2.2. Because the above boundary conditions may favor dislocation configurations that are approximately parallel to the x-axis, Fuchs and Rönnpagel [4.97] also investigated the expansion of an originally circular dislocation loop. In this case an additional stress is

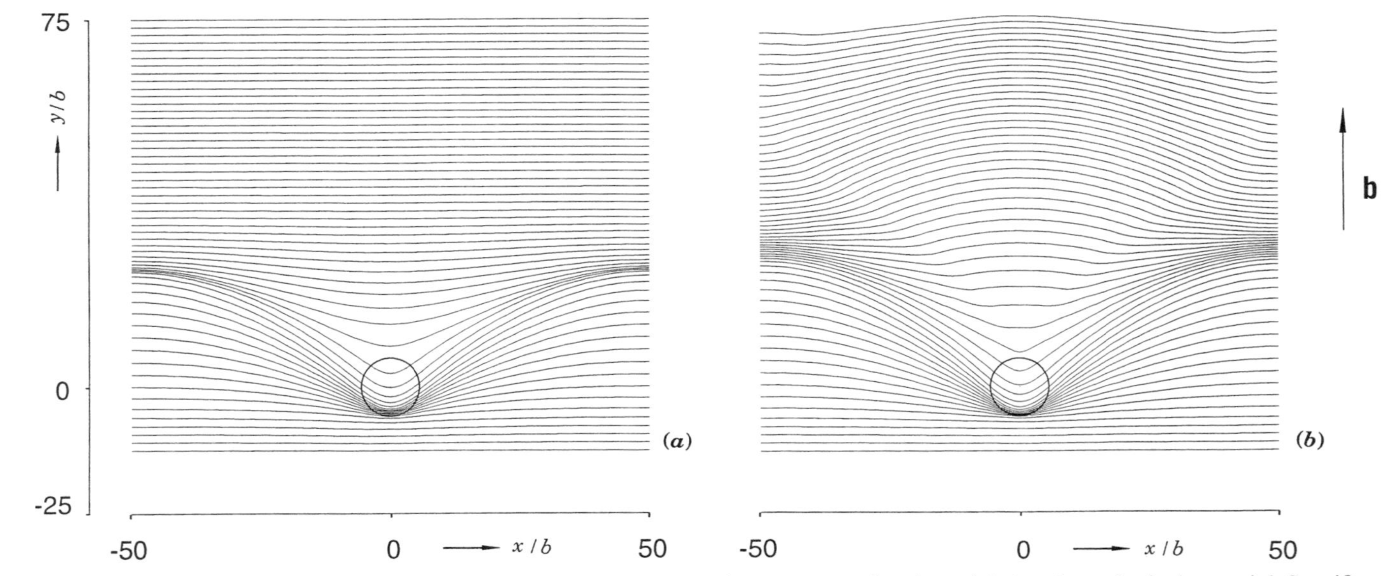

Fig. 4.34. Solutions of Eq. (4.73). An edge dislocation overcomes a long-range-ordered particle in a hypothetical material. Specific energy of the particle's antiphase boundary = 0.25 J/m^2, shear modulus of the matrix $\mu_s = 31$ GPa, Poisson's ratio of the matrix $\nu = 0.33$, drag coefficient = 1.7×10^{-5} Pa·s, $b = 0.25$ nm, particle radius $\rho = 8b$, coordinates of the center of the particle = $(0, 0, -\rho/\sqrt{2})$, density ρ_m of the matrix = 8.9×10^3 kg/m^3, inertial mass m_{eff} of the dislocation = $\rho_m b^2/2$. τ_{ext} increases at a constant rate. The area between two successive dislocation configurations is always the same. (*a*) τ_{inert} is disregarded. (*b*) τ_{inert} is allowed for. (Courtesy of D. Rönnpagel, Braunschweig.)

needed to compensate for the overall curvature of the loop; normally τ_{back} allows only for the stress required by the dislocation to bow out between two particles. In some cases (e.g., in Figure 4.34) τ_{ext} was increased continuously.

Rönnpagel et al.'s treatment of the motion of dislocations through three-dimensional obstacle arrangements is a highly significant advancement because various specific details of the hardening mechanisms can be allowed for. On the other hand, these details render it impossible to tabulate general results. Therefore only two examples are mentioned here; other details are presented in Chapter 5, where the various strengthening mechanisms are analyzed. In Figures 4.34–4.36 the boundary conditions are $\partial y/\partial x = 0$ for $x = \pm x_0$. Figure 4.37 shows the expansion of a pair of dislocation loops.

In Figure 4.34 an edge dislocation is seen to overcome a single particle. Because of its long-range order it repels the dislocation. τ_{ext} is slowly, but

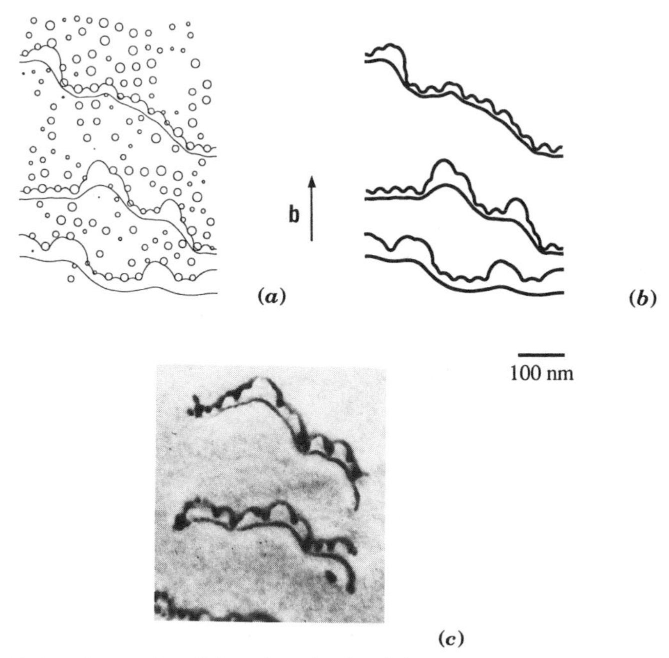

Fig. 4.35. Pairs of two edge dislocations in the nickel-base superalloy NIMONIC PE16, which is strengthened by long-range-ordered γ'-particles (Section 5.5). The dislocations glide upwards. (*a*) Solutions of Eq. (4.73), average γ'-particle radius $r = 7.5\,\text{nm}$, volume fraction $f = 0.089$, antiphase boundary energy $= 0.26\,\text{J/m}^2$, $\mu_s = 46\,\text{GPa}$, $\nu = 1/3$, $\tau_{ext} = 140\,\text{MPa}$. (*b*) As (*a*). For comparison with (*c*) the γ'-particles are not shown. (*c*) TEM image of dislocations under full load. $r = 8\,\text{nm}$, $f = 0.089$. In (*a*) and (*b*) the glide plane coincides with the image plane, and in (*c*) the glide plane is inclined; this shortens the images of the dislocations. The scale marker always refers to the image plane. ((*a*) and (*b*) Courtesy of T. Pretorius, Braunschweig. (*c*) After Ref. 4.2.)

continuously, increased. In Figure 4.34a, τ_{inert}, which describes the effects of inertia, is disregarded. Figure 4.34b shows the results obtained if τ_{inert} is allowed for: Behind the particle the dislocation vibrates. Far behind the particle, Eq. (4.54) is applicable: v_n is proportional to τ_{ext}. Only one-half of the inertial mass given by Eq. (2.15) has been inserted into Eq. (4.73).

Figures 4.35 and 4.36 illustrate order strengthening [4.99]. The particles are long-range-ordered. It has been mentioned in Section 4.1 and will be detailed further in Section 5.5 that this leads to the formation of dislocation pairs; their leading and trailing dislocations are referred to by D1 and D2, respectively. Figure 4.5a shows $\tau_{\text{obst}}(x, y)$ of D1. Because the sharp increase of $\tau_{\text{obst}}(x, y)$ at the particle's periphery may cause numerical problems in solving Eq. (4.73), $\tau_{\text{obst}}(x, y)$ has been smoothed there over the length $2b$. The resulting configurations of D1 and D2 strikingly resemble those observed in the TEM under full load (Figure 4.35). The configurations simulated for D1 and D2 in an

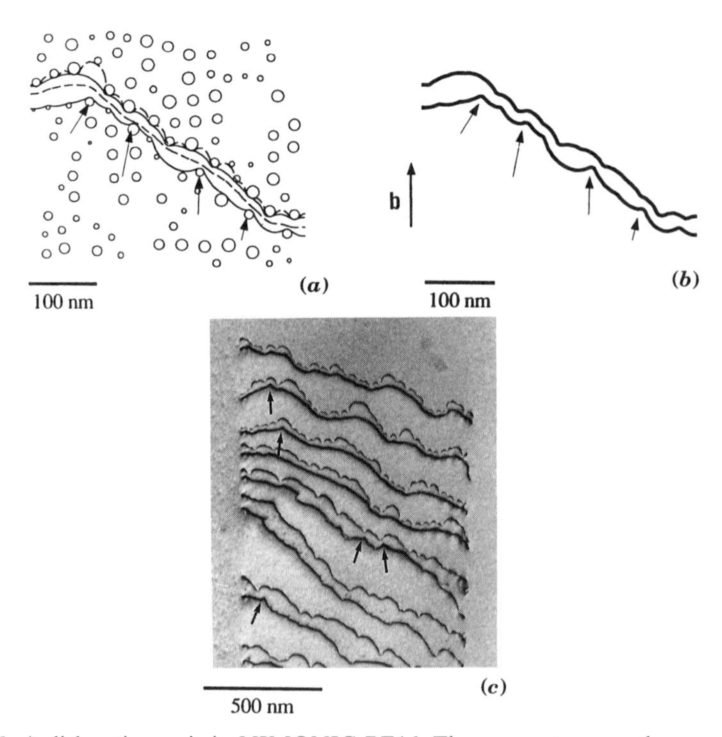

Fig. 4.36. A dislocation pair in NIMONIC PE16. The parameters are the same as in Figure 4.35a. The original glide direction is upwards. (*a*) Solutions of Eq. (4.73). --- $\tau_{\text{ext}} = 186\,\text{MPa}$ applied, —— τ_{ext} reduced to zero. The arrows indicate γ'-particles that prevent D2 from gliding backwards. (*b*) As (*a*), but only the configurations after unloading are shown. (*c*) Actual TEM image taken after unloading. For the scale markers see the caption of Figure 4.35 ((*a*) and (*b*) Courtesy of T. Pretorius, Braunschweig. (*c*) After Ref. 4.2.)

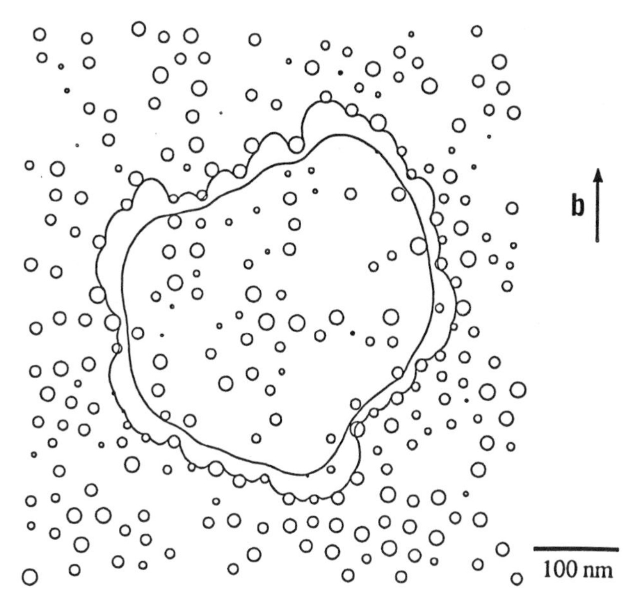

Fig. 4.37. Two simulated dislocation loops in NIMONIC PE16. The parameters are the same as in Figure 4.35. $\tau_{ext} = 182\,\text{MPa}$. (Courtesy of T. Pretorius, Braunschweig.)

unloaded specimen are shown in Figures 4.36a and 4.36b. First τ_{ext} had been slowly raised up to 186 MPa, and subsequently it was reduced to zero. Figure 4.36c gives the dislocation configurations seen in a stretched thin foil after unloading. Both D1 and D2 have relaxed. D2 would glide further backwards if it were not held up by the long-range-ordered particles. If D2 entered them, antiphase boundaries would be created. D2 forms sharp cusps at the particles. The cusps are marked by arrows. Again there is a very close agreement between the simulated dislocation configurations and those actually seen in the TEM. Figure 4.37 illustrates the expansion of two dislocation loops in a field of long-range-ordered particles.

The above examples prove the worth of Rönnpagel et al.'s computer simulations of the glide of dislocations through three-dimensional particle arrangements. It should be mentioned that Arsenault et al. [4.101] simulated the motion of a dislocation through a three-dimensional arrangement of solute atoms; that is, these authors analyzed solid solution strengthening.

4.3 SUMMARY

In Section 1.3 of the introduction it has already been stated that in relating the CRSS of a particle-strengthened metal or alloy to the properties of the shearable particles contained in it, one proceeds in two steps:

1. Derivation of the resolved shear stress τ_{obst} (Section 4.1), which the particles exert on the dislocation
2. Solving Eq. (2.18) or Eq. (2.20) (Section 4.2), each of which describes the motion of a dislocation through a field of particles.

In most cases, step 2 necessitates rather coarse approximations for τ_{obst}. Friedel [4.40, 4.41] (Section 4.2.1.3) solved Eq. (2.20) analytically. His most important simplifications are as follows: The obstacles are point obstacles of uniform strength, they are uniformly spaced, and the glide of the dislocations is steady. Friedel's result is given in Eqs. (4.26). Kocks [4.50, 4.51] and Hanson and Morris [4.39] eliminated the supposition that the obstacles are uniformly spaced. These authors analyzed planar random arrays of point obstacles statistically. Friedel's equations have been approximately maintained, but the CRSSs are 10–15% lower [Eqs. (4.38) and (4.43)] than those found by Friedel.

Foreman and Makin [4.37, 4.38] simulated the motion of a dislocation through a planar random array of point obstacles in a digital computer. The results were in satisfactory agreement with those derived by Kocks and Hanson and Morris analytically. Subsequently, Schwarz and Labusch [4.8] simulated the dislocation motion through planar random arrays of one-dimensional obstacles. Their interaction with dislocations was described by Eqs. (4.4) and (4.6). The CRSS was found to depend on the range of the interaction force profile [Eqs. (4.62) and (4.63)]. So far the most comprehensive computer simulations have been performed by Rönnpagel and associates [4.78, 4.97–4.99]. This group allowed for (i) the spatial variation of τ_{obst} produced by three-dimensional obstacle arrangements, (ii) the variation of the dislocation line tension with the bowing angle, and (iii) the mutual interaction between various dislocations gliding in the same glide plane. The complexity of the systems treated makes it difficult to tabulate general results. Some dislocation configurations obtained by these authors are shown in Figures 4.34–4.37.

Two important problems have not yet received the necessary attention:

1. The effect of particle volume fractions in excess of 0.2.
2. The mutual interaction between dislocations. Most authors limited their analyses to a single dislocation gliding in the slip plane. In the case of strengthening by ordered particles, a single pair of dislocations has been considered. Actually, however, many approximately parallel dislocations may glide simultaneously in a slip plane. τ_{mut} in Eq. (4.73) allows for this. In a pileup of n_d dislocations, τ_{mut} acting on the leading one is close to $[(n_d - 1)\tau_{ext}]$. Figures 4.6 and 5.24 show processions of dislocation pairs in the superalloy NIMONIC PE16.

Much of the material presented in this chapter is also applicable to solid solution hardening at 0 K. Because the interaction energy between a solute atom and a dislocation is small, thermal activation must not be disregarded at finite temperatures.

REFERENCES

4.1 S. Amelinckx, in *Dislocations in Solids*, Vol. 2 (edited by F. R. N. Nabarro), p. 67, North-Holland, Amsterdam, 1979.

4.2 E. Nembach, K. Suzuki, M. Ichihara, and S. Takeuchi, *Philos. Mag. A*, **51**, 607, 1985.

4.3 A. Pineau and F. Baudier, *Scripta Metall.*, **3**, 757, 1969.

4.4 B. Reppich, *Acta Metall.*, **23**, 1055, 1975.

4.5 A. Melander and P. Å. Persson, *Metal Sci.*, **12**, 391, 1978.

4.6 E. Nembach, *phys. stat. sol. (a)*, **78**, 571, 1983.

4.7 E. Nembach and G. Neite, *Prog. Mater. Sci.*, **29**, 177, 1985.

4.8 R. B. Schwarz and R. Labusch, *J. Appl. Phys.*, **49**, 5174, 1978.

4.9 R. B. Schwarz, *Phys. Rev. B*, **21**, 5617, 1980.

4.10 J. W. Morris, Jr. and C. K. Syn, *J. Appl. Phys.*, **45**, 961, 1974.

4.11 V. Gerold and H. Haberkorn, *phys. stat. sol.*, **16**, 675, 1966.

4.12 N. Büttner, K.-D. Fusenig, and E. Nembach, *Acta Metall.*, **35**, 845, 1987.

4.13 K.-D. Fusenig and E. Nembach, *Acta Metall. Mater.*, **41**, 3181, 1993.

4.14 E. Nembach, K. Suzuki, M. Ichihara, and S. Takeuchi, *Mater. Sci. Eng. A*, **101**, 109, 1988.

4.15 U. Messerschmidt, F. Appel, and H. Schmid, *Philos. Mag. A*, **51**, 781, 1985.

4.16 N. Büttner and E. Nembach, *Acta Metall.*, **30**, 83, 1982.

4.17 E. Nembach and M. Martin, *Acta Metall.*, **28**, 1069, 1980.

4.18 E. Orowan, *Z. Phys.*, **89**, 634, 1934.

4.19 C. N. Reid, *Deformation Geometry for Materials Scientists*, Pergamon Press, Oxford, 1973.

4.20 H. Alexander and P. Haasen, in *Solid State Physics*, Vol. 22 (edited by F. Seitz, D. Turnbull, and H. Ehrenreich), p. 27, Academic Press, New York 1968.

4.21 H. Alexander, in *Dislocations in Solids*, Vol. 7 (edited by F. R. N. Nabarro), p. 113, North-Holland, Amsterdam, 1986.

4.22 E. Orowan, *Proc. Phys. Soc.*, **52**, 8, 1940.

4.23 J. L. Snoek, *Physica*, **8**, 734, 1941.

4.24 A. H. Cottrell and B. A. Bilby, *Proc. Phys. Soc. A*, **62**, 49, 1949.

4.25 H. W. Paxton and A. T. Churchman, *Acta Metall.*, **1**, 473, 1953.

4.26 A. H. Cottrell, *Dislocations and Plastic Flow in Crystals*, Clarendon Press, Oxford, 1953.

4.27 H. Suzuki, in *Dislocations and Mechanical Properties of Crystals* (edited by J. C. Fisher, W. G. Johnston, R. Thomson, and T. Vreeland, Jr.), p. 361, John Wiley & Sons, New York, Chapman & Hall, London, 1957.

4.28 H. Suzuki, in *Proceedings of the 5th International Conference on the Strength of Metals and Alloys*, Vol. 3, (edited by P. Haasen, V. Gerold, and G. Kostorz), p. 1595, Pergamon Press, Toronto, 1979.

4.29 U. F. Kocks, in *Physics of Strength and Plasticity* (edited by A. S. Argon), p. 143, MIT Press, Cambridge, Massachusetts, 1969.

4.30 N. F. Mott and F. R. N. Nabarro, *Proc. Phys. Soc.*, **52**, 86, 1940.

4.31 N. F. Mott, *J. Inst. Metals*, **72**, 367, 1946.

4.32 F. R. N. Nabarro, *Proc. Phys. Soc.*, **58**, 669, 1946.

4.33 N. F. Mott and F. R. N. Nabarro, *Report of a Conference on Strength of Solids*, p. 1, The Physical Society, London, 1948.

4.34 F. R. N. Nabarro, *Symposium on Internal Stresses in Metals and Alloys*, p. 237, Institute of Metals, London, 1948.

4.35 N. F. Mott, in *Imperfections in Nearly Perfect Crystals* (edited by W. Shockley, J. H. Hollomon, R. Maurer, and F. Seitz), p. 173, John Wiley & Sons, New York, Chapman & Hall, London, 1952.

4.36 F. R. N. Nabarro, *J. Less-Common Metals*, **28**, 257, 1972.

4.37 A. J. E. Foreman and M. J. Makin, *Philos. Mag.*, **14**, 911, 1966.

4.38 A. J. E. Foreman and M. J. Makin, *Can. J. Phys.*, **45**, 511, 1967.

4.39 K. Hanson and J. W. Morris, Jr., *J. Appl. Phys.*, **46**, 983, 1975.

4.40 J. Friedel, *Les Dislocations*, Gauthier-Villars, Paris, 1956.

4.41 J. Friedel, *Dislocations*, Pergamon Press, Oxford, 1964.

4.42 R. L. Fleischer and W. R. Hibbard, Jr., in *The Relation Between the Structure and Mechanical Properties of Metals*, Vol. I, p. 262, Her Majesty's Stationery Office, London, 1963.

4.43 R. L. Fleischer, in *The Strengthening of Metals* (edited by D. Peckner), p. 93, Reinhold, New York, Chapman & Hall, London, 1964.

4.44 E. Orowan, *Symposium on Internal Stresses in Metals and Alloys*, p. 451, Institute of Metals, London, 1948.

4.45 J. Glazer and J. W. Morris, Jr., *Philos. Mag. A*, **56**, 507, 1987.

4.46 A. J. Ardell and J. C. Huang, *Philos. Mag. Lett.*, **58**, 189, 1988.

4.47 J. Glazer and J. W. Morris, Jr., *Philos. Mag. Lett.*, **62**, 33, 1990.

4.48 A. Orlová, K. Milicka, and J. Cadek, *Mater. Sci. Eng.*, **50**, 221, 1981.

4.49 L. M. Brown and R. K. Ham, in *Strengthening Methods in Crystals* (edited by A. Kelly and R. B. Nicholson), p. 9, Applied Science Publishers, London, 1971.

4.50 U. F. Kocks, *Philos. Mag.*, **13**, 541, 1966.

4.51 U. F. Kocks, *Can. J. Phys.*, **45**, 737, 1967.

4.52 U. F. Kocks, *Acta Metall.*, **14**, 1629, 1966.

4.53 J. E. Dorn, P. Guyot, and T. Stefansky, in *Physics of Strength and Plasticity* (edited by A. S. Argon), p. 133, MIT Press, Cambridge, Massachusetts, 1969.

4.54 J. W. Morris, Jr. and D. H. Klahn, *J. Appl. Phys.*, **44**, 4882, 1973.

4.55 J. W. Morris, Jr. and D. H. Klahn, *J. Appl. Phys.*, **45**, 2027, 1974.

4.56 K. Hanson, S. Altintas, and J. W. Morris, Jr., *Nucl. Metall.*, **20**, 917, 1976.

4.57 W. Feller, *An Introduction to Probability Theory and its Applications*, Vol. 1, 2nd Edition, 10th printing, John Wiley & Sons, New York, 1965.

4.58 R. Labusch, *J. Appl. Phys.*, **48**, 4550, 1977.

4.59 R. K. Ham, *Trans. JIM Suppl.*, **9**, 52, 1968.

4.60 R. Labusch, *phys. stat. sol.*, **41**, 659, 1970.

4.61 R. Labusch, *Acta Metall.*, **20**, 917, 1972.

4.62 F. R. N. Nabarro, *Philos. Mag.*, **35**, 613, 1977.

4.63 R. J. Arsenault, S. Patu, and D. M. Esterling, *Met. Trans. A*, **20A**, 1411, 1989.

4.64 G. deWit and J. S. Koehler, *Phys. Rev.*, **116**, 1113, 1959.

4.65 L. M. Brown, *Philos. Mag.*, **10**, 441, 1964.

4.66 H. Schmid and H. O. K. Kirchner, *Philos. Mag. A*, **58**, 905, 1988.

4.67 D. J. Bacon and R. O. Scattergood, *J. Phys. F: Metal Phys.*, **4**, 2126, 1974.

4.68 U. F. Kocks, A. S. Argon, and M. F. Ashby, *Prog. Mater. Sci.*, **19**, 1, 1975.

4.69 A. Melander and P. Å. Persson, *Acta Metall.*, **26**, 267, 1978.

4.70 F. C. Frank and W. T. Read, Jr., *A Symposium on the Plastic Deformation of Crystalline Solids*, p. 44, Pittsburgh, 1950.

4.71 F. C. Frank and W. T. Read, Jr., *Phys. Rev.*, **79**, 722, 1950.

4.72 A. J. E. Foreman, *Philos. Mag.*, **15**, 1011, 1967.

4.73 D. J. Bacon, U. F. Kocks, and R. O. Scattergood, *Philos. Mag.*, **28**, 1241, 1973.

4.74 R. O. Scattergood and D. J. Bacon, *Philos. Mag.*, **31**, 179, 1975.

4.75 K. Fuchs, *Proc. R. Soc. London A*, **157**, 444, 1936.

4.76 W. C. Overton, Jr. and J. Gaffney, *Phys. Rev.*, **98**, 969, 1955.

4.77 G. A. Alers, J. R. Neighbours, and H. Sato, *J. Phys. Chem. Sol.*, **13**, 40, 1960.

4.78 D. Rönnpagel, *Proceedings of the 8th Risø International Symposium on Metallurgy and Materials Science, Constitutive Relations and Their Physical Basis* (edited by S. I. Andersen, J. B. Bilde-Sørensen, N. Hansen, T. Leffers, H. Lilholt, O B. Pedersen, and B. Ralph), p. 503, Risø National Laboratory, Roskilde, Denmark, 1987.

4.79 B. Reppich, W. Kühlein, G. Meyer, D. Puppel, M. Schulz, and G. Schumann, *Mater. Sci. Eng.*, **83**, 45, 1986.

4.80 A. J. Ardell, in *Intermetallic Compounds*, Vol. 2 (edited by J. H. Westbrook and R. L. Fleischer), p. 257, John Wiley & Sons, New York, 1994.

4.81 A. J. Ardell, *Met. Trans. A*, **16A**, 2131, 1985.

4.82 B. M. Strunin and A. B. Popov, *phys. stat. sol. (a)*, **34**, 761, 1976.

4.83 R. Labusch and R. B. Schwarz, *Nucl. Metall.*, **20**, 650, 1976.

4.84 V. I. Alshits and V. L. Indenbom, in *Dislocations in Solids*, Vol. 7 (edited by F. R. N. Nabarro), p. 43, North-Holland, Amsterdam, 1986.

4.85 E. Nadgornyi, *Prog. Mater. Sci.*, **31**, 1, 1988.

4.86 R. W. Balluffi and A. V. Granato, in *Dislocations in Solids*, Vol. 4 (edited by F. R. N. Nabarro), p. 1, North-Holland, Amsterdam, 1979.

4.87 T. A. Parkhomenko and V. V. Pustovalov, *phys. stat. sol. (a)*, **74**, 11, 1982.

4.88 H. Kojima and T. Suzuki, *Phys. Rev. Lett.*, **21**, 896, 1968.

4.89 H. Kojima, T. Moriya, and T. Suzuki, *J. Phys. Soc. Jpn.*, **38**, 1032, 1975.

4.90 M. Suenaga and J. M. Galligan, *Scripta Metall.*, **5**, 829, 1971.

4.91 G. Kostorz, *phys. stat. sol. (b)*, **58**, 9, 1973.

4.92 F. R. N. Nabarro and A. T. Quintanilha, in *Dislocations in Solids*, Vol. 5 (edited by F. R. N. Nabarro), p. 193, North-Holland, Amsterdam, 1980.

4.93 J. Ohser, private communication.

4.94 V. Gerold and K. Hartmann, *Proceedings of the International Conference on the Strength of Metals and Alloys, Trans. JIM, Suppl.*, **9**, 509, 1968.

4.95 E. Nembach, *Scripta Metall.*, **20**, 763, 1986.

4.96 F. Rabe and P. Haasen, *Proceedings of the 8th International Conference on the Strength of Metals and Alloys*, Vol. 1 (edited by P. O. Kettunen, T. K. Lepistö, and M. E. Lehtonen), p. 561, Pergamon Press, Oxford, England, 1988.

4.97 A. Fuchs and D. Rönnpagel, *Mater. Sci. Eng.*, **A164**, 340, 1993.

4.98 D. Rönnpagel, T. Streit, and T. Pretorius, *phys. stat. sol. (a)*, **135**, 445, 1993.

4.99 T. Pretorius and D. Rönnpagel, *Proceedings of the 10th International Conference of the Strength of Materials*, (edited by H. Oikawa, K. Maruyama, S. Takeuchi, and M. Yamaguchi), p. 689, JIM, 1994.

4.100 J. P. Hirth and J. Lothe, *Theory of Dislocations*, 2nd edition, John Wiley & Sons, New York, 1982.

4.101 R. J. Arsenault, S. Patu, and D. M. Esterling, *Met. Trans. A*, **20A**, 1419, 1989.

CHAPTER 5

The Critical Resolved Shear Stress of Particle-Hardened Materials. Particle Shearing: The Different Particle–Dislocation Interaction Mechanisms

In Chapter 4, models of strengthening by coherent shearable particles were discussed from a general point of view. In the present chapter the various hardening mechanisms will be analyzed: Five interactions between penetrable particles and dislocations will be distinguished. They are listed here in the order of increasing complexity of their treatment. The section in which the respective mechanism is discussed is given in parentheses. It is emphasized again (see beginning of Chapter 4) that the particle–matrix interface is assumed to be free of dislocations. Thermal activation will be disregarded throughout.

Chemical Strengthening (Section 5.1). As a dislocation shears a coherent particle, an additional particle–matrix interface is created. This is illustrated in Figures 1.4b and 5.1. The externally applied stress must supply the energy stored in the interface. Because the particle is coherent, the specific energy is low (Section 3.2.1.1) and the ensuing hardening effect is very small. An alternative name for this strengthening mechanism is "surface" strengthening. This particle–dislocation interaction is of the energy storing type defined in Section 4.1.

Modulus Mismatch Strengthening (Section 5.2). Within the framework of the theory of linear isotropic elasticity, the energy density in the strain field of a dislocation is proportional to the shear modulus of the material [Eqs. (2.3)]. If that of the matrix and that of the particle differ from each other, the latter attracts or repels the dislocation, depending on whether the modulus of the particle is smaller or larger than that of the matrix. This interaction is elastic.

Stacking Fault Energy Mismatch Strengthening (Section 5.3). This mechanism is analogous to modulus mismatch strengthening. A difference in the specific stacking fault energy between particle and matrix leads to an elastic particle–dislocation interaction.

Lattice Mismatch Strengthening (Section 5.4). If the lattice constant of the coherent particle differs from that of the matrix, the particle is surrounded by a stress field, via which it interacts elastically with dislocations. This mechanism is often referred to as "coherency" strengthening.

Order Strengthening (Section 5.5). If the matrix and the particles have the same crystal structure, but only the latter ones are long-range-ordered, a perfect matrix dislocation is only a partial dislocation in the particles: It creates antiphase boundaries in them. In such materials, dislocations glide in pairs: The trailing dislocation eliminates the antiphase boundaries. This has already been mentioned in Sections 4.1 and 4.2.2.3 and is illustrated in Figures 4.6 and 4.35–4.37. In contrast to chemical strengthening where the planar fault is confined to the periphery of the particle, the antiphase boundary extends over the entire particle–glide-plane intersection. Order strengthening is the relevant mechanism in technical nickel-base superalloys. Long-range-ordered particles are obstacles of the energy-storing type (Section 4.1.1), because the two dislocations of a pair interact rather independently of each other with the particles. In the case of stacking fault energy mismatch strengthening, the coupling of the two partial dislocations is much stronger.

In the following sections the five above-named strengthening mechanisms will be discussed in the light of the general models presented in the preceding chapter. The coordinate system is chosen as before: The glide plane is the plane characterized by $z = 0$, and the dislocation is approximately parallel to the x-axis and glides towards $+y$. First the resolved shear stress τ_{obst}, respectively, the force profile $F(y)$ and/or $F_0 = \text{Max}\{|F(y)|\}$ are derived. Subsequently these parameters are inserted into the formulae presented in Section 4.2 for the CRSS. τ_{obst}, $F(y)$, and F_0 have been defined in Sections 4.1, 4.1.1, and 4.1.2, respectively. It is of course an idealization to assume that the particles of a given system interact with dislocations via only one mechanism. Order strengthening is always accompanied by chemical hardening; moreover, the long-range-ordered particles may have a lattice and/or modulus mismatch. In such cases, τ_{obst} and F represent the sums over all interaction stresses and forces, respectively [5.1–5.5]. Examples for superpositions of force profiles are given in Figures 4.5d and 5.7, respectively, for the following combinations: (a) order with lattice mismatch strengthening and (b) lattice mismatch with modulus mismatch strengthening. It is the sum over the stresses and forces, which enters Eqs. (2.18) and (2.20) and the formulae given in Section 4.2. This has already been emphasized in Section 4.1. One must not calculate CRSSs $\tau_{p\,ord}$, $\tau_{p\,chem}$, and $\tau_{p\varepsilon}$ for order, chemical, and lattice mismatch hardening first and then sum over these CRSSs. There are several instances for this inappropriate approach in the literature [e.g., Refs. 5.6 and 5.7].

One hopes of course that one of the several mechanisms via which a particle may interact with dislocations actually outweights the others. One measures the CRSS $\tau_{p\,\exp}$ as a function of f and r; the subscript "exp" stands for "experiment." r and f are the particles' mean radius and their volume fraction, respectively. On the basis of the supposed hardening mechanism, the theoretical function $\tau_{p\,\mathrm{theo}}(r, f, P_i)$ is established. "theo" of $\tau_{p\,\mathrm{theo}}$ indicates "theory." The respective models have been presented in Section 4.2. P_i stands for the various parameters involved in the model—for example, for the constrained lattice mismatch in the case of lattice mismatch strengthening and for the numerical constants C_i appearing in Eqs. (4.62) and (4.63). Because the particles will have a distribution of radii and they may be sheared off-center by dislocations, the appropriate effective radius has to be inserted into the formulae. To test whether the chosen model describes the data $\tau_{p\,\exp}(r, f)$ well, they are compared with $\tau_{p\,\mathrm{theo}}(r, f, P_i)$. If the agreement is good, the supposed interaction mechanism and the model are confirmed. This procedure requires the precise knowledge of the parameters P_i. Often one chooses an alternative approach: One fits the function $\tau_{p\,\mathrm{theo}}(r, f, P_i)$ to the data $\tau_{p\,\exp}(r, f)$. One or several of the P_i are treated as adjustable parameters. If $\tau_{p\,\mathrm{theo}}$ represents the data well AND (logic AND) if the parameters P_i turn out to have reasonable values, the model on the basis of which $\tau_{p\,\mathrm{theo}}(r, f, P_i)$ had been derived can be accepted. Because various hardening mechanisms may lead to identical functional variations of $\tau_{p\,\mathrm{theo}}$ with r and f, the correct functional dependence of τ_p on r and f alone is no absolute proof of the applicability of the respective model. This point has been stressed by Kelly [5.8], and its relevance is demonstrated by the following example. Friedel's equation (4.26d) states that for any interaction mechanism for which the force F_0 is proportional to r, the CRSS is proportional to $[(rf)^{1/2}]$. The slight variation of the dislocation line tension S with the dislocation's outer cutoff radius is disregarded (Section 4.2.1.4.3). The term $(rf)^{1/2}$ appears, for example, in Eqs. (5.52b) and (5.67a), which apply to lattice mismatch and order strengthening, respectively.

The problem that particles of a given type interact with dislocations via several different mechanisms has to be distinguished from the problem that there are different types of particles simultaneously present, each type having a different interaction mechanism. There may be, for example, coherent long-range-ordered γ'-particles and incoherent oxide particles in superalloys. This example has been mentioned in Section 1.5. The problem that various types of obstacles to the glide of dislocations are simultaneously present in a specimen will be dealt with in Chapter 7.

Before the five above-named strengthening mechanisms are subjected to detailed analyses, some of the points covered in Chapter 4 are recalled.

In the derivation of Friedel's [5.9] equations (4.26), the obstacles to the glide of the dislocation have been assumed to be uniformly distributed in the glide plane. This was assumption 3 in Section 4.2.1.3.1. Random distributions have been discussed in Sections 4.2.1.4.1 and 4.2.2. The randomness lowered the CRSS by about 10%. This is expressed by the factor 0.887 in Eq. (4.43) and by the coefficient $C_1 \approx 0.9$ in Eqs. (4.62) and (4.63). When in the following Friedel's equations are applied, this statistical factor is normally disregarded. This has the consequences that in the case of very small normalized ranges η_0 [Eq. (4.58g)], τ_{pF} calculated on the basis of Friedel's equations (4.26) exceeds τ_{pL} derived from Schwarz and Labusch's [5.10] equations (4.62) or (4.63). In Sections 5.4.2 and 5.5.3.1 the factor C_1 will be introduced into equations based on Friedel's equations (4.26). Thus one has the option to allow for the randomness of the obstacle arrangement by inserting Schwarz and Labusch's result for C_1. If the randomness is to be disregarded, C_1 equals unity. In Section 3.3.1.5 it has been shown that the particles in a material that has been subjected to Ostwald ripening are not distributed at random; the distribution function of interparticle spacings is rather narrow.

Friedel's as well as Schwarz and Labusch's (Section 4.2.2.2) equations apply to underaged (Section 4.2.1.3.2) materials: The particles are sheared and dislocation bowing is assumed to be only slight. This implies that (i) the particle spacing L_c along the dislocation exceeds $L_{\min}$ (Section 3.1) in the critical moment and (ii) the maximum force exerted by a particle on a dislocation is smaller than the dislocation line tension [relation (4.9c)].

As in Chapter 4, only one dislocation is assumed to glide in a plane. Processions of dislocations such as those shown in Figure 4.6 are not treated in general. Some exceptions are (i) a pair of partial dislocations in the case of stacking fault strengthening (Section 5.3), (ii) a pair of dislocations in materials strengthened by long-range-ordered particles (Section 5.5), and (iii) Eq. (4.73), which allows for a procession of dislocations.

Since with the only exception presented in Section 5.4.3, the motion of the dislocation is assumed to be overdamped, all discussions of particle strengthening will be based on the relations in Sections 4.2.1, 4.2.2.1, and 4.2.2.2.1. An example for dynamic dislocation effects in a particle-strengthened material will be analyzed in Section 5.4.3.

In many of the equations presented in this chapter, the statistical factors ω_r, ω_d, and ω_q appear. They have been defined in Eqs. (3.3)–(3.5). Their numerical values depend on the distribution function g of the particle radii. ω_r, ω_d, and ω_q of g_0 [Eq. (3.6)] and of g_{WLS} [Eq. (3.7)] are listed in Table 3.2. In all equations concerning the CRSS, unity is inserted for the statistical factors ω_n defined in Eq. (3.5b).

5.1 CHEMICAL STRENGTHENING

When a dislocation shears a coherent particle, additional lunar-shaped particle–matrix interface is created. This is sketched in Figure 5.1. By virtue of the coherency, the specific energy Γ involved is low: 0.01–0.2 J/m^2 [5.11, 5.12]. Possible anisotropies of Γ will be disregarded in the following. Obstacles that lead to chemical strengthening are of the energy storing type defined in Section 4.1.1.

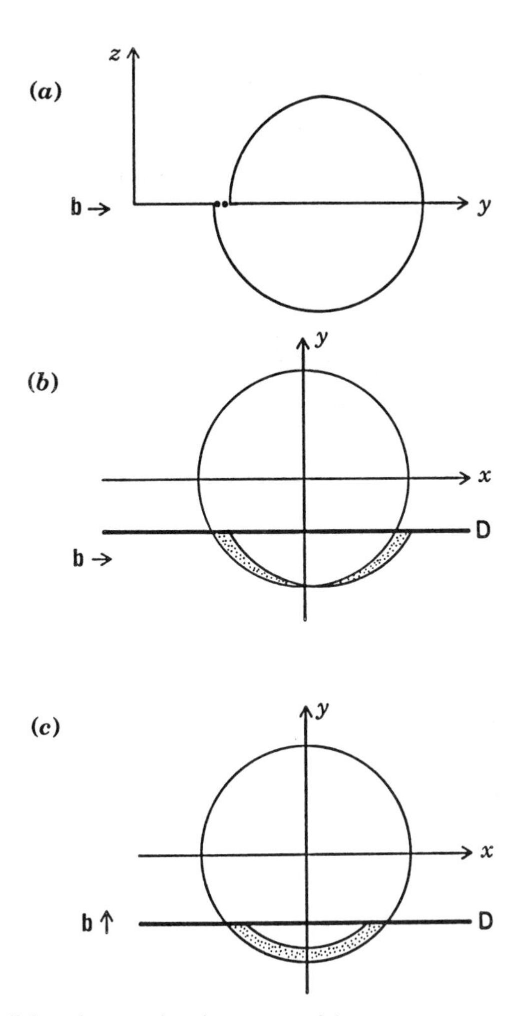

Fig. 5.1. A dislocation cutting into a particle creates new particle–matrix interface, shown schematically. The new interface is dotted. (*a*) Edge dislocation normal to the image plane. (*b*) Screw dislocation (D). The image plane is the glide plane. (*c*) As (*b*) for an edge dislocation (D).

Kelly and Fine [5.13] interpreted particle hardening of two aluminum base alloys as being due to chemical hardening:

1. Aluminum-rich aluminum–copper alloys containing copper-rich plate-like Guinier–Preston zones on {100}-planes; such zones have been described in Section 1.1.
2. Aluminum-rich aluminum–silver alloys containing silver-rich spherical particles.

Kelly and Fine derived estimates of the critical resolved shear stresses (CRSSs) of these alloys from the energy difference ΔE between a copper and a silver atom in the particle and in the aluminum-rich matrix. ΔE was related to the heat of reversion of the alloy. The CRSSs obtained in this way were of the correct magnitude. Today, however, particle hardening of aluminum-rich aluminum–silver alloys is believed to be caused by the difference of the stacking fault energies of the two phases involved [5.14, 5.15]. This will be detailed in Section 5.3. Brown and Ham [5.16] interpreted strengthening of copper by beryllium-rich zones as being due to chemical strengthening.

If the precipitates are plate-like, one expects that their interaction force with dislocations depends on the direction of the Burgers vector **b** relative to the plane of the precipitate. The force will be high if **b** is normal to the plate [5.17].

For spherical particles the profile $F_\Gamma(y)$ of the particle–dislocation interaction force is calculated on the basis of the straight-line approximation introduced in Section 4.1.1. The subscript Γ of F_Γ refers to chemical hardening. Because F_Γ is related to the area A of the new particle–matrix interface, A is calculated first:

$$F_\Gamma(y) = -\Gamma\, \partial A(y)/\partial y$$

In Figure 5.1b a straight screw dislocation has reached the position characterized by y. The area $A(y)$ is dotted. It is given by

$$A(y) = 2\int_{-\rho'}^{y} \{(\rho'^2 - y'^2)^{1/2} - [(\rho'^2 - y'^2)^{1/2} - b]\}\, dy' = 2b(y + \rho') \qquad (5.1)$$

ρ' is the radius of the intersection of the glide plane with the particle; ρ' is assumed to be much larger than the length b of the Burgers vector. Figure 5.1c illustrates the situation for an edge dislocation:

$$A(y) = 2\left\{\int_{-\rho'}^{y} (\rho'^2 - y'^2)^{1/2}\, dy' - \int_{-\rho'+b}^{y} (\rho'^2 - [y' - b]^2)^{1/2}\, dy'\right\}$$

$$A(y) = 2\left\{\int_{-\rho'}^{y} (\rho'^2 - y'^2)^{1/2}\, dy' - \int_{-\rho'}^{y-b} (\rho'^2 - y'^2)^{1/2}\, dy'\right\} \approx 2b(\rho'^2 - y^2)^{1/2} \qquad (5.2a)$$

This holds for $y \leqslant 0$. The results for $y > 0$ is

$$A(y) \approx 2b[2\rho' - (\rho'^2 - y^2)^{1/2}] \qquad (5.2b)$$

For the force F_Γ of the screw dislocation, Eq. (5.1) yields

$$F_\Gamma(y) = -2b\Gamma \qquad (5.3a)$$

Because $F_\Gamma(y)$ of the edge dislocation diverges at $y = \pm\rho'$ (i.e., when the dislocation just touches the particle), $F_\Gamma(y)$ is approximated by an average:

$$F_\Gamma(y) = -\Gamma[A(y=0) - A(y=-\rho')]/\rho' = -\Gamma 2b\rho'/\rho'$$

$$F_\Gamma(y) = -2b\Gamma \tag{5.3b}$$

This is the same result that has been obtained for the screw dislocation. Brown and Ham [5.16] pointed out that edge dislocations will penetrate into the particles by the formation of kinks [5.18]. Within the above approximations, F_Γ is independent of y, of ρ', and of the character of the dislocation.

It follows from Eqs. (5.3) that $F_{\Gamma_0} = \text{Max}\{|F_\Gamma(y)|\}$ of screw as well as of edge dislocations equals $2b\Gamma$:

$$F_{\Gamma_0} = 2b\Gamma \tag{5.4}$$

The breaking angle $\phi_c/2$ defined in Eq. (4.8b) is given by

$$\cos(\phi_c/2) = b\Gamma/S \tag{5.5}$$

where S is the line tension of the dislocation.

The range of Γ of coherent particles has been given above: 0.01–0.2 J/m^2 [5.11, 5.12]. The resulting forces F_{Γ_0} are small. This is demonstrated here for the copper–cobalt system. Γ of coherent cobalt-rich precipitates in copper is relatively high: around 0.2 J/m^2 [5.19–5.21]. Thus with $b = 0.25$ nm, one obtains for this system: $F_{\Gamma_0} = 10^{-10}$ N. The edge dislocation line tension parameter $K_S(\theta_d = 90°)$ of copper equals 2 GPa [5.22]; K_S has been introduced in Eq. (2.6). Inserting 1000 for the ratio of the outer to the inner dislocation cutoff radius yields $S = 9.3 \times 10^{-10}$ N and $\cos(\phi_c/2) = 0.055$. Thus $\phi_c/2$ is 87°. Since Γ of most coherent particles will be smaller than 0.2 J/m^2, $\phi_c/2$ will be very close to 90°. The cobalt-rich particles have a constrained lattice mismatch ε of 0.015. It yields the maximum particle–dislocation interaction force $F_{\varepsilon 0}$. For particles of 2 nm radius, $F_{\varepsilon 0}$ equals 10^{-9} N (Section 5.2.2). Evidently the ratio $F_{\varepsilon 0}/F_{\Gamma_0}$ is 10. All numerical data refers to room temperature. The CRSS of the copper–cobalt system is actually governed by ε (Sections 5.4.2.1.8 and 5.4.2.2.2). Because F_{Γ_0} and $F_{\varepsilon 0}$ have been derived for undissociated edge dislocations, both forces have been overestimated by nearly the factor 2. This will be discussed further in Section 5.4.2.2. Due to the disregard of the dissociation of dislocations, the CRSS derived below for chemical strengthening is too high by about the factor 2.8. But even this overestimated CRSS of a chemically strengthened material turns out to be negligible.

To obtain an estimate for the CRSS τ_p, F_{Γ_0} of Eq. (5.4) is inserted into Friedel's equation (4.26d):

$$\tau_{pF} = \frac{1}{b} \frac{(2b\Gamma)^{3/2}}{(2S\pi\omega_q)^{1/2}} \frac{f^{1/2}}{r} \tag{5.6}$$

where f and r are the volume fraction and the average radius of the particles; the statistical factor ω_q has been defined in Section 3.1.

In a material subjected to Ostwald ripening (Section 3.2.1.2), f stays virtually constant, but r increases with time. According to Eq. (5.6) the CRSS of a material that derives its strength from chemical hardening is highest at the beginning of the ripening treatment. There is no range of r in which τ_{pF} increases with r. Therefore chemical strengthening will only be of interest in the early stages of the precipitation processes. In the case of modulus mismatch, stacking fault energy mismatch, lattice mismatch, and order strengthening, there are ranges of age hardening where $\partial\tau_p/\partial r$ is positive. According to Friedel's equation (4.26d), τ_{pF} increases with r provided that F_0 increases with r more rapidly than does $r^{2/3}$. In the case of chemical strengthening, F_{Γ_0} is independent of r [Eq. (5.4)] and there is no age hardening but, instead, age softening. For the copper-cobalt system the above equation [Eq. (5.6)] yields for $f = 0.02$ and $r = 2\,\text{nm}$: $\tau_{pF} = 4.3\,\text{MPa}$. The dissociation of dislocations into Shockley partials will reduce the latter result further. At ambient temperature the actual CRSS of copper single crystals containing such cobalt-rich particles is around 40 MPa [5.23, 5.24]; it is governed by the particles' lattice mismatch (Sections 5.4.2.1.8 and 5.4.2.2.2). Evidently, chemical strengthening is negligible even in the copper–cobalt system, in which Γ is relatively high.

Reppich [5.25] has pointed out that since $\phi_c/2$ is close to 90°, the normalized range η_0 of $F(y)$ may be large. η_0 has been defined in Eq. (4.58g). Inserting the above numerical values of the copper–cobalt system ($f = 0.02, r = 2\,\text{nm}$) and assuming that the non-normalized range w_y equals r yields $\eta_0 = 0.39$. For finite values of η_0, the non-normalized CRSS τ_{pL} is given by Eqs. (4.58) and (4.62), and the coefficients C_1 and C_2 are listed in Table 4.1. The subscript L of τ_{pL} indicates that Schwarz and "Labusch's" [5.10] model is applied, which has been presented in Section 4.2.2.2.1.

$$\tau_{pL} = C_1 \frac{1}{b} \frac{(2b\Gamma)^{3/2}}{(2S\pi\omega_q)^{1/2}} \frac{f^{1/2}}{r} \left[1 + C_2 \frac{f^{1/2}}{(\pi\omega_q)^{1/2}} \left(\frac{S}{b\Gamma}\right)^{1/2}\right] \tag{5.7}$$

Inserting the above parameters of the copper–cobalt system yields $\tau_{pL} = 1.2 \cdot \tau_{pF} = 5.2\,\text{MPa}$. Allowing for η_0 raises the CRSS, but it still remains small. τ_{pL} increases monotonically with Γ. Because the CRSS decreases during ripening, all numerical evaluations have been made for rather small particles: r was only 2 nm.

So far chemical strengthening has not been proven to be the principal hardening mechanism of any system. Gerold [5.26] rejected Harkness and Hren's [5.27] interpretation that hardening of aluminum–zinc alloys is due to chemical strengthening by Guinier–Preston (Section 1.1) zones. Gerold found fault with the authors' derivations and with their value for Γ. It is far too high: $0.32\,\text{J/m}^2$. Actually the CRSS of these alloys is believed to be governed by the zones' lattice mismatch. This will be discussed in Section 5.4.2.1.8. Knoch and Reppich [5.28] investigated the CRSS of magnesium oxide single crystals hardened by magnesia ferrite particles. These authors assumed that order strengthening (Section 5.5) and chemical strengthening contribute to the CRSS and that the specific energy of antiphase boundaries in the particles equals Γ; both specific energies were found to be high: $0.82\,\mathrm{J/m^2}$.

5.2 MODULUS MISMATCH STRENGTHENING

A difference in the shear moduli μ_s and μ_p of the solid solution matrix and of the particles, respectively, gives rise to an elastic interaction force between particles and dislocations. This force will be discussed from a general point of view in Section 5.2.1. In Section 5.2.2, some actual alloy systems will be analyzed and the CRSS will be calculated.

5.2.1 General Derivation of the Particle–Dislocation Interaction Force

There have been two different approaches for obtaining the particle–dislocation interaction force. Their bases are (i) the dislocation line energy and (ii) the Peach–Koehler equation (2.9a). The most important method is the first one. One starts with Eqs. (5.8), which relate the energy density e_{st} stored in the strain field of a straight, infinitely long, undissociated dislocation to the properties of the material [5.18, 5.29]; elastic isotropy is assumed:

$$\text{Edge dislocation:} \qquad e_{\mathrm{st}}(y, x', y', z', \mu) = \frac{\mu b^2}{8\pi^2(1-\nu)^2} \frac{1 - 2\nu \sin^2\vartheta}{R^2} \tag{5.8a}$$

$$\text{Screw dislocation:} \qquad e_{\mathrm{st}}(y, x', y', z', \mu) = \frac{\mu b^2}{8\pi^2} \frac{1}{R^2} \tag{5.8b}$$

The subscript "st" of e_{st} stands for "strain." The geometry is sketched in Figure 5.2. The dislocation glides in the plane $z = 0$, and it is parallel to the x-axis; its position is characterized by the coordinate y. Equations (5.8) give e_{st} at the point with the coordinates (x', y', z'). μ and ν are the shear modulus and Poisson's ratio of the material. R and ϑ are defined by

$$R = [(y' - y)^2 + z'^2]^{1/2} \tag{5.8c}$$

$$\vartheta = \arctan\left(\frac{z'}{y' - y}\right) \tag{5.8d}$$

Equations (5.8a) and (5.8b) hold only outside of the dislocation core of radius R_c. For $R < R_c$, atomistic calculations are required. In a range close to R_c, anharmonic calculations may be appropriate. The elastic anisotropy of the material can be allowed for to some extent by the choice of the modulus. In view of the glide geometry, one may insert μ_F for face-centered cubic (fcc)

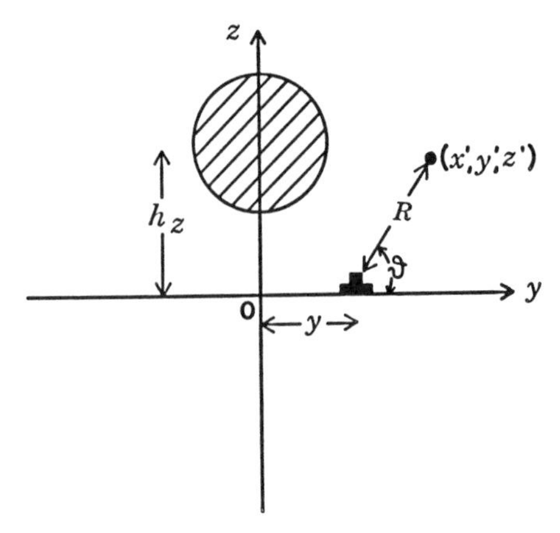

Fig. 5.2. Geometry for Eqs. (5.8). The particle (hatched) has the coordinates $(0, 0, h_z)$.

alloys. μ_F is the modulus on $\{111\}$-planes in $\langle 110\rangle$-directions [5.30]:

$$\mu_F = \frac{3C_{44}(C_{11} - C_{12})}{C_{11} - C_{12} + 4C_{44}} \tag{5.9a}$$

The C_{ik} are the elastic stiffnesses. The subscript F of μ_F stands for "fraction." An alternative is the average modulus μ_R, where R refers to "root":

$$\mu_R = [C_{44}(C_{11} - C_{12})/2]^{1/2} \tag{5.9b}$$

A third choice is μ_K, which is derived from the dislocation line energy parameter $K_E(\theta_d = 0°)$ of screw dislocations. K_E has been defined in Eq. (2.2). Because $K_E(\theta_d = 0°)$ is calculated anisotropically, the argument "aniso" is quoted in K_E:

$$\mu_K = 4\pi K_E(\theta_d = 0°, \text{aniso}) \tag{5.9c}$$

The subscript K of μ_K reminds one of K_E. The three moduli μ_F, μ_R, and μ_K of copper differ by $\pm 15\%$. This will be shown in Section 5.2.2. Equation (5.9c) is formulated for screw dislocations. Of course, one may as well do so for edge dislocations, but then Poisson's ratio v appears on the right-hand side of Eq. (5.9c). The choice of v is somewhat arbitrary. An example will be given in Eq. (6.5b). If a particle is embedded in the matrix, the energy density e_{st} is locally altered because μ_p will differ from μ_s. Moreover, there will be some relaxation of the strain field next to the particle–matrix interface.

If the relaxations at the particle–matrix interface are neglected, the interaction energy $E_\mu(y, -h_z)$ between a particle centered at $(0, 0, h_z)$ and an undissociated straight dislocation is obtained by integrating Eqs. (5.8) over the volume of the particle:

$$E_\mu(y, -h_z) = \int_p [e_{st}(\mu_p) - e_{st}(\mu_s)]\, dV \tag{5.10}$$

In accordance with the definitions given in Section 4.1, $-h_z$ is quoted as argument of E_μ and of F_μ introduced below. The subscript μ of E_μ and F_μ indicates the "modulus" effect. Only μ_p and μ_s are given in the argument of e_{st}, and the coordinates (x', y', z') and y have been suppressed. In the evaluation of E_μ, all authors except Weeks et al. [5.31] neglected the effect of the relaxation at the particle–matrix interface. Therefore e_{st} changes abruptly at the particle–matrix interface. The dislocation is assumed to stay straight in spite of its interaction with the particle. This is the straight-line approximation introduced in Section 4.1.1. Differentiating E_μ with respect to y yields the particle–dislocation interaction force $F_\mu(y, -h_z)$:

$$F_\mu(y, -h_z) = -\partial E_\mu/\partial y \tag{5.11}$$

Weeks et al. [5.31] have published analytical solutions for two cases:

1. $R \gg \rho$, where ρ is the radius of the spherical particle.
2. A spherical bubble of radius ρ lies symmetrically on a screw dislocation.

The treatment of case 1 is straightforward, because e_{st} approximately equals $e_{st}(y, x' = 0,\ y' = 0,\ z' = h_z, \mu_p)$ throughout the entire particle. The more interesting case 2 yielded the approximate result:

$$E_\mu(y = 0, h_z = 0) = -\frac{\mu_s b^2}{2\pi}\,\rho\left[\frac{\pi^2}{12} + \ln(\rho/R_i)\right] \tag{5.12}$$

where R_i is the dislocation's inner cutoff radius. In the derivation of Eq. (5.12) the relaxation at the bubble–matrix interface has been taken into account. It changed E_μ by about 25%. Weeks et al. allowed for the interaction with the dislocation core by subtracting $[0.2\mu_s b^2 \rho]$ from Eq. (5.12). The maximum interaction force $F_{\mu 0}$ was approximated by $|E_\mu(y = 0, h_z = 0)/\rho|$. The interaction is attractive.

Knowles and Kelly [5.32] carried out the integration of Eq. (5.10) numerically. These authors treated a spherical particle that interacts with a straight, undissociated screw dislocation. Graphic differentiation yielded $F_{\mu 0} = \mathrm{Max}\{|F_\mu(y, h_z = 0)|\}$. The core contribution to E_μ was allowed for by assigning the core a constant energy density. The authors tabulated their results.

Melander and Persson [5.33] have published analytical expressions for the force $F_\mu(y, -h_z)$ that a cuboidal particle exerts on a straight, undissociated screw dislocation. The cube axes are parallel to the axes of the coordinate system. Thus the dislocation is parallel to one edge and one face of the cube:

$$F_\mu(y, -h_z) = \frac{\Delta\mu b^2}{8\pi^2}(2c) \cdot \left[-\frac{\arctan\left(\frac{h_z - c}{\sqrt{(y-c)^2+b^2}}\right)}{\sqrt{(y-c)^2+b^2}} + \frac{\arctan\left(\frac{h_z - c}{\sqrt{(y+c)^2+b^2}}\right)}{\sqrt{(y+c)^2+b^2}} + \frac{\arctan\left(\frac{h_z + c}{\sqrt{(y-c)^2+b^2}}\right)}{\sqrt{(y-c)^2+b^2}} - \frac{\arctan\left(\frac{h_z + c}{\sqrt{(y+c)^2+b^2}}\right)}{\sqrt{(y+c)^2+b^2}} \right] \tag{5.13}$$

with

$$\Delta\mu = \mu_p - \mu_s \tag{5.14}$$

where $2c$ is the cube length.

Later on, Melander and Persson applied Eq. (5.13) also to spherical particles. They replaced the sphere of radius ρ by a cube of the same volume. c corresponding to ρ is given by

$$c = \rho(\pi/6)^{1/3} \tag{5.15}$$

By introducing the term b^2 in the roots in the denominators of Eq. (5.13), the authors avoided the divergencies at $y = \pm c$. The core models CM1–CM4 presented below serve the same purpose.

In Figure 5.3, $F_\mu/(\Delta\mu b^2)$ of Eq. (5.13) is plotted versus y/b for three choices of h_z. The entire cube length $2c$ equals $2(\pi/6)^{1/3} \cdot 20b = 32.2b$. A sphere of the same volume has the radius $20b$. $F_\mu(y, -h_z)$ is an antisymmetric function of y and a symmetric one of h_z. $|F_\mu(y, -h_z)|$ peaks for $h_z = 0$, the maximum value $F_{\mu 0}$ is reached at the particle–matrix interface:

$$F_{\mu 0} = \frac{|\Delta\mu| b^2}{4\pi^2}(2c)\left[-\frac{\arctan\left(\frac{c}{\sqrt{4c^2+b^2}}\right)}{\sqrt{4c^2+b^2}} + \frac{1}{b}\arctan\left(\frac{c}{b}\right)\right] \tag{5.16}$$

Figure 5.4 shows $F_{\mu 0}/(|\Delta\mu| b^2)$ as a function of c/b. $F_{\mu 0}$ varies approximately linearly with c/b.

The merit of Melander and Persson's derivations lies in the fact that an analytical expression has been obtained for $F_\mu(y, -h_z)$. The drawback is of course that only screw dislocations have been dealt with. The interaction

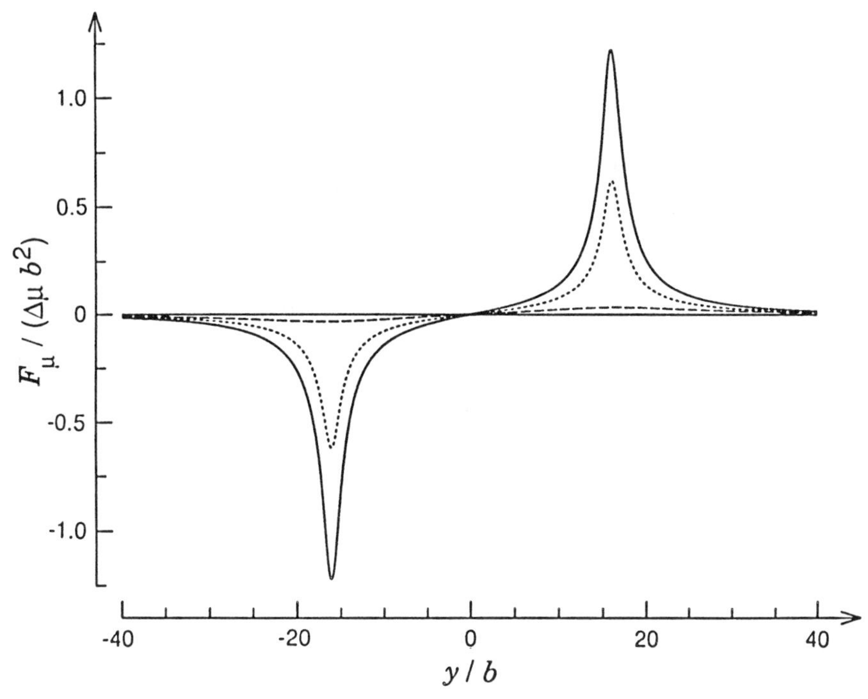

Fig. 5.3. $F_\mu(y/b, -h_z)/(\Delta\mu b^2)$ of Eq. (5.13) versus y/b. The entire cube length $2c$ equals $32.2b$. —— $h_z = 0$, --- $h_z = c$, – – – $h_z = 1.5c$.

energies of edge dislocations are estimated to exceed those of screw dislocations by about the factor $1/(1 - \nu)$. This follows from Eqs. (2.3b) and (5.8). Melander and Persson, however, argued that in the aluminum-rich aluminum–zinc–magnesium alloys for which they studied modulus strengthening the CRSS is governed by the mobility of screw dislocations. This will be discussed further in the next section. The authors' approximation to replace a spherical particle by a cuboidal one of the same volume is likely to lead to forces $F_{\mu 0}$ that are too high. One face and one edge of the cube were supposed to be parallel to the dislocation. $F_\mu(y, -h_z)$ peaks when the dislocation just contacts the cube or the sphere. The length along which contact is established is larger for the cube than for the sphere; hence R in Eqs. (5.8) and (5.10) assumes small values in a larger volume for the cube than for the sphere. This leads to higher interaction energies and forces for the cube than for the sphere. The force profiles $F_\mu(y, -h_z)$ calculated by Melander and Persson for cubes have sharper maxima than the profiles derived by Nembach [5.4] for spheres; see below. This is evident in Figures 5.3 and 5.5.

Nembach [5.4] evaluated the integral in Eq. (5.10) numerically. The interaction energies E_μ between spherical particles and straight, undissociated

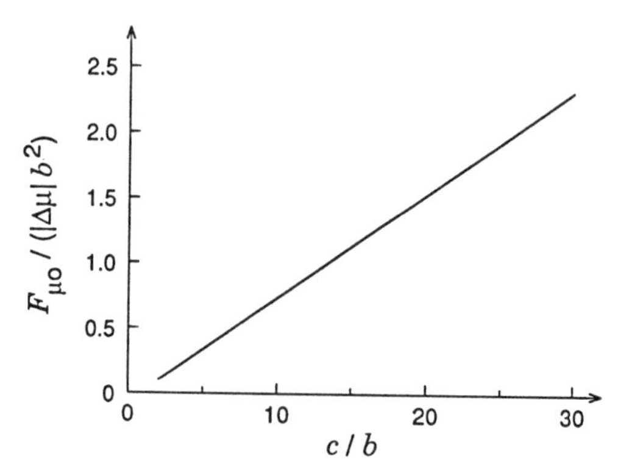

Fig. 5.4. $F_{\mu 0}/(|\Delta\mu|b^2)$ of Eq. (5.16) versus c/b. c is half the cube length.

screw and edge dislocations have been obtained. Equation (5.10) is rewritten as follows:

$$E_\mu(y, -h_z) = \frac{\Delta\mu}{\mu_s} \int_p e_{\text{st}}(\mu_s)\, dV \tag{5.17}$$

Once $E_\mu(y, -h_z)$ has been established, $F_\mu(y, -h_z)$ follows from Eq. (5.11). Because $F_\mu(y, h_z = 0)$ peaks when the dislocation core is in contact with the particle–matrix interface, the core energy is of exceeding importance. Weeks et al. [5.31], Knowles and Kelly [5.32], and Melander and Persson [5.33] treated the core rather cursorily. In Weeks et al.'s equation (5.12), the inner cutoff radius R_i appears and the term $0.2\mu_s b^2 \rho$ has subsequently been subtracted. Knowles and Kelly assigned a constant energy density to the core. This is similar to the core model CM2 presented below. Melander and Persson introduced the term b^2 in the denominators of Eq. (5.13). Nembach [5.4] demonstrated the importance of the core energy by comparing the results of the four different empirical core models CM1–CM4. Two alternatives have been inserted for the core radius R_c: $1b$ and $2b$. R_c is larger than the inner cutoff radius R_i; this has been shown in Section 2.2. R has been defined in Eq. (5.8c).

CM1—for $R \leqslant R_c$: $R = \infty$. The dislocation is considered as hollow; that is, $e_{\text{st}}(R \leqslant R_c) = 0$. Because this can be allowed for by replacing R in Eqs. (5.8a) and (5.8b) by infinity, this model is characterized by "for $R \leqslant R_c$: $R = \infty$."

CM2—for $R \leqslant R_c$: $R = R_c$. e_{st} is assumed to be constant within the core: $e_{\text{st}}(R \leqslant R_c) = e_{\text{st}}(R = R_c)$. This model will be referred to as "for $R \leqslant R_c$: $R = R_c$."

CM3—for $R \leqslant \infty$: $R = R + R_c$. R in Eqs. (5.8a) and (5.8b) is replaced by $R + R_c$ for all values of R. This explains the short title of this model: "for $R \leqslant \infty$: $R = R + R_c$." CM3 is suggested by the Peierls–Nabarro dislocation model [5.18, 5.34–5.37].

CM4—for $R \leqslant R_c$: $R = R + R_c$. This model is similar to CM3, but $R + R_c$ is substituted for R only in the core. Therefore CM4 is characterized by "for $R \leqslant R_c$: $R = R + R_c$."

For CM1 and CM4, the energy density e_{st} does not vary continuously across the boundary of the core: e_{st} decreases there abruptly. $\partial e_{st}/\partial R$ of CM2 changes abruptly at $R = R_c$. The energy E_{core} stored in the core can be compared with the energy E_{strain} stored in the strain field of the dislocation between R_c and R_o. R_o is the outer cutoff radius. For $R_o = 20R_c$, the ratios E_{core}/E_{strain} equal 0.00, 0.17, 0.10, and 0.06 for CM1–CM4, respectively. This is reasonable.

Some results of Nembach's numerical evaluations are shown in Figure 5.5. $F_\mu/(\Delta\mu b^2)$ is plotted versus y/b, in Figure 5.5a for edge and in Figure 5.5b for screw dislocations. The particle radius ρ is $20b$, R_c equals $2b$, and h_z is zero; that is, the center of the particle lies in the glide plane. In all evaluations, Poisson's ratio of the matrix and of the particles has been assumed to equal 1/3. F_μ of edge dislocations is about 33% higher than F_μ of screw dislocations. This reflects the well-known difference of the line energies and is also to be expected on the basis of Eqs. (5.8). The maximum of $|F_\mu(y, -h_z)|$ is at $|y| \approx [\rho - R_c]$ and $h_z = 0$. For the core models CM1 (for $R \leqslant R_c$: $R = \infty$) and CM4 (for $R \leqslant R_c$: $R = R + R_c$), $|F_\mu(y, -h_z)|$ of edge dislocations has secondary maxima next to the highest ones (Figure 5.5a). They are caused by the abrupt changes of e_{st} at the boundary of the core. For screw dislocations, there is only a shoulder visible for CM1 (Figure 5.5b). If the core is completely outside of the particle, CM1, CM2, and CM4 yield identical results; therefore the respective curves coincide in Figures 5.5a and 5.5b. CM2 (for $R \leqslant R_c$: $R = R_c$) yields the strongest forces F_μ. The variation of $F_\mu(y, -h_z)$ with R_c and with h_z is demonstrated in Figure 5.5c. Small core radii lead to strong interaction forces because $e_{st}(\mu_s)$ appearing in the integral of Eq. (5.17) becomes large if R assumes small values [Eqs. (5.8)]. $|F_\mu|$ decreases as $|h_z|$ increases. Because $|F_\mu|$ of edge dislocations exceeds $|F_\mu|$ of screw dislocations, only edge dislocations will be considered further.

Except for the mentioned secondary maxima, all curves presented in Figure 5.5 are similar in shape. From them $F_{\mu 0} = \mathrm{Max}\{|F_\mu(y, -h_z)|\}$ is derived. $F_{\mu 0}$ of edge dislocations has been related to the particle radius ρ by an empirical expression:

$$F_{\mu 0}(\rho/b)/(|\Delta\mu| b^2) = \alpha_1(\rho/b)^{\beta_1} \tag{5.18}$$

The parameters α_1 and β_1 depend on the core model and on the core radius. α_1 and β_1 are listed in Table 5.1 for CM1—CM4 and $R_c = 1b$ and $R_c = 2b$.

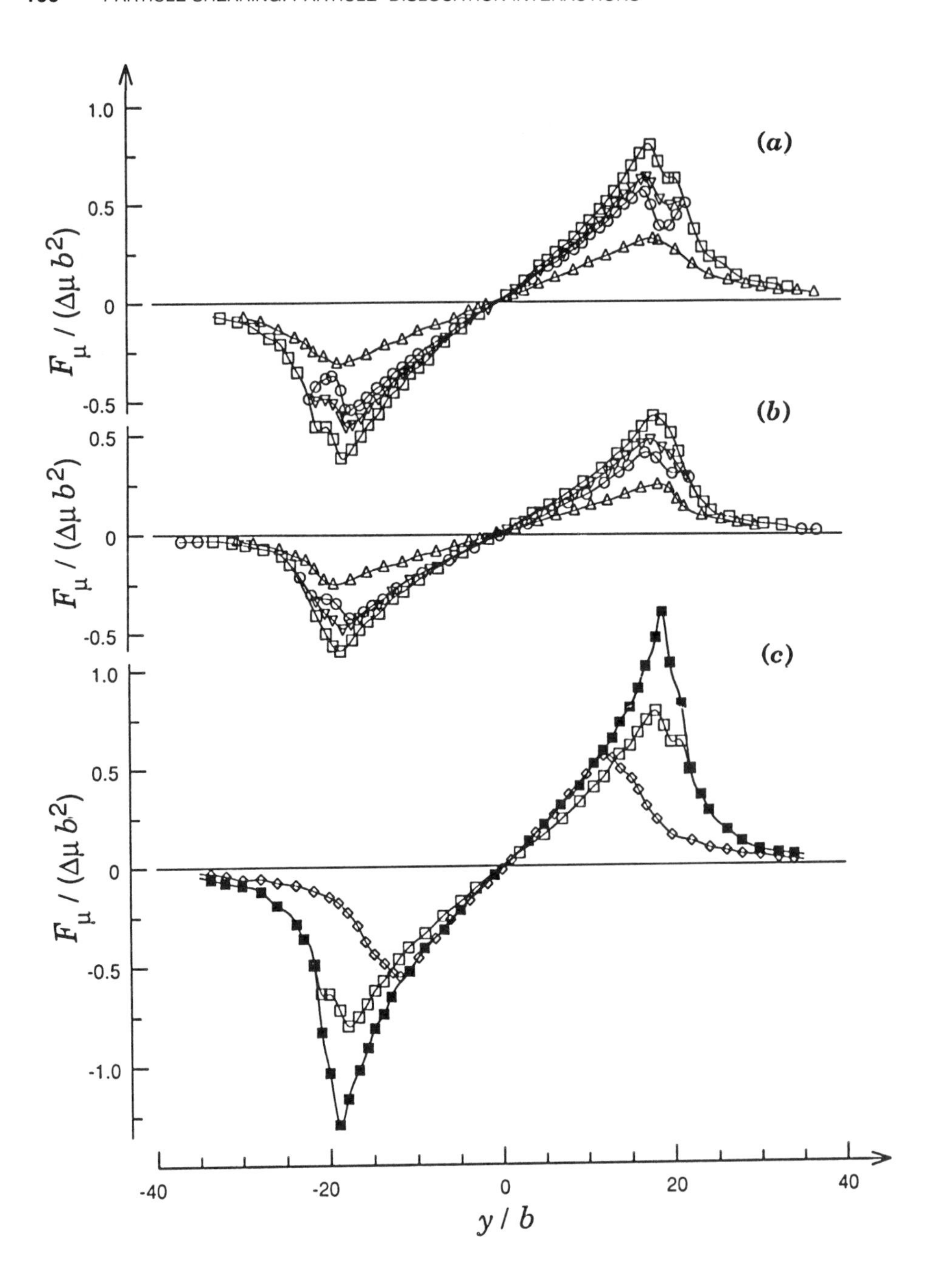
(a)
(b)
(c)
1.0
0.5
0
-0.5
-1.0
$F_\mu / (\Delta\mu b^2)$
-40
-20
0
20
40
y / b

Table 5.1. Parameters α_1 and β_1 of Eq. (5.18), $F'_{\mu 0} = F_{\mu 0}(\rho/b = 8, R_c = 2b)/(|\Delta\mu| b^2)$, ω of Eq. (5.19), and α_2 and β_2 of Eq. (5.20)[a]

	$R_c = 1b$		$R_c = 2b$				$R_c = 1b$		$R_c = 2b$	
Core Model	α_1	β_1	α_1	β_1	$F'_{\mu 0}$	ω	α_2	β_2	α_2	β_2
CM1— for $R \leqslant R_c$: $R = \infty$	0.096	0.756	0.0353	0.905	0.23	0.50	−2.86	4.56	−5.26	6.81
CM2— for $R \leqslant R_c$: $R = R_c$	0.162	0.688	0.0722	0.792	0.37	0.55	−2.15	3.63	−4.29	5.52
CM3— for $R \leqslant \infty$: $R = R + R_c$	0.0477	0.813	0.0175	0.951	0.13	0.49	−8.59	6.96	−12.3	9.39
CM4— for $R \leqslant R_c$: $R = R + R_c$	0.115	0.736	0.0460	0.864	0.28	0.52	−2.20	4.04	−4.00	6.04

[a]All data apply to edge dislocations. The particle radius ρ is limited to the range $8b$–50b.

Source: After Ref. 5.4.

Equation (5.18) is applicable in the range $8b \leqslant \rho \leqslant 50b$. $F_{\mu 0}(\rho/b = 8,\ R_c = 2b)/(|\Delta\mu| b^2)$ and the ratio ω have also been compiled in Table 5.1:

$$\omega = \frac{F_{\mu 0}(\rho/b = 8, R_c = 2b)}{F_{\mu 0}(\rho/b = 8, R_c = 1b)} \tag{5.19}$$

For all four core models, ω is close to 0.5. ω increases to about 0.65 for $\rho = 50b$. Melander and Persson's equation (5.16) yields the following for the maximum of the interaction force between a cube, which has the same volume as a sphere of radius $8b$, and a screw dislocation: $F_{\mu 0}/(|\Delta\mu| b^2) = 0.45$. This is in reasonable agreement with the data presented for $[F_{\mu 0}(\rho/b = 8, R_c = 1b)/(|\Delta\mu| b^2)] = [F'_{\mu 0}/\omega]$ of edge dislocations in Table 5.1. The difference between Melander and Persson's and Nembach's results is smaller than expected because there are two effects that partially cancel each other. Screw dislocations lead to lower

Fig. 5.5. $F_\mu(y/b,\ -h_z)/(\Delta\mu b^2)$ versus y/b for core models CM1–CM4. The particle radius ρ equals $20b$. The data points represent individual numerical results. R is defined in Figure 5.2 and in Eq. (5.8c). R_c is the radius of the dislocation core. ○ CM1—for $R \leqslant R_c$: $R = \infty$. □ ■ ◇ CM2—for $R \leqslant R_c$: $R = R_c$. △ CM3—for $R \leqslant \infty$: $R = R + R_c$. ▽ CM4—for $R \leqslant R_c$: $R = R + R_c$. In (*a*) and (*b*) the curves for CM1, CM2, and CM4 coincide for $|y/b| > (\rho + R_c)/b = 22$; in (*c*) the analogue holds for □ and ■. (*a*) Edge dislocation, CM1–CM4: $R_c = 2b$, $h_z = 0$. (*b*) Screw dislocation, CM1–CM4: $R_c = 2b$, $h_z = 0$. (*c*) Edge dislocation, CM2: □ $R_c = 2b$, $h_z = 0$; ■ $R_c = 1b$, $h_z = 0$; ◇ $R_c = 2b$, $h_z = \rho/\sqrt{2}$. (After Ref. 5.4.)

interaction forces than edge dislocations do, but the maximum interaction force between a dislocation and a cube exceeds that between a dislocation and a sphere. This has been explained above.

The range w_y of $F_\mu(y, h_z = 0)$ is defined as its full width at half-maximum. For $8b \leqslant \rho \leqslant 50b$, the function $w_y(\rho/b)$ of edge dislocations can be approximated by

$$w_y(\rho/b) = b[\alpha_2 + \beta_2 \ln(\rho/b)] \tag{5.20}$$

The coefficients α_2 and β_2 have been derived from curves like those shown in Figure 5.5a. α_2 and β_2 are given in Table 5.1. For $8b \leqslant \rho \leqslant 50b$, the ratio $[w_y(\rho, R_c = 1b)/w_y(\rho, R_c = 2b)]$ is around 0.7. This holds for all four core models.

Core model CM1 (for $R \leqslant R_c$: $R = \infty$) yields interaction forces $F_{\mu 0}$ that are close to the average $F_{\mu 0}$ of all four core models. Therefore most of the numerical examples presented in Section 5.2.2 will be based on this model.

The length b of the Burgers vector affects F_μ and $F_{\mu 0}$ in two ways: (i) On the basis of Eqs. (5.8), one expects that both forces are proportional to $|\Delta\mu|b^2$. (ii) Since all lengths, including the core radius R_c, have been defined in units of b, b enters the square brackets of Eq. (5.16) and appears as $(\rho/b)^{\beta_1}$ in Eq. (5.18). All the above evaluations of F_μ and $F_{\mu 0}$ have been meant for undissociated dislocations. If the width d_s of the stacking fault ribbon of dissociated dislocations is not much smaller than the range w_y of each partial dislocation, all the above equations for F_μ and $F_{\mu 0}$ have to be revised. For $d_s \gg w_y$, the length b of the Burgers vector of the perfect dislocation has to be replaced by the length of the Burgers vector of a partial dislocation. In the case of fcc alloys, this reduces b to roughly one-half. This renormalization of b must, however, only be made in the term $|\Delta\mu|b^2$ and neither in R_c nor in ρ/b [Eqs. (5.18) and (5.20)]. Though it is common practice to express R_c in units of b, it is by no means self-evident that the dissociation of a dislocation reduces its core radius in proportion. In this respect it would have been better to express R_c and ρ in units of the lattice constant and not in units of b. Thus $F_{\mu 0}$ is proportional to b^2 and the dissociation reduces $F_{\mu 0}$ of fcc alloys to about one-fourth of its original value. This reduction is of course different if one allows for effects of the dissociation on R_c.

It is stressed again that Melander and Persson's [5.33] equations (5.13) and (5.16) as well as Nembach's equations (5.18) and (5.20) involve the straight-line approximation (Section 4.1.1).

Russel and Brown [5.38] based their model of modulus mismatch strengthening on the dislocation line energy E_l. They supposed, however, that the line tension S equals E_l and that both S and E_l are equal to $[\mu_s b^2/2]$. Equilibrium of forces at the particle–matrix interface requires

$$E_{ls} \sin \alpha_s = E_{lp} \sin \alpha_p \tag{5.21}$$

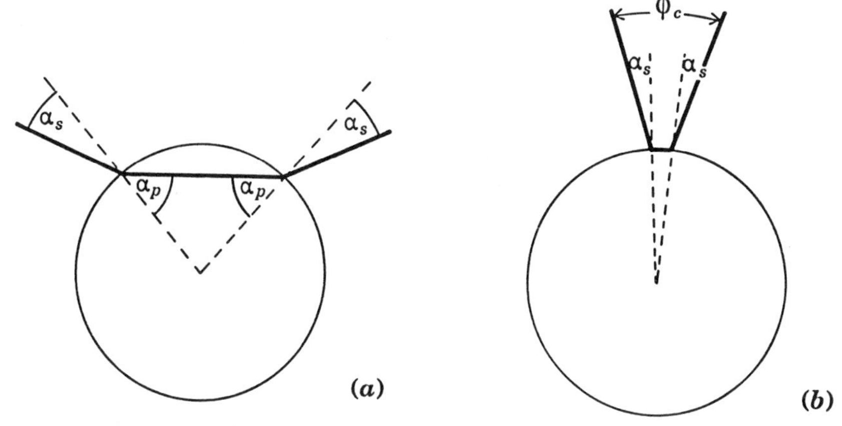

Fig. 5.6. Schematic depiction of a dislocation (thick lines) cutting through the interface of a spherical particle. α_s and α_p are the angles between the dislocation and the normal (dashed lines) of the particle–matrix interface (After Ref. 5.38.)

This is illustrated in Figure 5.6a for $E_{lp} < E_{ls}$. This relation is assumed to hold in all further evaluations: The particles are supposed to be softer than the matrix. α_s and α_p are the angles between the dislocation and the normal of the particle–matrix interface. The subscripts s and p refer to the "solid solution" matrix and to the "particle," respectively. Equation (5.21) is analogous to Snell's law of refraction [5.39]. Figures 5.6a and 5.6b are idealizations: Actually the dislocation does not consist of straight segments, but it is curved. The critical configuration is sketched in Figure 5.6b. In this case α_p is 90° and the critical angle ϕ_c equals $2\alpha_s$:

$$E_{ls}\sin(\phi_c/2) = E_{lp}$$
$$\phi_c/2 = \arcsin(E_{lp}/E_{ls}) \tag{5.22}$$

The main problem is to relate E_{lp} to the radius ρ of the particle. The authors defined E_{ls}^{∞} and E_{lp}^{∞}. These are the line energies in the respective infinite media. Russel and Brown made use of the logarithmic dependence of the dislocation energy on the dimensions of the crystal and chose the following expression for the ratio E_{lp}/E_{ls}:

$$\frac{E_{lp}}{E_{ls}} = \frac{E_{lp}^{\infty}\ln(\rho/R_i)}{E_{ls}^{\infty}\ln(R_o/R_i)} + \frac{\ln(R_o/\rho)}{\ln(R_o/R_i)} \tag{5.23}$$

R_o and R_i are the outer and inner cutoff radii of the dislocation, respectively. Evidently, Eq. (5.23) is correct in the two limiting cases: ρ is very small (i.e., $\rho \approx R_i$) and ρ is large (i.e., $\rho \approx R_o$). For a void, the ratio $E_{lp}^{\infty}/E_{ls}^{\infty}$ vanishes. The only justification given by the authors for Eq. (5.23) was that "it seems appropriate to choose" it. In Section 5.2.2, the CRSS will be derived from $\phi_c/2$ given by Eqs. (5.22) and (5.23).

Though it is not always as evident as in Eq. (5.18), all of the above results for F_μ are sensitive to the properties of the dislocation core: R_i enters Eq. (5.23) and b^2 appears in the denominators of Eqs. (5.13) and (5.16). The reason is as follows. Because in the present approximation the relaxations at the particle–matrix interface are disregarded, the particle produces no strain field. The dislocation feels the particle only because the latter one modifies the former one's line energy. The interaction energy E_μ is given by Eqs. (5.8), (5.10), and (5.17). Since the energy density e_{st} falls off with $1/R^2$ [Eqs. (5.8)], the integrand in Eq. (5.17) diverges for R approaching zero and the introduction of a cutoff becomes necessary. In the case of lattice-mismatch hardening, the particle gives rise to a stress field and F_ε can be calculated on the basis of the Peach–Koehler equation (2.9a). There is an analogy between modulus mismatch and lattice mismatch on the one side and diamagnetism and paramagnetism on the other side. Therefore some authors [5.29, 5.40] refer to the modulus and the lattice mismatch interaction as the diaelastic and the paraelastic interaction, respectively.

Comninou and Dundurs' [5.41] derivation of F_μ differed radically from all others presented so far. These authors calculated the interaction force between a spherical particle having a modulus mismatch and a straight, undissociated screw dislocation on the basis of the Peach–Koehler equation (2.9a). The authors primarily treated the long-range interaction. Gavazza and Barnett [5.42] have published similar, but more elaborate, calculations. Their results are, however, rather unhandy series.

5.2.2 Derivation of the CRSS and Discussion of Experimental Data

In the preceding section, $F_\mu(y, -h_z)/(\Delta\mu b^2)$ and $F_{\mu 0}/(|\Delta\mu| b^2)$ have been derived. To evaluate modulus mismatch strengthening of actual alloy systems, $\Delta\mu$ has to be known. Often it is impossible to establish μ_p because the particles are metastable. In such cases, μ_p is considered as an adjustable parameter that has to be derived from experimental CRSS data.

It is evident from Figure 5.5 and from Table 5.1, that core model CM1 (for $R \leqslant R_c$: $R = \infty$) leads to particle–dislocation interaction forces F_μ of average strength. Therefore primarily this model will be referred to below. $2b$ is inserted for the core radius R_c. Since $F_{\mu 0}$ of edge dislocations is larger than $F_{\mu 0}$ of screw dislocations and the opposite holds for the line tensions, edge dislocations are expected to lead to higher CRSSs than screw dislocations do. For that reason, only edge dislocations will be considered. There will be, however, one exception: Melander and Persson's analysis of the aluminum–zinc–magnesium system refers to screw dislocations.

From Figures 5.3 and 5.5c it appears that $F_\mu(y, -h_z)$ decreases drastically if $|h_z|$ is raised above zero. But even those particles that do not intersect the glide plane still interact with dislocations. In the case of lattice mismatch hardening, the analogous problem is encountered. It will be analyzed in Section 5.4. Because of this long-range interaction the derivation of the CRSS requires an intricate averaging procedure. To begin with, Melander and Persson's [5.33] simple suggestion is followed: All particles that contact the glide plane are assumed to exert the force $F_{\mu 0} = \text{Max}\{|F_{\mu}(y, h_z = 0)|\}$ on the dislocation. A possible distribution of the particle radii is neglected in the derivation of $F_{\mu 0}$; $F_{\mu 0}$ is calculated for the average radius r. The interaction with those particles that do not contact the glide plane is entirely disregarded. Thus when Friedel's equations (4.26) are applied, $F_{\mu 0}$ will be inserted for F_0 and $L_{\min}$ follows from Eq. (3.3c). This is a rather crude estimate, but it allows one to put different strengthening mechanisms into perspective. When modulus and lattice mismatch strengthening are compared with each other, both evaluations will involve this approximation. Therefore the relative importance of these two mechanisms can be judged. The two following comparisons are based on the straight-line approximation (Section 4.1.1). They are meant for room temperature. The dislocations are assumed to be undissociated. Melander and Persson's more elaborate averaging procedure will be presented towards the end of this section.

Modulus mismatch strengthening can be assessed on the basis of the following numerical comparisons. Two systems will be dealt with:

1. Copper containing cobalt-rich precipitates (Section 3.3.5)
2. The commercial nickel-base superalloy NIMONIC PE16 (Section 5.5.1) containing long-range-ordered γ'-precipitates

Actually the CRSSs of these two systems are governed by lattice mismatch (copper–cobalt, Sections 5.4.2.1.8 and 5.4.2.2.2) and by order strengthening (NIMONIC PE16, Section 5.5.4), respectively. The evaluations are given for core model CM1 (for $R \leqslant R_c$: $R_c = \infty$), because its maximum force $F_{\mu 0}$ is close to the average $F_{\mu 0}$ of all four models.

Copper–Cobalt System. This system is the textbook example for lattice mismatch strengthening. This will be shown in Sections 5.4.2.1.8 and 5.4.2.2.2. The relevant parameters are the constrained lattice mismatch ε, the resulting force profile $F_{\varepsilon}(y, -h_z)$, and its maximum $F_{\varepsilon 0} = \text{Max}\{|F_{\varepsilon}(y, -h_z)|\}$. The cobalt-rich particles have the same fcc crystal structure as the copper-rich matrix. Because their stiffnesses are not known, the mean of the Reuss [5.43] average and of the Voigt [5.44] average of the shear moduli of hexagonal cobalt is inserted for μ_p [5.45, 5.46]:

$$\mu_p = 82\,\text{GPa}$$

The analogous mean is used for the matrix [5.45, 5.47]:

$$\mu_s = 48\,\text{GPa}$$

$$\Delta\mu = 35\,\text{GPa}$$

$$\Delta\mu/\mu_s = 0.73$$

μ_F, μ_R, and μ_K [Eqs. (5.9)] of the matrix are: 31 GPa, 43 GPa, and 42 GPa, respectively. Evidently the choice of the mean of the Reuss and Voigt averages is reasonable. The line tension S of edge dislocations is $9.3 \cdot 10^{-10}$ N (Section 5.1). The ratio of the cutoff radii has been assumed to equal 1000. All further data refer to $r = 8b$, $f = 0.02$, core model CM1, edge dislocations, $h_z = 0.0$, $R_c = 2b$, and $b = 0.25$ nm:

$$F_{\mu 0}/(\Delta\mu b^2) = 0.23$$

$$F_{\mu 0} = 5.2 \times 10^{-10}\,\mathrm{N}$$

$$\cos(\phi_c/2) = F_{\mu 0}/(2S) = 0.28$$

$$\phi_c/2 = 74^\circ$$

$$w_y = 8.9b$$

$$\eta_0 = 0.19$$

The breaking angle ϕ_c and the normalized range η_0 have been defined in Eqs. (4.8) and (4.58g), respectively.

The CRSS is estimated on the basis of Friedel's [5.9] equations (4.26) (τ_{pF}) and of Schwarz and Labusch's [5.10] equations (4.58j) and (4.62) (τ_{pL}). τ_{pL} differs from τ_{pF} by the factor $[C_1(1 + C_2\eta_0)]$. C_1 and C_2 equal 0.93 and 0.55, respectively (Table 4.1). As stated above, all particles that contact the glide plane are assumed to exert $F_{\mu 0}$. All other particles are entirely disregarded. The analogous approximation will be made for $F_{\varepsilon 0}$. The results for the modulus effect are

$$\tau_{pF} = 49\,\mathrm{MPa} = \tau_{pF}(\mu)$$

$$\tau_{pL} = 1.03\tau_{pF} = 50\,\mathrm{MPa} = \tau_{pL}(\mu)$$

Evidently, τ_{pF} and τ_{pL} are nearly the same. In order to put the above results concerning the modulus mismatch into the right perspective, they are compared with the corresponding data referring to the lattice mismatch of the same system. r and f are again $8b$ and 0.02, respectively, and h_z equals $r/\sqrt{2}$. $F_{\varepsilon 0}$ follows from Eq. (5.43) with $\mu_s = \mu_F$. This yields the following for the lattice mismatch; all parameters refer to edge dislocations:

$$\varepsilon = 0.0152$$

$$F_{\varepsilon 0} = 9.8 \times 10^{-10}\,\mathrm{N}$$

$$F_{\mu 0}/F_{\varepsilon 0} = 0.53$$

$$\cos(\phi_c/2) = F_{\varepsilon 0}/(2S) = 0.53$$

$$\phi_c/2 = 58^\circ$$

$$w_y \approx r = 8b \text{ (estimate)}$$
$$\eta_0 = 0.13$$
$$\tau_{pF} = 126\,\text{MPa} = \tau_{pF}(\varepsilon)$$
$$\tau_{pL} = 0.99\tau_{pF} = 125\,\text{MPa} = \tau_{pL}(\varepsilon)$$

Again τ_{pL} is very close to τ_{pF}. τ_{pL} is slightly smaller than τ_{pF} because the statistical factor C_1 in Eq. (4.62) has been disregarded in the derivation of Friedel's equations (4.26). C_1 allows for the randomness of the particle arrangement in the glide plane. Evidently $\tau_{pF}(\varepsilon)$ equals about 2.5 times $\tau_{pF}(\mu)$. It must be stressed that the dissociation of the edge dislocation into two Shockley partial dislocations has been disregarded. In the above numerical comparison between modulus and lattice mismatch strengthening, r has been chosen to be $8b$. If r had been larger (e.g., $12b$), $\phi_c/2$ of lattice mismatch hardening would have been far below 60°, which is the lower limit set in relations (4.9) for "slight bending" of the dislocation. Friedel's equations (4.26), which yield τ_{pF}, have been based on the assumption that dislocations bow out only slightly. If the dissociation of the dislocations into Shockley partials is taken into consideration (see below), $\phi_c/2$ exceeds 50° even for $r = 12b$ (Section 5.4.2.2.2).

In Figure 5.7, $F_\varepsilon(y/b, -h_z = \rho/\sqrt{2}, \rho = 20b)$, $F_\mu(y/b, -h_z = \rho/\sqrt{2}, \rho = 20b)$, and the sum $F_{\varepsilon+\mu}(y/b, -h_z = \rho/\sqrt{2}, \rho = 20b) = F_\varepsilon(y/b, -h_z = \rho/\sqrt{2}, \rho = 20b) + F_\mu(y/b, -h_z = \rho/\sqrt{2}, \rho = 20b)$ are shown for an undissociated edge dislocation in copper–cobalt. $|F_\varepsilon|$ has its absolute maximum value for $|h_z| = \rho/\sqrt{2}$ (Section 5.4.1), and $|F_\mu|$ has its absolute maximum value for $h_z = 0.0$. Because F_ε vanishes for $h_z = 0.0$ (Section 5.4.1), $h_z = -\rho/\sqrt{2}$ has been chosen. F_μ is plotted for CM2 (for $R \leqslant R_c$: $R = R_c$) because this core model yields relatively large interaction forces $|F_\mu|$. This choice compensates at least in part for the choice of $h_z = -\rho/\sqrt{2}$, which favors F_ε. Evidently $\text{Max}\{|F_{\varepsilon+\mu}|\}$ exceeds $F_{\varepsilon 0}$ by 45%. If the dissociation of the dislocation had been allowed for, this percentage would be much lower. Because $|F_\mu(y/b)|$ and $|F_\varepsilon(y/b)|$ assume their maximum values at the particle's periphery, the difference $[\text{Max}\{|F_{\varepsilon+\mu}|\} - F_{\varepsilon 0}]$ is relatively large. In order to give $F_\mu(y)$ and $F_\varepsilon(y)$ the same sign, h_z has been chosen to be negative.

It remains to discuss the effects of the dissociation of dislocations into Shockley partials. If the width d_s of the stacking fault ribbon between the two partials is much larger than the particle radius, $F_{\mu 0}$ is reduced by nearly the factor 4, whereas the corresponding factor of $F_{\varepsilon 0}$ is close to 2. The reason is that $F_{\mu 0}$ is proportional to b^2 [Eqs. (5.16) and (5.18)], whereas $F_{\varepsilon 0}$ is proportional to b [Eq. (5.43)]. Since according to Eq. (4.26a) τ_{pF} is proportional to $F_0^{3/2}$, the corresponding reductions of $\tau_{pF}(\mu)$ and $\tau_{pF}(\varepsilon)$ are by the factors ≈ 8 and ≈ 2.8, respectively. In the copper–cobalt system, d_s is about 3.4 nm $\approx 13b$ [5.48]. Hence in the above numerical comparison, in which r equaled $8b$, the dissociation should not have been neglected. It

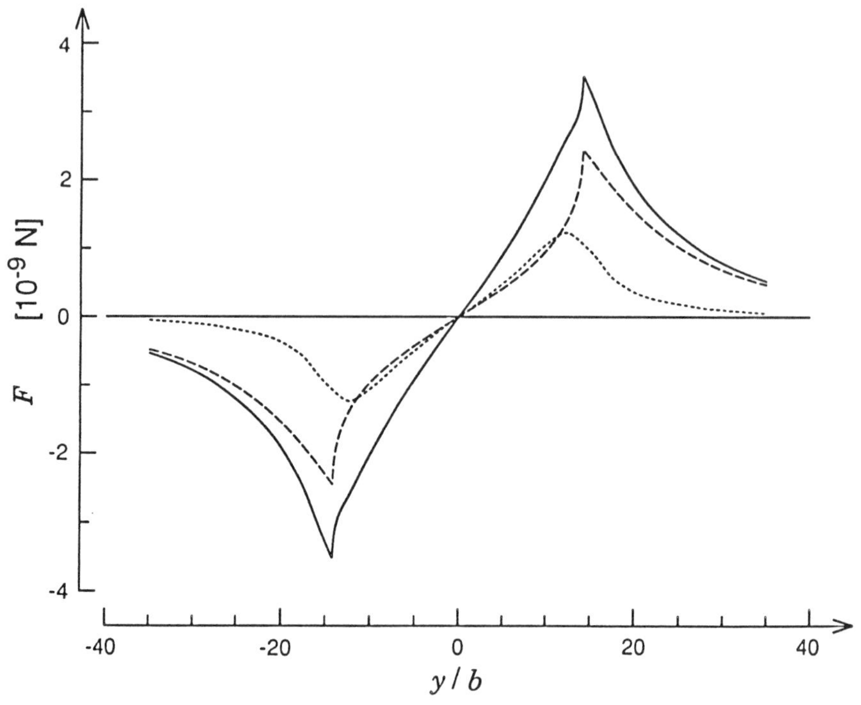

Fig. 5.7. Forces F exerted by a cobalt-rich particle on an undissociated edge dislocation in a copper-rich copper–cobalt alloy versus y/b. Particle radius $\rho = 20b$, $b = 0.25$ nm, $h_z = -\rho/\sqrt{2}$. $\varepsilon = 0.015$ ($a_s > a_p$), $\mu_s = 31$ GPa (for F_ε), $\Delta\mu = 35$ GPa, core model CM2 (for $R \leqslant R_c$, $R = R_c$), $R_c = 2b$. --- F_μ, — — — F_ε, —— $F_{\varepsilon+\mu} = F_\varepsilon + F_\mu$.

raises the ratio $[\tau_{pF}(\varepsilon)/\tau_{pF}(\mu)]$ from the above-mentioned value 2.5 to nearly $(2.5 \times 8/2.8) \approx 7.1$.

NIMONIC PE16. The analogous evaluations for the commercial nickel-base superalloy NIMONIC PE16, which is hardened by long-range-ordered γ'-precipitates, are as follows: μ_K of Eq. (5.9c) is inserted for μ_s as well as for μ_p. The stiffnesses C_{ik} and the dislocation parameters K_E and K_S (Chapter 2) of the matrix and of the γ'-particles of this alloy have been published by Pottebohm et al. [5.49] and by Wallow et al. [5.50]. S of edge disloctions in the matrix equals 1.5×10^{-9} N for $R_o/R_i = 1000$. The other parameters are:

$$\mu_p = 77\,\text{GPa}$$

$$\mu_s = 65\,\text{GPa}$$

$$\Delta\mu = 11\,\text{GPa}$$

$$\Delta\mu/\mu_s = 0.17$$

All further data refer to $r = 8b$, $f = 0.02$, core model CM1 (for $R \leqslant R_c$: $R = \infty$), $h_z = 0.0$, $R_c = 2b$, $b = 0.25$ nm:

$$F_{\mu 0}/(\Delta\mu b^2) = 0.23$$

$$F_{\mu 0} = 1.7 \times 10^{-10}\,\mathrm{N}$$

$$\cos(\phi_c/2) = 0.054$$

$$\phi_c/2 = 87^\circ$$

$$\eta_0 = 0.44$$

$$\tau_{pF} = 7.0\,\mathrm{MPa} = \tau_{pF}(\mu)$$

$$\tau_{pL} = 1.16\tau_{pF} = 8.1\,\mathrm{MPa} = \tau_{pL}(\mu)$$

The corresponding results for order strengthening of NIMONIC PE16 are the following. Pairing of dislocations (Section 5.5) is disregarded in this comparison. The force $F_{\gamma 0}$ due to the long-range order of the particle equals $2r\gamma$ (Section 5.5.3), and γ is the specific antiphase boundary energy:

$$\gamma = 0.25\,\mathrm{J/m^2}$$

$$F_{\gamma 0} = 10 \times 10^{-10}\,\mathrm{N}$$

$$F_{\mu 0}/F_{\gamma 0} = 0.17$$

$$\cos(\phi_c/2) = 0.33$$

$$\phi_c/2 = 71^\circ$$

$$w_y \approx r = 8b$$

$$\eta_0 = 0.16$$

$$\tau_{pF} = 104\,\mathrm{MPa} = \tau_{pF}(\gamma)$$

$$\tau_{pL} = 1.06\tau_{pF} = 110\,\mathrm{MPa} = \tau_{pL}(\gamma)$$

Because the γ'-particles are obstacles of the energy-storing type, C_1 and C_2 in Eq. (4.62) equal 0.96 and 0.64, respectively (Table 4.1).

Evidently $F_{\gamma 0}$ due to the long-range order of the particles exceeds $F_{\mu 0}$ due to the modulus mismatch sixfold, and $\tau_{pF}(\gamma)$ is close to 15 times $\tau_{pF}(\mu)$. $\tau_{pF}(\gamma)$ has, however, been overestimated because pairing of dislocations has been disregarded (Section 5.5). For the copper–cobalt system as well as for NIMONIC PE16, modulus strengthening appears to be rather weak. In Figure 5.8 the forces $F_\gamma(y/b, -h_z = -\rho/\sqrt{2}, \rho = 20b)$, $F_\mu(y/b, -h_z = -\rho/\sqrt{2}, \rho = 20b)$, and $F_{\gamma+\mu}(y/b, -h_z = -\rho/\sqrt{2}, \rho = 20b) = [F_\gamma(y/b, -h_z = -\rho/\sqrt{2}, \rho = 20b) + F_\mu(y/b, -h_z = -\rho/\sqrt{2}, \rho = 20b)]$ are plotted versus y/b. Though F_γ and

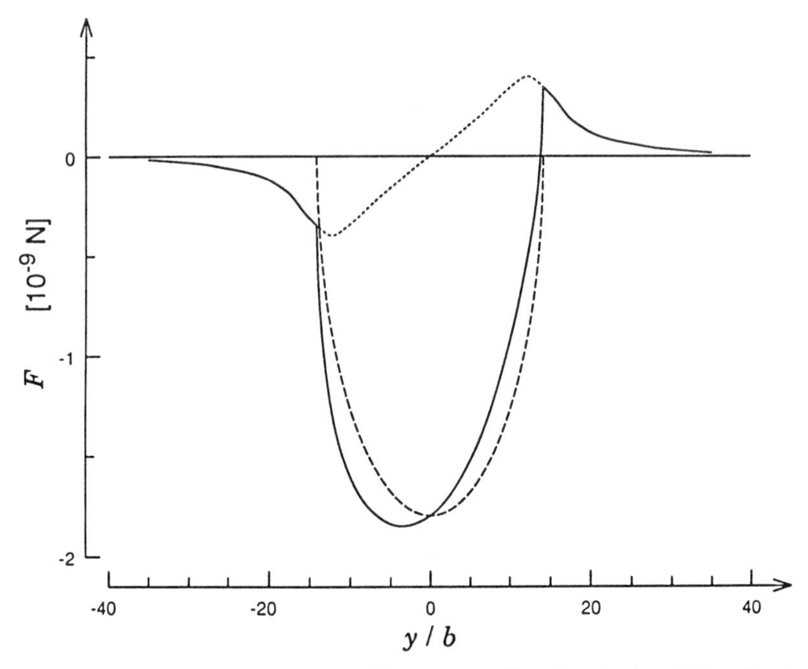

Fig. 5.8. Forces F exerted by a γ'-particle on an undissociated edge dislocation in the nickel-base superalloy NIMONIC PE16 versus y/b. Particle radius $\rho = 20b$, $b = 0.25\,\text{nm}$, $h_z = \rho/\sqrt{2}$, $\gamma = 0.25\,\text{J/m}^2$, $\Delta\mu = 11\,\text{GPa}$, core model CM2 (for $R \leqslant R_c$: $R = R_c$), $R_c = 2b$. --- F_μ, – – – F_γ, —— $F_{\gamma+\mu} = F_\gamma + F_\mu$. Outside of the particle—that is, for $|y| > \rho/\sqrt{2}$—F_γ vanishes and $F_{\gamma+\mu}$ equals F_μ.

F_μ peak for $h_z = 0$, h_z has been chosen to be $\rho/\sqrt{2}$ in order to be consistent with Figure 5.7. The maximum force $\text{Max}\{|F_\gamma + F_\mu|\}$ is very close to $\text{Max}\{|F_\gamma|\}$; that is, the modulus difference raises the maximum interaction force by only 3%. From Figure 5.8 it is also evident that the full width at half-maximum of F_γ and $F_{\gamma+\mu}$ are nearly the same; that is, the modulus difference does not alter w_y either. In Section 5.5.1 it will be shown that the dissociation of the dislocation into Shockley partial dislocations has hardly any effect on $F_{\gamma 0}$. It has been stated above that the dissociation reduces $F_{\mu 0}$ to about one-fourth of its original value.

For copper–cobalt and for NIMONIC PE16 [$f = 0.02$, $8b \leqslant r \leqslant 50b$, and core model CM1 (for $R \leqslant R_c$: $R = \infty$) with $R_c = 2b$], the normalized range η_0 of $F_\mu(y, h_z = 0, r)$ is between 0.03 and 0.44. Therefore the difference between $\tau_{pF}(\mu)$ and $\tau_{pL}(\mu)$ never amounts to more than 16%. The maximum difference is encountered for $r = 8b$ of NIMONIC PE16. Equations (4.26) (Friedel) and (4.62) (Schwarz–Labusch) are based on the assumption that dislocation bowing is only slight. Hence $F_{\mu 0}$ must not exceed the dislocation line tension S [relation (4.9c)].

In many cases, it is accurate enough to describe modulus strengthening by Friedel's equation (4.26d). $F_{\mu 0}$ is taken from Eq. (5.18):

$$\tau_{pF} = \frac{1}{br(2S\pi\omega_q)^{1/2}} (|\Delta\mu|b^2)^{3/2}\alpha_1^{3/2}(r/b)^{3\beta_1/2}f^{1/2}$$

$$\tau_{pF} = \frac{\alpha_1^{3/2}}{(2S\pi\omega_q)^{1/2}} |\Delta\mu|^{3/2} \frac{r^{(3\beta_1/2-1)}f^{1/2}}{b^{(3\beta_1/2-2)}} \tag{5.24}$$

where r is the average particle radius and f is the volume fraction. The statistical factor ω_q has been defined in Eq. (3.5d). All particles that contact the glide plane have been assumed to exert the force $\alpha_1(r/b)^{\beta_1}|\Delta\mu|b^2$ on the edge dislocation. Those particles that do not contact the glide plane are assumed to exert no force [5.33]. Because $F_{\mu 0}$ of edge dislocations exceeds that of screw dislocations and the opposite holds for the dislocation line tensions S, edge dislocations lead to the higher CRSS. Hence their parameters have to be inserted into Eq. (5.24). The above numerical data for $\tau_{pF}(\mu)$ and $\tau_{pL}(\mu)$ have been calculated on the basis of Eq. (5.24).

The exponent $3\beta_1/2 - 1$ of r in Eq. (5.24) is positive for all four core models CM1–CM4 and for both core radii R_c: $R_c = 1b$ or $R_c = 2b$. The minimum exponent is 0.032; it is that of CM2 with $R_c = 1b$. Evidently there is age hardening. If during an Ostwald ripening treatment, r increases with time t and f stays constant, the CRSS also increases with t. At later stages of aging when the critical particle spacing L_c along the dislocation is close to L_{min} (Section 3.1), Friedel's equation (4.26d) may have to be replaced by Eq. (4.16a) and finally the Orowan process will operate. This is the usual aging sequence described at the end of Section 4.2.1.3.2.

Knowles and Kelly [5.32] inserted their numerical results for $F_{\mu 0}$ into Eq. (4.16a), which yields an upper bound for the CRSS. These authors' result for the function $\tau_p(r)$ has a maximum for r equal to several times the inner cutoff radius R_i; that is, age softening starts at rather small particle radii r. The authors applied their model to α-iron strengthened by body-centered cubic (bcc) copper-rich zones, whose lattice mismatch they judged to be negligible. They concluded that their theory was in good agreement with the data. $\partial\tau_p/\partial r$ of the experimental data was negative throughout. r ranged from $9.4b$ to $41b$ and f was 0.012. τ_p was defined as the increase in shear strength due to the particles. Knowles and Kelly found in transmission electron microscopy (TEM) studies of deformed iron-rich iron–copper alloys that dislocations tend to thread through the low-modulus copper-rich particles. This evidence does support the view that they do attract dislocations, but this attraction may as well be due to some other interaction mechanism. The same authors analyzed data published by Mima et al. [5.51] for copper-rich copper–iron alloys hardened by iron-rich particles and claimed that their model of modulus strengthening described the data satisfactorily. Matsuura et al. [5.52] and Wendt and Wagner [5.53], who also investigated the latter system, concluded, however, that its CRSS is governed by the lattice mismatch of the iron-rich particles. Mime et al. had reached the same conclusion. In Matsuura et al.'s alloys the volume fraction f was around 0.01. The yield stress had a maximum at $r \approx 5$–$7\,\text{nm}$. For $r = 17\,\text{nm}$, Orowan loops were observed in the transmission electron microscope.

Russel and Brown [5.38] inserted their equations [Eqs. (5.22) and (5.23)] for $\phi_c/2$ into Eqs. (4.56) and (4.26c) and obtained

$$\tau_p = 0.8\,\frac{\mu_s b}{L_{\min}}\left[1 - \left(\frac{E_{lp}}{E_{ls}}\right)^2\right]^{1/2} \qquad \text{for } \phi_c/2 \leqslant 50^\circ \tag{5.25a}$$

and

$$\tau_p = \frac{\mu_s b}{L_{\min}}\left[1 - \left(\frac{E_{lp}}{E_{ls}}\right)^2\right]^{3/4} \qquad \text{for } \phi_c/2 > 50^\circ \tag{5.25b}$$

It must be remembered that E_{lp} is supposed to be smaller than E_{ls}; that is, the particles are supposed to be softer than the matrix. Equation (5.23) relates the ratio E_{lp}/E_{ls} to the average particle radius r; there ρ has to be replaced by r. Irrespective of the character of the dislocation, $\mu_s b^2/2$ has been inserted for the line tension S and no distinction has been made between the line energy E_l and S. $\tau_p(r)$ of Eq. (5.25a) as well as $\tau_p(r)$ of Eq. (5.25b) reach maxima for r around $2R_i$. This result is almost independent of R_o and of the ratio $E_{lp}^{\infty}/E_{ls}^{\infty}$. R_o and R_i are the dislocation's outer and inner cutoff radii, respectively. Russel and Brown investigated iron-rich iron–copper alloys, which had also been studied by Knowles and Kelly. Russel and Brown inserted 0.6 for the ratio $E_{lp}^{\infty}/E_{ls}^{\infty}$ and $1000R_i$ for R_o. The result for the maximum CRSS $\tau_{p\,\max}$ is

$$\tau_{p\,\max} = 0.041\mu_s b f^{1/2}/R_i \tag{5.26}$$

The authors fitted this equation to their own data and to those published by other groups for the same alloy system. Most data were well represented by this equation. R_i turned out to be $2.5b$. The authors also plotted the yield stress in the overaged state versus $1/L_{\min}$ and found satisfactory agreement with Eq. (5.25a). This is illustrated in Figure 5.9. The slope of the plot "yield stress versus $1/L_{\min}$" was within the expected bounds. The main uncertainty was the Taylor [5.58] factor. The ordinate intercept represents the yield stress of pure iron polycrystals. The authors assumed that this stress can be added to the stress derived from Eq. (5.25a); that is, they supposed that the exponent k in Eqs. (4.57) and (5.58) equals unity.

Ibrahim and Ardell [5.54, 5.55] measured the CRSS of copper-rich copper–gold–cobalt (Cu_3Au–Co) alloys containing cobalt-rich precipitates; the atomic fraction of gold was 0.25. The CRSS decreased monotonically during aging. The authors concluded that their data were better represented by Knowles and Kelly's than by Russel and Brown's model. Ibrahim and Ardell studied deformed copper–gold–cobalt specimens by TEM: no Orowan loops (Figure

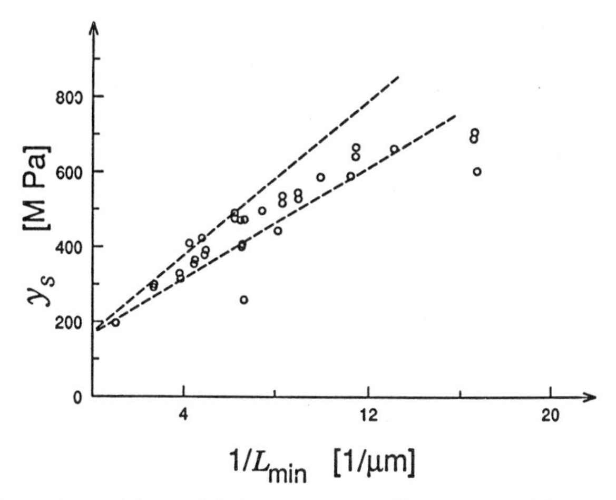

Fig. 5.9. Yield stress y_s of iron-rich iron–copper alloys versus $1/L_{\min}$. Data taken by different groups have been compiled by Russel and Brown [5.38]. The dashed lines indicate theoretical upper and lower limits derived from Eq. (5.25a). (After Ref 5.38.)

6.1d) were observed. The authors did not interpret this as evidence that the Orowan process did not operate, because the cobalt-rich particles may have been too small to sustain the pressure exerted on them by dislocation loops. In this context, Reppich [5.25] drew attention to the cross-slip processes proposed by Gleiter [5.56, 5.57].

Knowles and Kelly's and Russel and Brown's models of modulus strengthening yield functions $\tau_p(r)$ that peak at very small particle radii r. The attraction of these two models is that they can explain overaging without having to resort to the Orowan process, which is supposed to be accompanied by a high work-hardening rate. If the latter one is absent and/or if there are no Orowan loops left behind, it is tempting to interpret a negative slope $\partial\tau_p/\partial r|_{f=\text{const.}}$ of the respective system on the basis of these two models. There are, however, other mechanisms that lead to particle shearing AND (logic AND) negative slopes $\partial\tau_p/\partial r$. Examples are Eqs. (5.6) (chemical strengthening) and (5.30d) (stacking fault energy mismatch strengthening).

Melander and Persson [5.33] combined Eqs. (5.13) and (5.16) with Hansen and Morris' [5.59] statistical derivation (Section 4.2.1.4.1) of the CRSS. In this way it became possible to allow for a distribution of the particle–dislocation interaction forces. Two entirely different effects contribute to this distribution:

1. There may be different types of particles present or/and they may be of various sizes.

2. The separation h_z of the particle center from the glide plane varies from particle to particle.

If all characteristics of the particles are known, Melander and Persson's approach leads to the particles' contribution τ_p to the total CRSS. The authors applied their analysis to polycrystals of an aluminum-rich aluminum–zinc–magnesium alloy, which contained Guinier–Preston zones and particles of the η-phase. Only the latter ones are impenetrable to dislocations, and the zones are sheared. Unless the aging time t at 393 K exceeded 12 h, the contribution of the η-particles to the CRSS was negligible. The experimental derivation of τ_p involved the following two additional operations: (a) Because polycrystals had been studied, their proof stresses had to be converted to CRSSs by dividing the former ones by the Taylor factor [5.58]. (b) Before the measured proof stress of the particle-hardened specimens was converted, the lower yield stress of the solid solution of the alloy was subtracted. The authors plotted τ_p versus t and not versus some length, because it is impossible to represent the zones *and* the η-particles by a single length. The results are shown in Figure 5.10. The average radius of the zones was about 1 nm at $t = 3$ h and about 4 nm at $t = 798$ h. The strong increase of τ_p between $t = 12$ h and $t = 48$ h is caused by the appearance of particles of the η-phase. After 192 h, overaging starts.

The authors assumed that the interaction of dislocations with the zones is governed by their modulus mismatch. Their modulus μ_p was considered as an adjustable parameter. Equating the experimental and the theoretical CRSS for $t = 12$ h yielded 43 GPa for μ_p. The authors inserted 26 GPa for μ_s of the matrix. This agrees with the Voigt–Reuss averages [5.43–5.45, 5.60]. For $t > 12$ h, the contribution of the η-particles to the CRSS has to be taken into consideration. According to Figure 5.10, the theoretical CRSSs calculated for $t < 12$ h and for $t > 12$ h are in satisfactory agreement with the experimental data; that is, inserting 43 GPa for μ_p leads to a satisfactory overall agreement between theory and experiment.

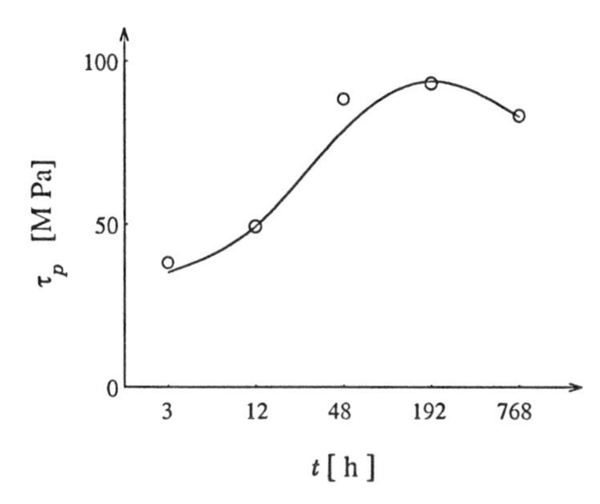

Fig. 5.10. CRSS τ_p of aluminum-base aluminum–zinc–magnesium alloys versus the duration t of the heat treatment at 393 K. The curve has been calculated by Melander and Persson. (After Ref. 5.33.)

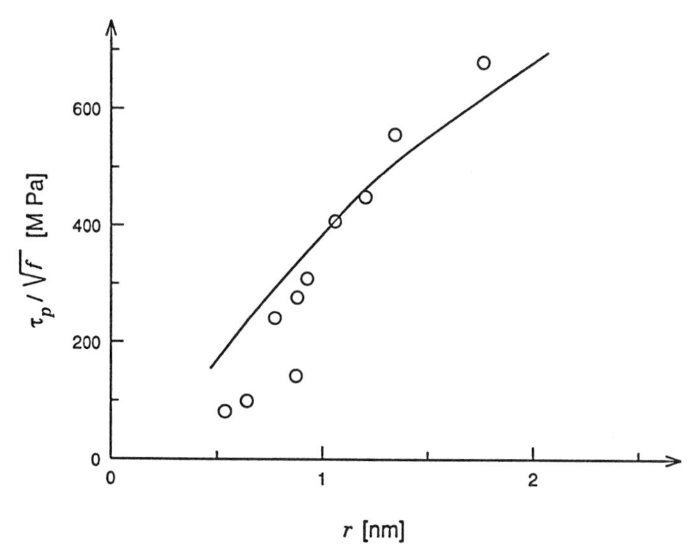

Fig. 5.11. Melander and Persson's [5.33] analysis of Dünkeloh et al.'s [5.61] data: $\tau_p/\sqrt{f}$ of aluminum-base aluminum–zinc–magnesium alloys versus the radius r of the Guinier–Preston zones. $\tau_t - \tau_s$ (Section 7.2.1) has been inserted for τ_p. f is the volume fraction of the zones. The curve represents the result of Melander and Persson's [5.33] model for $f = 0.02$. (After Ref. 5.33.)

Melander and Persson subjected experimental data published by Dünkeloh et al. [5.61] for similar aluminum–zinc–magnesium alloys to the same analysis. The results are shown in Figure 5.11. Again the model represents the data satisfactorily. τ_p increases monotonically with r, and there is no overaging. Haberkorn and Gerold [5.62, 5.63] interpreted precipitation hardening of *binary* aluminum–zinc alloys as being governed by the particles' lattice mismatch (Section 5.4.2.1.8).

The stacking fault energy of aluminum is known to be high [5.64] and the shear modulus is low. Therefore the stacking fault ribbon is very narrow and it may be appropriate to disregard the dissociation of dislocations in aluminum-base alloys.

Melander and Persson based their derivations on the assumption that screw dislocations control the CRSS. They justified this by the following two arguments:

1. The dislocations observed in deformed specimens of their aluminum-base alloy were essentially of screw character.

2. The critical angle $\phi_c/2$ derived from Eq. (5.22) is nearly independent of the character of the dislocation, because Poisson's ratio is nearly the same

for all materials. If therefore Eq. (5.22) is combined with Eq. (4.26c) or with Eq. (4.56), the CRSS τ_p is proportional to the dislocation line tension S. In Eqs. (5.25), $\mu_s b^2/2$ has been inserted for S. Because S of screw dislocations actually exceeds that of edge dislocations, the combination of Eq. (5.22) with Eqs. (4.26c) or (4.56) leads to higher CRSSs for screw than for edge dislocations.

The strength of iron-rich iron–copper alloys and of aluminum-rich aluminum–zinc–magnesium alloys derives probably from the modulus mismatch between particles and matrix. In the aluminum alloys discussed, this applies only to the coherent Guinier–Preston zones, but not to the impenetrable η-particles. In other systems, modulus strengthening may be an ancillary strengthening mechanism. The more so if the principal particle–dislocation interaction force peaks at the particle–matrix interface. This is, for example, the case for lattice mismatch strengthening (Figure 5.7). From this point of view it seems to be rather unlikely that modulus strengthening may enhance order strengthening: According to Figure 5.8 the respective interaction forces peak at the periphery and at the center of the particle. Nevertheless, Messerschmidt and Bartsch [5.65] discussed the possibility that the modulus mismatch of long-range-ordered δ'-precipitates in aluminum–lithium alloys noticeably contributes to the total particle–dislocation interaction.

5.3 STACKING FAULT ENERGY MISMATCH STRENGTHENING

If the specific stacking fault energy γ_s of the solid solution matrix differs from the corresponding energy γ_p of the particles, they interact elastically with dislocations. Let the respective widths of the stacking fault ribbons be d_s and d_p. Both widths should lie between $2b$ and $L_{\min}$, where b is the length of the Burgers vector and $L_{\min}$ is the square lattice spacing of the particles (Section 3.1). Too narrow stacking fault ribbons are better described as extended cores; some of the problems encountered in analyzing particle–core interactions have been touched upon in Section 5.2.1. In this section γ_s is assumed to be larger than γ_p. This is by no means a limitation of the generality; this assumption just allows one to draw a clear picture. As can be seen in Figure 5.12, the stacking fault ribbon widens in the particles, which attract the dislocation. The leading partial dislocation will be referred to as P1, and the trailing one will be called P2. The resolved shear stress $\tau_{\text{obst}}(x, y)$, which has been defined in Section 2.5, depends on γ_s, γ_p, ρ', and the elastic stiffnesses C_{ik} of both phases. ρ' is the radius of the circular intersection of the glide plane with the spherical particle of radius ρ. ρ' is related to ρ by

$$\rho' = (\rho^2 - h_z^2)^{1/2} \tag{5.27}$$

h_z is supposed to be smaller than ρ. The geometry is the same as always, and

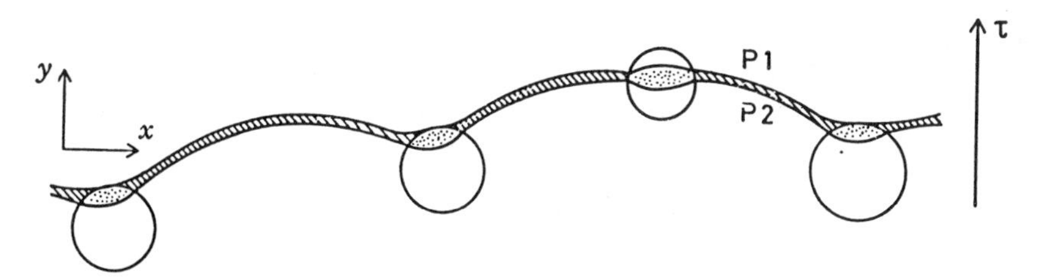

Fig. 5.12. A dissociated dislocation interacting with particles whose stacking fault energy is lower than that of the matrix. The stacking faults in the matrix are hatched, and in the particles they are dotted.

it is sketched in Figure 5.2. The particle is centered at $(0, 0, h_z)$. The dislocation glides in the plane $z = 0$.

Hirsch and Kelly [5.66] derived estimates for the maximum interaction force $F_{\gamma 0}$ that a particle exerts on a dissociated dislocation. The subscript γ of $F_{\gamma 0}$ reminds one of stacking fault energy mismatch hardening. The authors analyzed a variety of combinations of d_s, d_p, and ρ'. $F_{\gamma 0}$ is expressed in the following form:

$$F_{\gamma 0} = l_0 \Delta\gamma \tag{5.28a}$$

with

$$\Delta\gamma = \gamma_s - \gamma_p \tag{5.28b}$$

Equation (5.28a) defines the length l_0. The authors obtained for the case that ρ' is smaller than d_s and d_p:

$$l_0 = 2\rho' \tag{5.29a}$$

For ρ' much larger than d_p, they found that l_0 is proportional to $\sqrt{\rho'}$:

$$l_0 \propto \sqrt{\rho'} \tag{5.29b}$$

In their derivation of the CRSS the authors followed Mott [5.67]. This has been criticized by Brown and Ham [5.16], because a particle that has a stacking fault energy mismatch is no diffuse obstacle in the sense of Section 4.2.1.4.2. The normalized range η_0 [Eq. (4.58g)] of such a particle is actually small. This will be shown below. Hirsch and Kelly's result for the CRSS is similar to Eq. (4.65), which applies to the Mott limit, in which η_0 is much larger than unity.

So far $F_{\gamma 0}$ and l_0 have been given in terms of ρ'; hence $F_{\gamma 0}$ and l_0 are functions of h_z [Eq. (5.27)]. To obtain an estimate of the CRSS τ_p, an averaging procedure is required which relates τ_p to the average particle radius r. Gerold and Hartmann's [5.14] averaging procedure will be presented below. For the time being, the average ρ' is assumed to be proportional to r. Depending on whether the particles are strong or weak obstacles, τ_p has to be derived either from Eq. (4.56) or from Eqs. (4.26) (Friedel's equations). This has been shown by Brown and Ham [5.16] (Section 4.2.2.1). The results for the two above limiting cases [relations (5.29)] are as follows; only the variation of τ_p with r and f is indicated:

1. *Small Particles.* ρ' is smaller than the widths d_s and d_p. In this case $F_{\gamma 0}$ is given by Eqs. (5.29a) and (5.28a): $F_{\gamma 0}$ is proportional to r. This leads to:

   Strong obstacles, Eq. (4.56): τ_p is proportional to $f^{1/2}$, but independent of r (5.30a)

   Weak obstacles, Eqs. (4.26): τ_p is proportional to $(rf)^{1/2}$ (5.30b)

2. *Very Large Particles.* ρ' is much larger than the width d_p of the stacking fault inside of the particles. $F_{\gamma 0}$ is proportional to $r^{1/2}$ [relations (5.29b), (5.28a)]. Thus for τ_p one obtains

   Strong obstacles, Eq. (4.56): τ_p is proportional to $(f/r)^{1/2}$ (5.30c)

   Weak obstacles, Eqs. (4.26): τ_p is proportional to $(f/r^{1/2})^{1/2}$ (5.30d)

   The two relations (5.30c) and (5.30d) indicate age softening: if r increases and f is kept constant, τ_p decreases.

The most elaborate treatment of stacking fault energy mismatch strengthening has been published by Gerold and Hartmann [5.14], who analyzed the case that ρ' exceeds d_p. Figure 5.13a shows the critical configuration. The authors approximated it by the one sketched in Figure 5.13b. P1 is straight, and it has traversed the particle and is about to break free from it. The position of P2 is described by $y_2(x)$. For $|x| \gg \rho'$, P1 and P2 are parallel. The authors derived the static equilibrium configuration of P2 from Eq. (4.73). In the present case it reduces to

$$\tau_{\text{mut}} = \tau_{\text{back}} + \tau_{\text{fault}} \tag{5.31}$$

The boundary conditions are as follows:

1. $\partial y_2/\partial x = 0$ is reached inside of the particle.
2. $y_2 = \rho' - d_s$ for $|x| \gg \rho'$.

Because the solution of Eq. (5.31) depends on the specific stacking fault energies γ_s and γ_p and on the stiffnesses C_{ik} of both phases, no general results

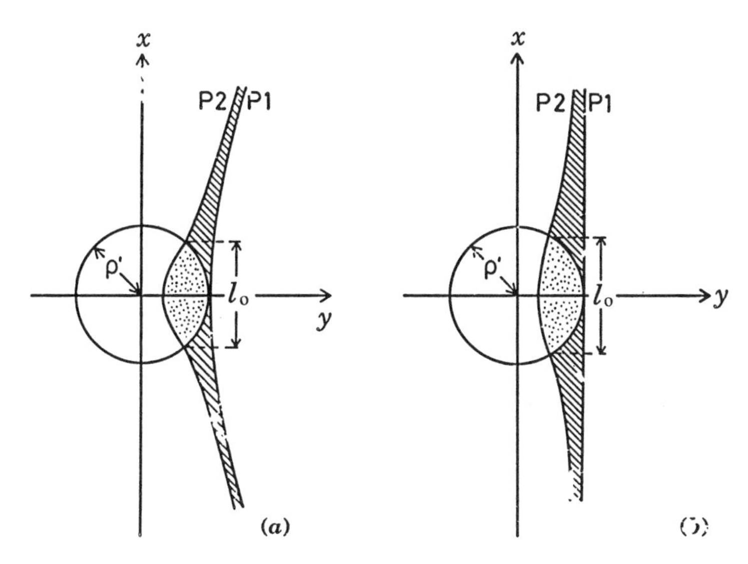

Fig. 5.13. A dissociated dislocation is about to break free from a particle whose stacking fault energy is lower than that of the matrix. The stacking faults are hatched/dotted in the matrix/particle. P1 and P2 are, respectively, the leading and trailing partial dislocations. (*a*) Actual configuration. (*b*) Gerold and Hartmann's [5.14] approximate configuration.

can be given. The authors solved Eq. (5.31) for the aluminum–silver system. Coherent spherical precipitates, whose atomic fraction of silver is between 0.35 and 0.54, form in the aluminum-rich matrix. The authors inserted 0.2 J/m^2 for γ_s and $\mu_F b^2/2 = 10^{-9}$ N for the line tension S of the perfect dislocation. b is the length of its Burgers vector. μ_F has been defined in Eq. (5.9a). τ_{back} in Eq. (5.31) involves the line tension S_p of the partial dislocation. S_p has been assumed to equal $S/3$. The quoted values for γ_s and S are too high. Actually γ_s is 0.135 J/m^2 [5.64] and S of edge dislocations is around 0.4×10^{-9} N. This will be shown below. To be consistent with Gerold and Hartmann's paper, their values for γ_s and S are kept for the time being. Ernst and Haasen [5.68] have published a detailed account of the precipitation process in aluminum–0.01silver.

Some of Gerold and Hartmann's results for l_0 are presented in Figure 5.14 for two choices of γ_p: 0.000 J/m^2 and 0.080 J/m^2. l_0/b is plotted versus ρ'/b. The authors went on to calculate the CRSS τ_p on the basis of Eq. (4.14). Since l_0, which governs $F_{\gamma 0}$ and thus L_c, is a function of ρ', which in turn depends on h_z [Eq (5.27)], an averaging procedure is required. L_c is the particle spacing measured along the dislocation when the latter one is about to break free. Equation (4.14) yields

$$\tau_p = \frac{1}{b} \left\langle \frac{l_0(\rho')\Delta\gamma}{L_c(\rho')} \right\rangle \tag{5.32}$$

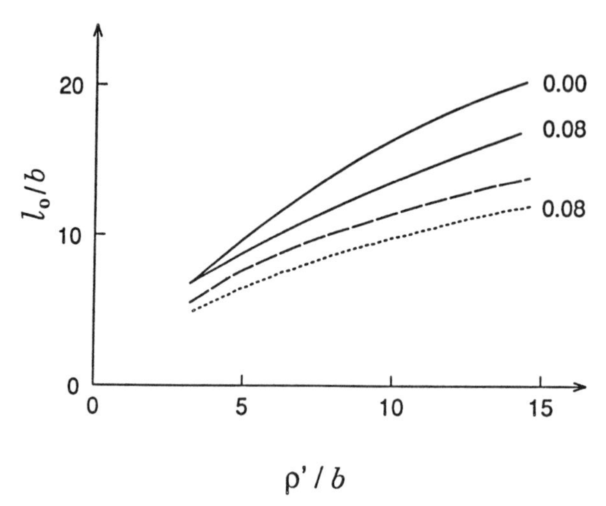

Fig. 5.14. l_0/b versus ρ'/b for aluminum-rich aluminum–silver alloys. The length l_0 and the radius ρ' are defined in Eqs. (5.28a) and (5.27), respectively. The stacking fault energy γ_s of the matrix is 0.2 J/m^2. γ_p of the particles is indicated by the numbers on the right-hand side in J/m^2. —— Gerold and Hartmann's [5.14] model. – – – Eq. (5.36a); this curve is independent of γ_p, $d_s = 1.73$b. --- Gerold and Hartmann's evaluation of Hirsch and Kelly's [5.66] model.

The angle brackets indicate the average. In Section 5.4.2 lattice mismatch strengthening will be dealt with; there similar averaging procedures will be required. Because Gerold and Hartmann assumed that all particles have the same radius r, the averaging procedure concerns only h_z. $1/L_c(\rho')$ is taken from Friedel's equation (4.27a):

$$\frac{1}{L_c(\rho')} = \frac{1}{L_{\min}} \left(\frac{l_0(\rho')\Delta\gamma}{2S} \right)^{1/2}$$

Equations (3.1a) and (3.3) are rewritten:

$$L_{\min} = \frac{r\sqrt{\pi\omega_q}}{\sqrt{f}} = \frac{1}{\sqrt{n_s}}$$

where n_s is the number of particles intersecting unit area of the glide plane. If $|h_z|$ of a particle is close to r, its radius ρ' and consequently its value for l_0 are close to zero. Therefore n_s is renormalized: Only those particles are counted for which $|h_z|$ does not exceed a limit h_1. Hence n_s is replaced by $n_s(h_1/r)$ and $L_{\min}$ is raised by the factor $(r/h_1)^{1/2}$. The limit h_1 will be obtained by an optimization procedure. Inserting for $L_c(\rho')$ from above yields the following for the average in Eq. (5.32):

$$\left\langle \frac{l_0(\rho')}{L_c(\rho')} \right\rangle = \left\langle l_0(\rho') \left(n_s \frac{h_1}{r} \frac{l_0(\rho')\Delta\gamma}{2S} \right)^{1/2} \right\rangle$$

and with Eq. (3.3b) one obtains

$$\left\langle \frac{l_0(\rho')}{L_c(\rho')} \right\rangle = \left\langle l_0(\rho') \left(\frac{l_0(\rho')\Delta\gamma}{2S\pi} f \frac{h_1}{r^3} \right)^{1/2} \right\rangle \tag{5.33}$$

The statistical factor ω_q [Eq. (3.3b)] has been suppressed. The derivation that the authors gave for Eq. (5.33) was analogous to that presented in Section 4.2.1.3.2 for Eq. (4.27a). Inserting Eq. (5.33) into Eq. (5.32) yields

$$\tau_p = \frac{1}{b} (\Delta\gamma)^{3/2} \left(\frac{fr}{2S\pi} \right)^{1/2} I(h_1/r) \tag{5.34a}$$

with

$$I(h_1/r) = \left\langle \left[\frac{l_0(\rho')}{r} \right]^{3/2} \left(\frac{h_1}{r} \right)^{1/2} \right\rangle = \left(\frac{h_1}{r} \right)^{1/2} \frac{1}{h_1} \int_0^{h_1} \left\{ \frac{l_0([r^2 - h_z^2]^{1/2})}{r} \right\}^{3/2} dh_z \tag{5.34b}$$

The function $l_0([r^2 - h_z^2]^{1/2})$ follows from curves like those shown in Figure 5.14. $I(h_1/r)$ has a broad maximum for $0.85 < h_1/r < 0.95$. The authors assumed that the actual CRSS τ_p is obtained by inserting the maximum I_{max} of $I(h_1/r)$ into Eq. (5.34a):

$$\tau_p = \frac{1}{b} (\Delta\gamma)^{3/2} \left(\frac{rf}{2S\pi} \right)^{1/2} I_{max} \tag{5.34c}$$

Since both the function $l_0(\rho')$ and the average in Eq. (5.34b) are nonlinear, the described averaging procedure is rather involved. Equation (5.34a) involves $(\Delta\gamma)^{3/2}$ and the average of $l_0^{3/2}$. This reflects the appearance of $F_0^{3/2}$ in Friedel's equation (4.26a); F_0 is the maximum particle–dislocation interaction force.

As already mentioned, the authors applied their model to the aluminum–silver system. The overall atomic fraction of silver was 0.018. The single crystals were subjected to Ostwald ripening treatments either at the temperature $T_H = 413$ K or at $T_H = 498$ K. The atomic fraction c_p of silver in the particles and their volume fractions f varied with T_H. The average radii r of the particles were measured by small-angle X-ray scattering. r ranged from $3b$ to $14b$. In Figure 5.15 the results for the total CRSS τ_t are plotted versus r/b. τ_t comprises τ_p referred to above and the CRSS τ_s of the matrix. Since τ_s is estimated to be only a minor fraction of τ_t, τ_p approximately equals τ_t and no distinction is made here between τ_t and τ_p. The authors measured the CRSS at 77 K and at 295 K. The CRSS decreased by about 10% as the deformation temperature was raised. Only the data taken at 295 K have been reproduced in Figure 5.15. The authors fitted Eq. (5.34c) to the data, treating γ_p as an adjustable parameter. The drawn-out curves in Figure 5.15 represent the results of the fitting routine. The curves follow the data satisfactorily. The results are as follows:

$$\text{For } T_H = 413\,\text{K:} \quad c_p = 0.54, \quad f = 0.020, \quad \gamma_p = 0.020\,\text{J/m}^2$$
$$\text{For } T_H = 498\,\text{K:} \quad c_p = 0.35, \quad f = 0.029, \quad \gamma_p = 0.060\,\text{J/m}^2$$

γ_s was assumed to equal $0.200\,\text{J/m}^2$. Evidently γ_p decreases as the silver concentration of the particles increases. In view of the current aluminum–silver phase diagram [5.69], in which a hexagonal phase appears at $c_p \approx 0.60$, the observed variation of γ_p with the silver content appears to be reasonable.

The slope $\partial\tau_p/\partial r$ is slightly negative for $r > 10b$ and $f = 0.029$; there is overaging even though the particles are sheared. It has been stated in Section 4.2.1.3.2 that this is to be expected if the maximum particle–dislocation interaction force F_0 increases with r less rapidly than $r^{2/3}$ does.

Nembach [5.70] has published a rather simple model of stacking fault energy mismatch hardening. Since this model is based on the straight-line approximation (Section 4.1.1), P1 and P2 are straight and their separation is invariably very close to d_s. The coordinate system is the same as in Figure 5.13b. P1 and P2 lie at y and at $y - d_s$, respectively. ρ' is defined in Eq. (5.27). The length L of P1 and P2 which reacts with one particle is considered. The length λ_i of Pi lies inside of a particle. As P1 and P2 advance rigidly by dy, there is the following change in energy:

$$dE_\gamma = \{(L - \lambda_1)dy\gamma_s + \lambda_1 dy\gamma_p\} - \{(L - \lambda_2)dy\gamma_s + \lambda_2 dy\gamma_p\}$$
$$dE_\gamma = -(\lambda_1 - \lambda_2)dy(\gamma_s - \gamma_p)$$

Thus for the particle–dislocation interaction force $F_\gamma(y, \rho')$ one obtains

$$F_\gamma = -\partial E_\gamma/\partial y$$
$$F_\gamma(y, \rho') = (\gamma_s - \gamma_p)(\lambda_1 - \lambda_2) = \Delta\gamma(\lambda_1 - \lambda_2) \tag{5.35a}$$

As before, γ_s is supposed to exceed γ_p. The lengths λ_i are given by

$$\lambda_1 = 2[\rho'^2 - y^2]^{1/2} \tag{5.35b}$$
$$\lambda_2 = 2[\rho'^2 - (y - d_s)^2]^{1/2} \tag{5.35c}$$

If a term in square brackets is negative, the respective length λ_i vanishes. Below, l will be written for the difference $\lambda_1 - \lambda_2$:

$$l = \lambda_1 - \lambda_2 \tag{5.35d}$$

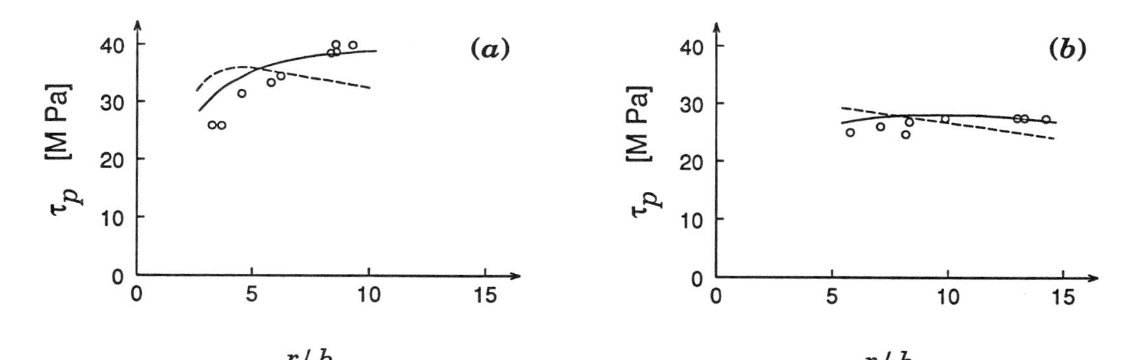

Fig. 5.15. Gerold and Hartmann's [5.14] data $\tau_p \approx \tau_t$ of aluminum–silver alloys versus r/b. r = average particle radius. The data have been taken at 295 K. —— Eq. (5.34a), $\gamma_s = 0.200$ J/m^2; – – – Eq. (5.38a), $\gamma_s = 0.135$ J/m^2. (*a*) $f = 0.02$, (*b*) $f = 0.029$. (After Ref. 5.70.)

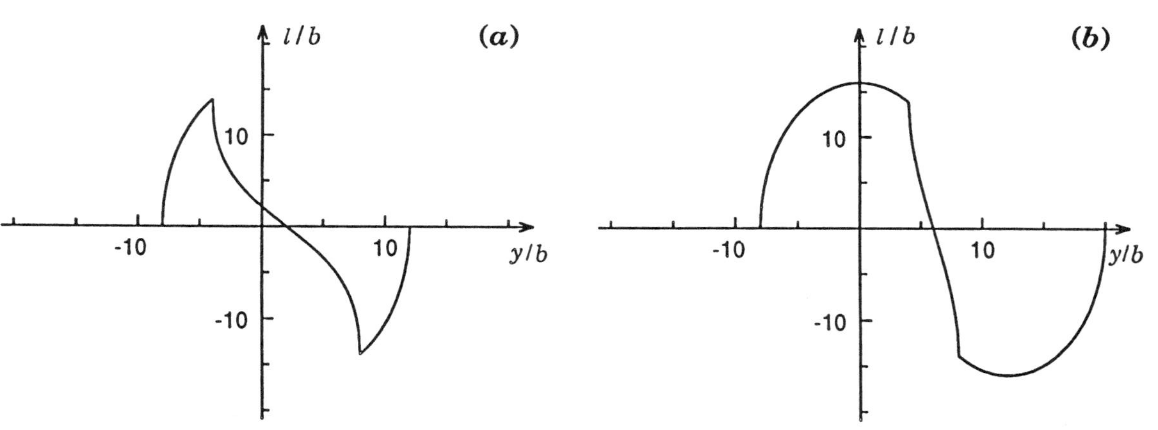

Fig. 5.16. The length l/b defined in Eq. (5.35d) versus the position y/b of the leading partial dislocation, $\rho' = 8b$. (*a*) $d_s = 4b$, (*b*) $d_s = 12b$.

$l(y, \rho' = 8b)$ is plotted in Figure 5.16. d_s equals $4b$ in Figure 5.16a and $12b$ in Figure 5.16b; that is, d_s is smaller, respectively larger, than ρ'. The maximum force

$$F_{\gamma 0}(\rho') = l_0(\rho')\Delta\gamma = \mathrm{Max}\{|F_y(y, \rho')|\}$$

and the positions y at which $|F_y|$ equals $F_{\gamma 0}$ depend on the relative size of d_s and ρ'. This implies that $F_{\gamma 0}$ varies with the radius ρ of the particle and with the position h_z of its center.

For $\rho' \geqslant d_s$, $|l|$ reaches its maximum l_0 when either P1 or P2 just touches the periphery of the particle and the other partial dislocation is inside of it:

$$l_0(\rho') = 2[\rho'^2 - (\rho' - d_s)^2]^{1/2}$$
$$l_0(\rho') = 2[2d_s\rho' - d_s^2]^{1/2} \tag{5.36a}$$

For $\rho' < d_s$, $|l|$ peaks when either of the partial dislocations is at $y = 0$:

$$l_0(\rho') = 2\rho' \tag{5.36b}$$

For $2\rho' \gg d_s$, Eq. (5.36a) is similar to relation (5.29b); Eq. (5.36b) is identical to Eq. (5.29a). Equations (5.36a) and (5.36b) can be combined into one relation for $F_{\gamma 0}(\rho', d_s)$:

$$F_{\gamma 0}(\rho', d_s) = 2[2\delta\rho' - \delta^2]^{1/2}\Delta\gamma \tag{5.36c}$$

with

$$\delta = \mathrm{Min}\{d_s, \rho'\} \tag{5.36d}$$

In Figure 5.14, l_0/b derived from Eq. (5.36a) for the aluminum–silver system is plotted versus ρ'/b. There also Gerold and Hartmann's results are reproduced. All curves have been calculated for $d_s = 1.73b$, and this value is based on $\gamma_s = 0.2\ \mathrm{J/m^2}$. Below, it will be shown that d_s is actually close to $2.6b$. d_s is below ρ'. Evidently l_0/b of Eq. (5.36a) is lower than l_0/b calculated by Gerold and Hartmann. In contrast to Hirsch and Kelly's and Gerold and Hartmann's results, l_0/b of Eqs. (5.36) is independent of γ_p. Because the widening of the stacking fault ribbon inside of the particle has been disregarded in the derivations of Eqs. (5.36), the particle–dislocation interaction energy has been underestimated.

Though the evaluation of the range $w_y(\rho')$ of $F_y(y, \rho')$ is straightforward, it involves some cumbersome arithmetic. Therefore only an approximation has been given for the full width at half-maximum of $F_y(y, \rho')$ defined in Eqs. (5.35)

[5.70]:

$$w_y(\rho') = \mathrm{Min}\{d_s, (\sqrt{3}\,\rho')\} \tag{5.37}$$

As shown above, the derivation of the CRSS τ_p from Eq. (5.32) involves an elaborate averaging procedure. Nembach [5.70] chose $F_{\gamma 0}(r_r, d_s)$ as the effective maximum interaction force. The function $F_{\gamma 0}$ is that of Eq. (5.36c). r_r, which is the average of ρ', has been defined in Section 3.1. There r_r has been written as $\omega_r r$ in Eq. (3.5a); r is the average particle radius. The statistical factor ω_r allows for a distribution of particle radii. Nembach inserted $F_{\gamma 0}(r_r, d_s)$ into Friedel's equation (4.26d) and found

$$\tau_{pF} = \frac{[2\Delta\gamma(2\delta\omega_r r - \delta^2)^{1/2}]^{3/2}}{br(2S\pi\omega_q)^{1/2}} f^{1/2} \tag{5.38a}$$

with

$$\delta = \mathrm{Min}\{d_s, \omega_r r\} \tag{5.38b}$$

f is the particles' volume fraction and ω_q is the statistical coefficient defined in Eq. (3.5d).

In the derivation of Eq. (5.38a), all particles that intersect the glide plane have been assumed to exert the force $F_{\gamma 0}(r_r, d_s)$ on the dislocation. This averaging procedure is similar to the one that led to Eq. (5.24) describing modulus strengthening. If the slight variation of $S^{1/2}$ with r is disregarded, the function $\tau_{pF}(r, f = \text{const.})$ of Eq. (5.38a) has a maximum for $r = r_{\max} = 2d_s/\omega_r$. If r exceeds $r_{\max}$, $\partial\tau_p/\partial r$ is negative. Thus there is overaging even though the particles are sheared. In the same approximation, for the range w_y one obtains

$$w_y = \mathrm{Min}\{d_s, (\sqrt{3}\,\omega_r r)\} \tag{5.39}$$

Equation (5.38a) with $\delta = d_s$ has been fitted to Gerold and Hartmann's data; $\Delta\gamma$ was treated as an adjustable parameter. In the evaluations the following parameters have been used: $\gamma_s = 0.135\ \mathrm{J/m^2}$ [5.64], shear modulus $\mu_F = 24.6\ \mathrm{GPa}$ [5.60], dislocation line tension parameter $K_S(\theta_d = 90°) = 0.91\ \mathrm{GPa}$ [5.70], ratio R_0/R_i of the cutoff radii $= L_{\min}/b$, $\omega_r = 0.82$, $\omega_q = 0.75$. γ_s and μ_F yield $d_s = 2.56b$. d_s refers to edge dislocations. The fitting routine led to the following:

For $T_H = 413$ K: $f = 0.020$, $\Delta\gamma = 0.114\ \mathrm{J/m^2}$, $\gamma_p = 0.021\ \mathrm{J/m^2}$ $(0.020\ \mathrm{J/m^2})$

For $T_H = 498$ K: $f = 0.029$, $\Delta\gamma = 0.086\ \mathrm{J/m^2}$, $\gamma_p = 0.049\ \mathrm{J/m^2}$ $(0.060\ \mathrm{J/m^2})$

The data in parentheses are Gerold and Hartmann's respective results for γ_p. As mentioned above, they have been based on too high values for γ_s and S. In spite of this, the agreement is remarkably good. The curves fitted to the data are shown in Figure 5.15. Gerold and Hartmann's curves represent the data somewhat better. It must, however, be remembered that their model involves elaborate derivations of l_0 (Figure 5.14) and averaging procedures [Eqs. (5.32)–(5.34)], all of which are specific to the system under investigation.

In the range of Gerold and Hartmann's data, the breaking angle $\phi_c/2$ [Eqs. (4.8)] obtained on the basis of Nembach's model and his above-mentioned values for γ_s, γ_p, d_s, and S is not less than 57°. Thus dislocation bowing is only slight. This justifies the application of Eq. (5.38a), which has been derived from Friedel's equation (4.26d), which in turn had been based on the assumption that dislocations bow out only slightly. In Section 4.1.2 a lower limit has been set for $\phi_c/2$: For $\phi_c/2 \geqslant 60°$, dislocations bow out only slightly. A similar evaluation shows that the normalized range η_0 [Eq. (4.58g)] of all data presented in Figure 5.15 is less than 0.14. The CRSS τ_{pL} of Eq. (4.62) differs from τ_{pF} of Eq. (4.26d) by 6% at most. For $\gamma_s = 0.135\,\mathrm{J/m^2}$, $d_s = 2.56b$, $\gamma_p = 0.021\,\mathrm{J/m^2}$, $f = 0.02$, $r = 8b$, and $b = 0.286\,\mathrm{nm}$, Eqs. (5.36a), (5.36c), (4.8), (4.58g), (5.38), and (4.62) yield

$$l_0(r_r) = 10.4b$$

$$F_{\gamma 0}(r_r) = 3.4 \times 10^{-10}\,\mathrm{N}$$

$$\cos(\phi_c/2) = 0.51$$

$$\phi_c/2 = 59°$$

$$\eta_0 = 0.04$$

$$\tau_{pF} = 34.0\,\mathrm{MPa}$$

$$\tau_{pL} = 0.95\tau_{pF} = 32.3\,\mathrm{MPa}$$

τ_{pL} is very close to τ_{pF}, but slightly below it. The reason is the same one as in Section 5.2.2 when $\tau_{pL}(\varepsilon)$ and $\tau_{pF}(\varepsilon)$ were compared with each other: C_1, which appears in Eq. (4.62) and which allows for the randomness of the arrangement of obstacles in the glide plane, is smaller than unity and η_0 is negligible. Because the two partial dislocations are strongly coupled, their total interaction with the particle having a stacking fault energy mismatch is of the elastic type (Section 4.1).

Rabe and Haasen [5.15] investigated the aluminum–silver system, but with a low atomic fraction of silver: 0.009. These authors found a very broad maximum of the function $\tau_t(r)$ for r around 1 nm. Close to the maximum the scatter of the τ_t data was extremely high. Hattenhauer et al. [5.71] deformed thin foils of such an alloy inside of a transmission electron microscope and observed the critical configurations of the dislocations under full load. The fact that only rather few precipitates appeared to interact with dislocations shed some doubt on the applicability of the above models to aluminum–silver alloys.

In Sections 5.1, 5.2.2, 5.4.2.1.8, and 5.4.2.2.2, copper-rich copper–cobalt alloys are discussed. They are strengthened by coherent cobalt-rich particles, which have a lattice mismatch ε. Because they have the fcc crystal structure, they are only metastable. This may render γ_p negative. γ_s is 0.041 J/m^2, and d_s of edge dislocations equals $13b$ [5.22, 5.48]. For $\gamma_p = 0.000$ J/m^2 and $\rho' = r = 8b$, Eq. (5.36c) yields $F_{\gamma 0} = 1.6 \times 10^{-10}$ N. Due to the dissociation of the dislocation, $F_{\varepsilon\,\mathrm{dis}\,0}$ caused by the lattice mismatch is reduced to 5.3×10^{-10} N. This will be shown in Section 5.4.1. "dis" of $F_{\varepsilon\,\mathrm{dis}\,0}$ indicates that "dissociated" dislocations are being dealt with. In Sections 5.1 and 5.2.2, $F_{\varepsilon 0}$ of undissociated dislocations has been quoted: 10^{-9} N. On the basis of Friedel's equations (4.26) the ratio $[\tau_p(\varepsilon, \mathrm{dis})/\tau_p(\gamma)]$ is estimated to be about 6.0.

The forces $|F_\gamma(y)|$ and $|F_{\varepsilon\,\mathrm{dis}}(y)|$ equal their respective overall maximum values $F_{\gamma 0}$ and $F_{\varepsilon\,\mathrm{dis}\,0}$ when the partial dislocations are in the following positions; ρ' is smaller than d_s, and γ_s exceeds γ_p:

$F_{\gamma 0}$: Either P1 or P2 passes through the center of the particle, $h_z = 0.0$

$F_{\varepsilon\,\mathrm{dis}\,0}$: Either P1 or P2 is at the periphery of the particle, and the other partial lies outside of it; h_z equals about $\pm\rho/\sqrt{2}$ (Section 5.4.1).

Therefore the maximum of $|F_\gamma + F_{\varepsilon\,\mathrm{dis}}|$ is very close to either $F_{\gamma 0}$ or $F_{\varepsilon 0}$, whichever is larger.

Because $F_{\gamma 0}$ increases with $\Delta\gamma$ [Eq. (5.28a)], low values of γ_p lead to high CRSSs. γ_s has two opposite effects on $F_{\gamma 0}$. It raises $F_{\gamma 0}$ via $\Delta\gamma$, but it lowers $F_{\gamma 0}$ because d_s is proportional to $1/\gamma_s$. This lowering of $F_{\gamma 0}$ occurs only if d_s is smaller than ρ' [Eq. (5.36a)]. These comparisons apply to the case considered above: $\gamma_s > \gamma_p$.

5.4 LATTICE MISMATCH STRENGTHENING

Coherent particles whose lattice constant a_p differs from that of the matrix (a_s) are surrounded by a stress field, which obviously exists independently of the presence of dislocations. This contrasts with modulus mismatch strengthening, which has been treated in Section 5.2. In the latter case the interaction between dislocations and particles arises only from the fact that the latter ones alter the energy stored in the stress field of the dislocations. At the end of Section 5.2.1 the terms *diaelastic* for the modulus mismatch interaction and *paraelastic* for the lattice mismatch interaction have been introduced. These terms demonstrate the analogy with magnetic interactions. The paraelastic particle–dislocation interaction is treated on the basis of the Peach–Koehler equation (2.9a).

The stress tensor $\sigma^{\mathrm{obst}\,\varepsilon}$, the resolved shear stress $\tau_{\mathrm{obst}\,\varepsilon}$, and the forces F_ε and $F_{\varepsilon 0}$ will be derived in Section 5.4.1. The subscript ε indicates that lattice mismatch strengthening is being dealt with. The resulting CRSS will be calculated in Section 5.4.2. The sign of the constrained lattice mismatch ε has been chosen such that ε is positive if a_s is larger than a_p; this follows from Eqs.

(3.16) and (3.32). The geometry is the standard one; it is sketched in Figure 5.2: The spherical particle of radius ρ is centered at $(0, 0, h_z)$; the glide plane is characterized by $z = 0$ and the dislocation is approximately parallel to the x-axis and glides towards $+y$. If both h_z and ε are positive, the particle attracts edge dislocations.

5.4.1 Derivation of $\sigma_{\text{obst}\,\varepsilon}$, $\tau_{\text{obst}\,\varepsilon}$, F_ε, and $F_{\varepsilon 0}$

Mott and Nabarro [5.72] and Eshelby [5.73] related the stress tensor $\sigma_{\text{obst}\,\varepsilon}$ to the radius ρ and the constrained lattice mismatch ε [Eqs. (3.16) and (3.32)] of the particle and to the shear modulus μ_s of the matrix; linear, isotropic theory of elasticity has been applied. Inside of the particle—that is, for $[x^2 + y^2 + (z - h_z)^2] \leqslant \rho^2$—$\sigma_{\text{obst}\,\varepsilon}$ is purely hydrostatic [5.74]:

$$\sigma_{\text{obst}\,\varepsilon}(x, y, z - h_z) = M^* \begin{pmatrix} 1 & 0 & 0 \\ 0 & 1 & 0 \\ 0 & 0 & 1 \end{pmatrix} \tag{5.40a}$$

Because $\sigma_{\text{obst}\,\varepsilon}$ of Eq. (5.40a) has no shear stress components, it does not affect the glide of dislocations. The factor M^* involves ε and elastic parameters. In Figure 1.3a a coherent misfitting spherical particle is shown. Its crystal structure as well as that of the matrix was originally cubic. Figure 1.3a gives the solution of the relevant elastic differential equations; isotropy has been assumed. Because there is no shear stress inside of the particle, its original, cubic structure and its original, spherical shape are maintained. Outside of the particle—that is, for $[x^2 + y^2 + (z - h_z)^2] > \rho^2$—Eq. (5.40b) holds [5.74]:

$$\sigma_{\text{obst}\,\varepsilon}(x, y, z - h_z) = \frac{2\mu_s \varepsilon \rho^3}{R^5} \begin{pmatrix} 3x^2 - R^2 & 3xy & 3x(z-h_z) \\ 3xy & 3y^2 - R^2 & 3y(z-h_z) \\ 3x(z-h_z) & 3y(z-h_z) & 3(z-h_z)^2 - R^2 \end{pmatrix} \tag{5.40b}$$

with

$$R^2 = [x^2 + y^2 + (z - h_z)^2] \tag{5.40c}$$

The resolved shear stress $\tau_{\text{obst}\,\varepsilon}$ follows from Eq. (4.1). $\tau_{\text{obst}\,\varepsilon}$ is given here in a form suitable for the standard glide geometry shown in Figure 5.2: The dislocation is parallel to the x-axis and glides in the plane $z = 0$ towards $+y$:

$$\tau_{\text{obst}\,\varepsilon} = -\frac{6\mu_s \varepsilon \rho^3}{R^5} [x h_z \cos\theta_d + y h_z \sin\theta_d] \tag{5.41a}$$

According to Eq. (4.1), the angle between the x-axis and the Burgers vector **b** should take the place of θ_d: θ_d had been defined as the angle between the direction **s** of the dislocation and **b** (Section 2.1). Because **s** is parallel to the x-axis, θ_d equals the angle between **b** and the x-axis. Equation (5.41a) is written separately for edge and screw dislocations:

Edge dislocation, **b** is normal to the x-axis; that is, $\theta_d = 90°$:

$$\tau_{\text{obst }\varepsilon} = -\frac{6\mu_s \varepsilon \rho^3}{R^5}\, y h_z \tag{5.41b}$$

Screw dislocation, **b** is parallel to the x-axis; that is, $\theta_d = 0°$:

$$\tau_{\text{obst }\varepsilon} = -\frac{6\mu_s \varepsilon \rho^3}{R^5}\, x h_z \tag{5.41c}$$

Equations (5.41) hold only outside of the particle; inside of it, $\tau_{\text{obst }\varepsilon}$ vanishes. This follows directly from Eq. (5.40a). $\tau_{\text{obst }\varepsilon}$ of edge dislocations depends on the position x along the dislocation only through R; therefore $\tau_{\text{obst }\varepsilon}$ is a symmetric function of x. $\tau_{\text{obst }\varepsilon}$ of screw dislocations, however, is an antisymmetric function of x. $\tau_{\text{obst }\varepsilon}$ of both types of dislocations is antisymmetric in h_z and ε. $\tau_{\text{obst}}(x, y, -h_z = -1.5\rho)$ is shown in Figure 4.3 for a cobalt-rich particle embedded in copper; ε is positive.

The straight-line approximation introduced in Section 4.1.1 yields for the edge dislocation [Eq. (4.2)] [5.63]; b is the length of the Burgers vector:

$$F_\varepsilon(y, -h_z, \rho) = b \int_{-\infty}^{\infty} \tau_{\text{obst }\varepsilon}(x, y, -h_z, \rho)\, dx$$

$$F_\varepsilon(y, -h_z, \rho) = -\frac{8\mu_s \varepsilon \rho^3 y h_z b}{(y^2 + h_z^2)^2}\left[1 - \phi\,\frac{2\rho^2 + y^2 + h_z^2}{2\rho^3}\right] \tag{5.42a}$$

with

$$\phi = [\rho^2 - y^2 - h_z^2]^{1/2} \qquad \text{for } [y^2 + h_z^2] \leqslant \rho^2 \tag{5.42b}$$

and

$$\phi = 0 \qquad \text{for } [y^2 + h_z^2] > \rho^2 \tag{5.42c}$$

Equation (5.42b) reflects the fact expressed by Eq. (5.40a) that there is no shear stress inside of the particle. Since according to Eq. (5.41c), $\tau_{\text{obst }\varepsilon}(x, y, -h_z, \rho)$ of screw dislocations is antisymmetric in x, the straight-line approximation yields zero for their force $F_\varepsilon(y, -h_z, \rho)$. This follows also from Figure 4.3b: the screw dislocation lies across the positive and the negative peak, whereas the

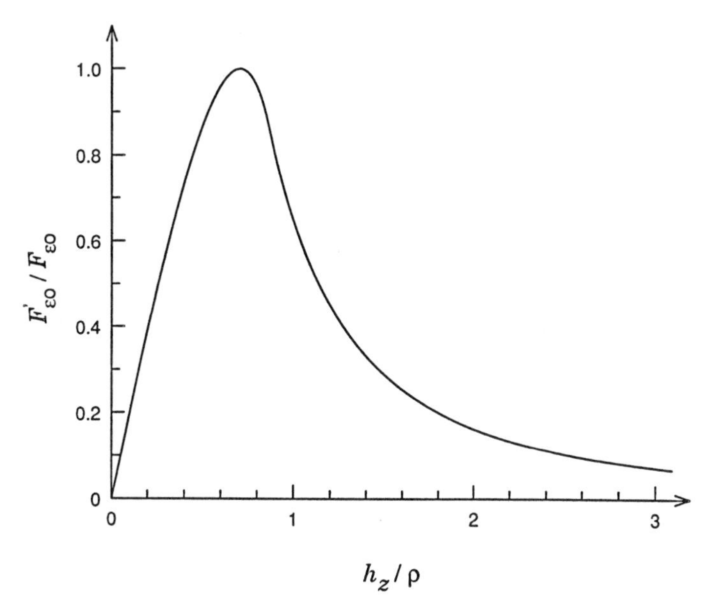

Fig. 5.17. $F'_{\varepsilon 0}/F_{\varepsilon 0}$ [Eqs. (5.44)] versus h_z/ρ. $F'_{\varepsilon 0}/F_{\varepsilon 0}$ are, respectively, the relative and the overall maxima of the particle–edge dislocation interaction force. The particle has a lattice mismatch. Its center is at $(0, 0, h_z)$, and its radius is ρ.

edge dislocation encounters the two peaks in succession (Figure 4.3a). With the exception of Eqs. (5.47), only edge dislocations will be considered in the remainder of this section.

$F_\varepsilon(y, -h_z = -1.5\rho, \rho = 8b)$ and $F_\varepsilon(y, -h_z = -\rho/\sqrt{2}, \rho = 20b)$ of edge dislocations are plotted in Figures 4.3c and 4.27a, respectively. ε is positive. The sign of F_ε is the opposite of that of the product $\varepsilon y h_z$. A particle whose lattice constant is smaller than that of the matrix has a positive misfit parameter ε and attracts edge dislocations if h_z is positive—that is, if the center of the particle lies on the compressive side of the edge dislocations. F_ε does not vanish even if $|h_z|$ exceeds ρ—even those particles that do not intersect the glide plane exert forces on edge dislocations. This is the same situation as in the case of the diaelastic interaction treated in Section 5.2.1. The long range of $F_\varepsilon(h_z)$ complicates the derivation of the CRSS, and it necessitates intricate averaging procedures.

The maximum $F'_{\varepsilon 0}(h_z, \rho) = \text{Max}\{|F_\varepsilon(y, -h_z = \text{const.}, \rho)|\}$ is plotted in Figure 5.17 versus h_z/ρ. $F'_{\varepsilon 0}$ is the maximum interaction force between a particle centered at $(0, 0, h_z)$ and an edge dislocation gliding in the plane $z = 0$. The overall maximum $F_{\varepsilon 0}(\rho)$ is reached for $|y| = |h_z| = \rho/\sqrt{2}$:

$$F_{\varepsilon 0}(\rho) = \text{Max}\{|F_\varepsilon(y, -h_z, \rho)|\} = \text{Max}\{F'_{\varepsilon 0}(h_z, \rho)\}$$
$$F_{\varepsilon 0}(\rho) = 4\mu_s|\varepsilon|b\rho \tag{5.43}$$

$F'_{\varepsilon 0}$ is normalized to $F_{\varepsilon 0}$ in Figure 5.17. Gerold and Haberkorn [5.63] wrote $F'_{\varepsilon 0}(h_z, \rho)$ as

$$F'_{\varepsilon 0}(h_z, \rho) = F_{\varepsilon 0}(\rho)\varphi(h_z/\rho) \tag{5.44a}$$

with

$$\varphi(q) = 2|q|\sqrt{1-q^2} \qquad \text{for } q^2 \leqslant 3/4 \tag{5.44b}$$

and

$$\varphi(q) = \frac{3\sqrt{3}}{8q^2} \qquad \text{for } q^2 > 3/4 \tag{5.44c}$$

φ peaks at $q = 1/\sqrt{2}$: $\varphi(q = 1/\sqrt{2}) = 1.0$, and φ vanishes at the particle's equator—that is, $\varphi(q = 0) = 0.0$.

So far the straight-line approximation has been applied. Gleiter [5.75], Wiedersich [5.74], and Gerold and Pham [5.76, 5.77] calculated the maximum force $F'_{\varepsilon 0}(h_z = \text{const.}, \rho, A)$ for flexible edge dislocations. The parameter A characterizes the amount of dislocation bowing:

$$A = \frac{2\mu_s|\varepsilon|b\rho}{2S} \tag{5.45}$$

The numerator of A equals $F_{\varepsilon 0}/2$ of the straight undissociated edge dislocation [Eq. (5.43)]. Hence, within the straight-line approximation, A is equal to $[\frac{1}{2}\cos(\phi_c/2)]$ defined in Eq. (4.8a); this holds only for $A \leqslant 0.5$. In order to avoid a mix-up of the parameters used for straight and for bent dislocations, $\phi_c/2$ is not used here. Moreover, for $A > 0.5$, $\phi_c/2$ is not defined and the operation of the Orowan process (Chapter 6) becomes possible. The quoted authors expressed the function $F'_{\varepsilon 0}(h_z, \rho, A)$ in the following form:

$$F'_{\varepsilon 0}(h_z, \rho, A) = F_{\varepsilon 0}(\rho) \quad \varphi(h_z/\rho, A) \tag{5.46}$$

This is a generalization of Eqs. (5.44). For A approaching zero, Eq. (5.46) reduces to Eqs. (5.44). The effects of finite values of A are as follows: (i) The height of the maximum of φ is reduced below unity. (ii) The maximum of φ as a function of $[h_z/\rho]$ is less sharp. (iii) φ is no more an antisymmetric function of $[h_z/\rho]$. If A is large, the edge dislocation bows out strongly and thus acquires a substantial screw component. The interaction of the latter with the misfitting particle is only weak. Provided that A does not exceed 0.25, the maximum of φ is reduced by no more than 10%. Thus the effect of A on φ and $F'_{\varepsilon 0}$ can be neglected as long as A stays below 0.25. In the following section, analytical expressions will be presented for the CRSS of materials strengthened by particles that have a lattice mismatch. These equations are based on the assumption that dislocations bow out only slightly. According to relation (4.9c), this requires that the maximum particle–dislocation interaction force does not exceed the dislocation line tension S. If $F_{\varepsilon 0}$ of the straight edge dislocation [Eq. (5.43)] is inserted into relation (4.9c), one finds that $[4\mu_s|\varepsilon|b\rho]$ has to be smaller than S:

$$4\mu_s|\varepsilon|b\rho < S$$

or

$$\frac{\mu_s|\varepsilon|b\rho}{S} < 1/4$$

Inserting this into Eq. (5.45) yields the result that A must be below 0.25. Thus the prerequisite of "slight dislocation bowing" and of "negligible effects of A on $F'_{\varepsilon 0}$" is the same: A must be smaller than 0.25. In view of the derivations of the CRSS, it is evidently consequent to limit A already at the present stage to 0.25. Therefore Eqs. (5.44) can be used throughout and Eqs. (5.45) and (5.46) are not needed.

Because $F_\varepsilon(y, -h_z, \rho)$ of the screw dislocation vanishes if the straight-line approximation is performed, Gerold and Haberkorn [5.63] inserted zero as lower limit of the integral in Eq. (4.2). The authors justified this on the ground that bending will impart screw-dislocations edge-components. The authors considered the result obtained in this way as an upper limit for the actual maximum $F'_{\varepsilon 0}(h_z = \text{const.}, \rho, \text{screw})$. "screw" is included in the argument in order to stress that screw dislocations are being dealt with:

$$F'_{\varepsilon 0}(h_z, \rho, \text{screw}) = 4\mu_s|\varepsilon|b\rho\psi(h_z/\rho) \tag{5.47a}$$

with

$$\psi(q) = q/2 \qquad \text{for } |q| \leqslant 1 \tag{5.47b}$$

and

$$\psi(q) = \frac{1}{2q^2} \qquad \text{for } q > 1 \tag{5.47c}$$

Since the introduction of $F'_{\varepsilon 0}(h_z, \rho, \text{screw})$ is rather arbitrary and the maximum of $|F'_{\varepsilon 0}(h_z, \rho, \text{screw})|$ equals only half the respective maximum of edge dislocations [Eq. (5.43)], $F'_{\varepsilon 0}(h_z, \rho, \text{screw})$ is not considered any further; below only edge dislocations are studied.

So far all calculations have been based on linear, isotropic elasticity. Because most materials are actually anisotropic, the question arises which shear modulus should be inserted for μ_s. In view of the glide geometry of fcc materials, μ_F defined in Eq. (5.9a) is a good choice: μ_F is the shear modulus in $\{111\}$-planes in $\langle 110 \rangle$-directions [5.30]. The effects of the elastic anisotropy on ε and $F_{\varepsilon 0}$ are demonstrated here for the copper–cobalt system, which has already been referred to in Sections 5.1, 5.2.2, and 5.3: Coherent, spherical cobalt-rich precipitates are embedded in a copper-rich matrix. The anisotropy factor $2C_{44}/(C_{11} - C_{12})$ of copper is relatively large: 3.2 [5.18, 5.47, 5.78]. In Sections 5.1, 5.2.2, and 5.3, ε and $F_{\varepsilon 0}$ have been calculated on the basis of μ_F. As an alternative, now μ_R defined in Eq. (5.9b) is inserted into Eqs. (3.32) and (5.43). For $\rho = 8b$, $b = 0.25$ nm, and room temperature [5.24], this yields

$$\mu_F = 30.9\,\text{GPa}$$
$$\mu_R = 42.6\,\text{GPa}$$
$$\varepsilon(\mu_F) = 0.0152$$
$$\varepsilon(\mu_R) = 0.0142$$
$$F_{\varepsilon 0}(\mu_F) = 9.8 \times 10^{-10}\,\text{N}$$
$$F_{\varepsilon 0}(\mu_R) = 12.6 \times 10^{-10}\,\text{N}$$

The bulk modulus χ_p of the cobalt-rich particles is 186 GPa. Evidently, $F_{\varepsilon 0}(\mu_R)$ exceeds $F_{\varepsilon 0}(\mu_F)$ by nearly 30%. The *elastic* anisotropy is not the only one relevant to lattice mismatch strengthening. If the particles are not spherical, there is a geometric anisotropy. In the latter case the orientation of the particles relative to the Burgers vector of the dislocation is important. Obvious examples are the Guinier–Preston zones in the aluminum-base alloys mentioned in Section 1.1. For further treatments of both types of anisotropies, reference is made to the literature [5.79–5.82].

Up to now, no distinction has been made between perfect and partial dislocations. The length b of the Burgers vector appearing in Eqs. (5.42a), (5.43), (5.45), and (5.47a) might be that of either type of dislocation. For the sake of clarity, however, throughout the discussions of lattice mismatch strengthening, b stands for the length of perfect Burgers vectors. Only if the width d_s of the stacking fault ribbon between the partial dislocations is far below the particle radius ρ, the dissociation can be disregarded. d_s refers to the matrix. Since the partials of a dissociated dislocation are strongly coupled, it is the total force exerted by the misfitting particle on all partials together which enters formulae for the CRSS. Of course the dissociation affects also the dislocation line tension. An example for this is given in Section 5.4.2.1.8. If dissociated dislocations are being dealt with, $F_{\varepsilon\,\text{dis}}$, $F_{\varepsilon\,\text{dis}\,0}$, and $F'_{\varepsilon\,\text{dis}\,0}$ are written instead of F_ε, $F_{\varepsilon 0}$, and $F'_{\varepsilon 0}$, respectively. The latter three forces are to be calculated for perfect Burgers vectors.

In fcc materials, the edge component of the Burgers vector of each partial of a dissociated edge dislocation has the length $b/2 = a_s/2^{3/2}$. Because a particle having a lattice mismatch exerts no force on a straight screw dislocation, the screw components of the straight partial dislocations of a dissociated edge dislocation experience no force either. Let P1 and P2 be the leading and the trailing partial, respectively, of a straight dissociated edge dislocation. P1 lies at y and P2 lies at $y - d_s$. Applying the straight-line approximation, the force profile $F_{\varepsilon\,\mathrm{dis}}(y, -h_z, \rho, d_s)$ follows from Eqs. (5.42). In order to emphasize the dependence on d_s, d_s is included as argument of $F_{\varepsilon\,\mathrm{dis}}$, $F'_{\varepsilon\,\mathrm{dis}\,0}$, and $F_{\varepsilon\,\mathrm{dis}\,0}$:

$$F_{\varepsilon\,\mathrm{dis}}(y, -h_z, \rho, d_s) = \tfrac{1}{2}\{F_\varepsilon(y, -h_z, \rho) + F_\varepsilon(y - d_s, -h_z, \rho)\} \tag{5.48}$$

Nembach [5.22] calculated $F_{\varepsilon\,\mathrm{dis}}$ numerically and determined the maximum $F'_{\varepsilon\,\mathrm{dis}\,0}(h_z, \rho, d_s) = \mathrm{Max}\{|F_{\varepsilon\,\mathrm{dis}}(y, -h_z = \mathrm{const.}, \rho, d_s)|\}$ at given h_z and the overall maximum $F_{\varepsilon\,\mathrm{dis}\,0}(\rho, d_s) = \mathrm{Max}\{|F_{\varepsilon\,\mathrm{dis}}(y, -h_z, \rho, d_s)|\} = \mathrm{Max}\{F'_{\varepsilon\,\mathrm{dis}\,0}(h_z, \rho, d_s)\}$. For $|h_z| < \rho$, $|F_{\varepsilon\,\mathrm{dis}}(y, -h_z = \mathrm{const.}, \rho, d_s)|$ assumes its maximum value $F'_{\varepsilon\,\mathrm{dis}\,0}(h_z, \rho, d_s)$ when both partial dislocations lie on the same side outside of the particle and one of them just touches its periphery. The values of h_z for which $|F_{\varepsilon\,\mathrm{dis}}(y, -h_z, \rho, d_s)|$ reaches its overall maximum $F_{\varepsilon\,\mathrm{dis}\,0}(\rho, d_s)$ depend on d_s. For $d_s = 0.5\rho$, this maximum is at $|h_z| \approx 0.79\rho$. For $d_s = 0.0$, the optimum position has been given above: $|h_z| = \rho/\sqrt{2} = 0.707\rho$. The quoted author wrote the function $F_{\varepsilon\,\mathrm{dis}\,0}(\rho, d_s)$ as

$$F_{\varepsilon\,\mathrm{dis}\,0}(\rho, d_s) = F_{\varepsilon 0}\beta_1(\rho/b)^{\beta_2} \tag{5.49}$$

The coefficients β_1 and β_2 depend on the ranges of ρ and d_s to which Eq. (5.49) is to be applied. β_1 and β_2 are listed in Table 5.2. If d_s is much larger than ρ, $F_{\varepsilon\,\mathrm{dis}\,0}/F_{\varepsilon 0}$ is close to 0.5. In Figure 5.18 the ratio $F_{\varepsilon\,\mathrm{dis}\,0}/F_{\varepsilon 0}$ is plotted versus ρ/b for two choices of d_s: $4b$ and the range $12b$–$16b$.

Table 5.2. Coefficients β_1 and β_2 of Eq. (5.49)

Range of r	d_s	β_1	β_2
0–∞	0	1.0	0.000
$6b$–$30b$	$4b$	0.515	0.175
$6b$–$30b$	$8b$	0.366	0.249
$6b$–$30b$	$12b$	0.330	0.251
$6b$–$30b$	$16b$	0.332	0.225
$6b$–$30b$	$20b$	0.346	0.192
$8b$–$20b$	$12b$–$16b$	0.328	0.239

Source: After Ref. 5.22.

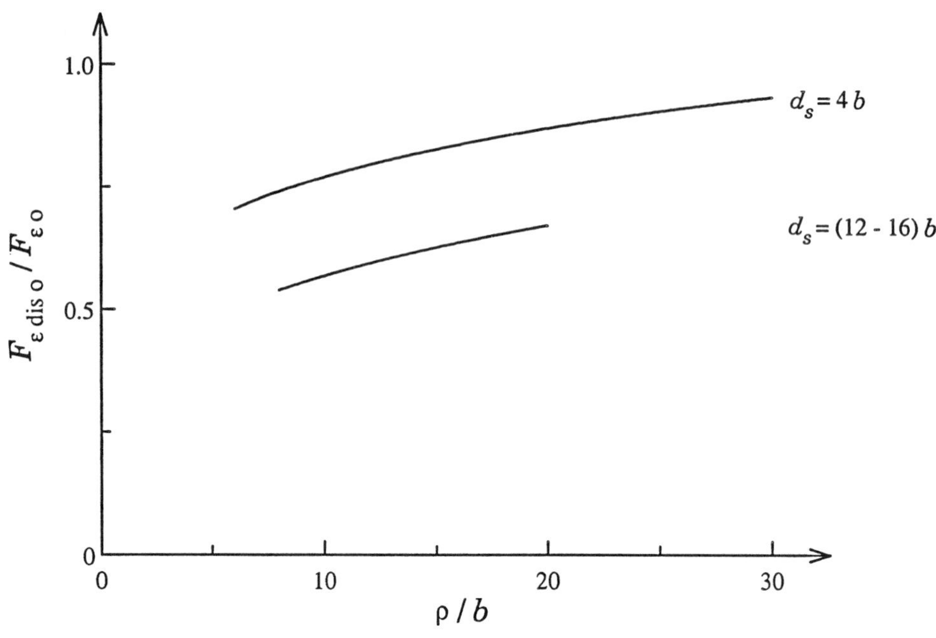

Fig. 5.18. $F_{\varepsilon\,\mathrm{dis}\,0}/F_{\varepsilon 0}$ versus ρ/b. These forces are defined in Eqs. (5.43) and (5.49). The width d_s of the stacking fault ribbon is indicated.

5.4.2 Derivation of the CRSS and Discussion of Experimental Data

There are two reasons that make one hope to develop quantitatively correct descriptions of lattice mismatch strengthening: (i) From the comparisons with other hardening mechanisms presented in Sections 5.1, 5.2.2, and 5.3, it appears that lattice mismatch strengthening is strong. This explains its technical importance. (ii) In contrast to modulus (Section 5.2.1) and stacking fault energy (Section 5.3) mismatch hardening, it is possible to calculate accurately the interaction forces $F'_{\varepsilon 0}(h_z, \rho, \varepsilon)$ [Eqs. (5.44)] and $F_{\varepsilon 0}(\rho, \varepsilon)$ [Eq. (5.43)] between a straight undissociated edge dislocation and a spherical particle that has a lattice mismatch—at least in the case of elastic isotropy. Because the interaction force between such a particle and a straight infinitely long *screw* dislocation vanishes, only *edge* dislocations have to be considered. The long range of their force $F'_{\varepsilon 0}(h_z, \rho, \varepsilon)$, however, poses a problem. According to Eqs. (5.44) and Figure 5.17, $F'_{\varepsilon 0}(h_z, \rho, \varepsilon)$ decreases in proportion to $1/h_z^2$ provided that h_z^2 exceeds $0.75\rho^2$. This long range necessitates intricate averaging procedures. The early one proposed by Mott and Nabarro [5.72] has been described in Section 4.2.1.1. The authors identified the CRSS with the average shear stress in the specimen. They related this stress to the interparticle spacing. The result was Eq. (4.13): The CRSS is independent of the particles' radius r and proportional to their volume fraction f. Both dependencies are at variance

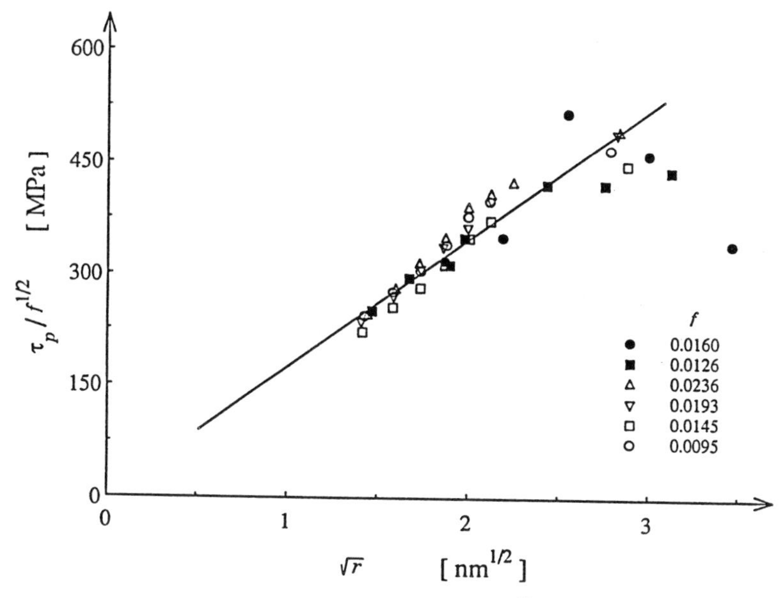

Fig. 5.19. $\tau_p/f^{1/2}$ of copper–cobalt alloys versus $\sqrt{r}$. Unfilled symbols: $T_D = 523$ K. (After Ref. 5.23.) Filled symbols: $T_D = 293$ K. (After Refs. 5.77 and 5.93.)

with the experimental observations: The CRSS of underaged specimens has been found to be approximately proportional to $(rf)^{1/2}$; this is evident in Figures 4.14 and 5.19. The reason why Mott and Nabarro's model led to a faulty variation of the CRSS with r and f is that the authors averaged τ_{obst} over the interparticle spacing, which depends on r and f. This will be discussed further in connection with Gleiter's [5.75] averaging procedure (Section 5.4.2.1.5).

An estimate of the CRSS can be obtained by averaging $F'_{\varepsilon 0}(h_z, \rho, \varepsilon)$ over an appropriate range $h_0 \leqslant h_z \leqslant h_1$ and inserting this average into Friedel's equation (4.26a). The choice of the limits h_0 and h_1 is, however, rather arbitrary. If h_0 approaches zero and h_1 approaches infinity, the average of $F'_{\varepsilon 0}$ vanishes; this follows from Eqs. (5.44) and from Figure 5.17:

$$\lim_{\substack{h_0 \to 0 \\ h_1 \to \infty}} \left\{ \frac{1}{h_1 - h_0} \int_{h_0}^{h_1} F'_{\varepsilon 0}(h_z, \rho)\, dh_z \right\} = 0.0$$

Since according to Eq. (4.26a) the dependence of the CRSS on the particle–dislocation interaction force is nonlinear, it is conceptually better to relate the CRSS to $F'_{\varepsilon 0}(h_z, \rho, \varepsilon)$ first and to take the average afterwards. In this approach, those particles whose centers lie between h_z and $[h_z + dh_z]$ are grouped into one class. The contributions of all classes are somehow added to obtain the total CRSS. Evidently one is faced here with the problem of superimposing CRSS contributions of different classes (i.e., types) of obstacles to the glide of the dislocations. This problem has been touched upon in Section 4.2.2.1 and will be discussed further in Section 7.1. Equation (4.57) describes such superpositions; it will be applied in Section 5.4.2.1.3 with the exponent k equal to 2.0.

In Section 5.4.2.1 the dissociation of the edge dislocation into two Shockley partial dislocations will be disregarded, but in Section 5.4.2.2 it will be taken into consideration. Though Gerold and Haberkorn [5.63] mentioned the effects of the dissociation on the CRSS in 1966, they were later on entirely disregarded, until in 1984 Nembach [5.22] studied them in detail.

Fuchs and Rönnpagel [5.83] did not postulate that the edge dislocations bow out only slightly. Most other authors made this assumption, which allows one to use Friedel's equations (4.26) or similar ones. The dislocations will be slightly bent if relation (4.9c) is fulfilled. Thus $F_{\varepsilon 0}$ and $F_{\varepsilon\,\mathrm{dis}\,0}$ must be smaller than the dislocation line tension S. Moreover, all authors except Fuchs and Rönnpagel assumed that all particles present in the specimen have the same radius; that is, the distribution function of particle radii was assumed to be g_0 defined in Eq. (3.6). In some cases the actual distribution can be approximately allowed for by inserting the appropriate statistical factor ω_q [Eq. (3.5d)] in the final relation for the CRSS.

5.4.2.1 *Undissociated Dislocations.* In this section the CRSS is derived for undissociated dislocations, and in Section 5.4.2.2 their dissociation will be taken into account. The particles are sheared. In Sections 5.4.2.1.1–5.4.2.1.6 the underaged state is considered, and the peak-aged one will be discussed in Section 5.4.2.1.7. These states of aging have been defined in Sections 1.4 and 4.2.1.3.2.

5.4.2.1.1 Rough Estimate of the CRSS. A rough estimate of the CRSS is obtained by approximating $F'_{\varepsilon 0}(h_z, r, \varepsilon)$ by

$$
\begin{aligned}
F'_{\varepsilon 0}(h_z, r, \varepsilon) &= F_{\varepsilon 0}(r, \varepsilon) = 4\mu_s|\varepsilon|br && \text{for } |h_z| \leqslant r \\
F'_{\varepsilon 0}(h_z, r, \varepsilon) &= 0.0 && \text{for } |h_z| > r
\end{aligned}
\qquad (5.50)
$$

Because all particles are assumed to have the same radius r, only r appears in Eqs. (5.50). Evidently those particles that contact the glide plane are assumed to exert $F_{\varepsilon 0}(r, \varepsilon)$ given in Eq. (5.43); the interaction with those particles that are not intersected by the glide plane is entirely disregarded. The analogous approximation had been made in Section 5.2.2 when modulus mismatch strengthening had been analyzed. Equation (5.24) was the result. In the terminology of Section 5.4.2, Eqs. (5.50) imply $h_0 = 0.0$ and $h_1 = r$; hence Eq. (3.3c), which relates the square lattice spacing $L_{\min}$ to r and the particles' volume fraction f, is applicable. $F'_{\varepsilon 0}(h_z, r, \varepsilon)$ of Eqs. (5.50) is inserted into

Friedel's equation (4.26a). The coefficient C_1 is, however, introduced into it. In Friedel's original derivation the particles were assumed to be uniformly spaced (Section 4.2.1.3.1), and this leads to $C_1 = 1.0$. If one wishes to allow for the randomness of the particle arrangement, ≈ 0.9 has to be inserted for C_1. This has been shown in Section 4.2.2.2.1.

$$\tau_p = \frac{C_1}{bL_{\min}} (4\mu_s|\varepsilon|br)^{3/2}/(2S)^{1/2}$$

$$\tau_p = \frac{C_1 f^{1/2}}{b(\pi\omega_q)^{1/2} r} \frac{8(\mu_s|\varepsilon|)^{3/2} b^{3/2} r^{3/2}}{(2S)^{1/2}}$$

$$\tau_p = \frac{8C_1}{(\pi\omega_q)^{1/2}} (\mu_s|\varepsilon|)^{3/2} (rfb)^{1/2}/(2S)^{1/2} \quad (5.51a)$$

For future reference the parameter α_ε and the function $\tau_{p0}(r, f, \varepsilon)$ are defined:

$$\tau_p = \alpha_\varepsilon \tau_{p0}(r, f, \varepsilon) \quad (5.52a)$$

with

$$\tau_{p0}(r, f, \varepsilon) = (\mu_s|\varepsilon|)^{3/2} (rfb)^{1/2}/(2S)^{1/2} \quad (5.52b)$$

The numerical factor α_ε of Eq. (5.51a) is given by

$$\alpha_\varepsilon = \frac{8C_1}{(\pi\omega_q)^{1/2}} \quad (5.51b)$$

Results for α_ε have been compiled in Table 5.3. Though the above derivations have been meant for particles of uniform size—that is, for the distribution

Table 5.3. Coefficient α_ε of Eq. (5.52a)[a]

		α_ε	
Source	C_1	g_0	g_{WLS}
Eq. (5.51a)	1.00	5.5	
	0.93		4.8
Gerold and Pham [5.76]	1.00	3.7	
	0.93	3.4	3.3
Brown and Ham [5.16]	1.00	4.0	
	0.93	3.7	
Jansson and Melander [5.86]	0.93	3.7	

[a]The distribution functions g_0 and g_{WLS} of particle radii have been defined in Eqs. (3.6) and (3.7), respectively.

function g_0 [Eq. (3.6)]—there is also an entry for the Wagner–Lifshitz–Slyozov [5.84, 5.85] distribution g_{WLS} [Eq. (3.7)] in Table 5.3. g_{WLS} has only been allowed for in the calculation of the coefficient ω_q, but all combinations of ω_n [Eq. (3.5b)] have been set equal to unity.

Via the outer cutoff radius, S depends logarithmically on r and f. If this is disregarded, τ_p is proportional to $(rf)^{1/2}$; this is in agreement with the experimental data shown in Figures 4.14 and 5.19. Equations (5.51a) and (5.52b) are based on Friedel's equation (4.26a). Unless the constrained lattice mismatch parameter ε is very small, it is not necessary to replace Eq. (4.26a) by Schwarz and Labusch's [5.10] equation (4.62). It has been shown in Section 5.2.2 and it will be detailed further in Section 5.4.2.1.8 that the normalized range η_0 [Eq. (4.58g)] of cobalt-rich particles in copper–cobalt alloys is so small that the Schwarz–Labusch correction factor $[C_1(1 + C_2\eta_0)]$ [Eq. (4.62)] is close to unity.

5.4.2.1.2 Gerold and Haberkorn's Model. Gerold and Haberkorn's [5.63] treatment of lattice mismatch hardening is similar to Gerold and Hartmann's [5.14] treatment of stacking fault energy mismatch strengthening (Section 5.3). Since Friedel's equation (4.26a) for the CRSS involves $F_0^{3/2}$, $[F'_{\varepsilon 0}(h_z, \rho)]^{3/2} = [F_{\varepsilon 0} \cdot \varphi(h_z, \rho)]^{3/2}$ is averaged over the range $h_0 \leqslant h_z \leqslant h_1$. $F_{\varepsilon 0}$, $F'_{\varepsilon 0}$, and φ are given in Eqs. (5.43) and (5.44). The limits h_0 and h_1 are obtained by maximizing the CRSS τ_p. L_{min} appearing in Eq. (4.26a) has to be replaced by L'_{min}, where L_{min} is given by Eq. (3.1a):

$$L'_{min} = L_{min} \left(\frac{r}{h_1 - h_0} \right)^{1/2} \tag{5.53}$$

This equation allows for the fact that the number of particles that interact with the dislocation is proportional to the factor $[(h_1 - h_0)/r]^{1/2}$. This has been shown in the derivations of Eqs. (5.33) and (5.34). Thus on the basis of Friedel's equation (4.26a) one obtains

$$\tau_p = \frac{C_1}{bL_{min}} F_{\varepsilon 0}^{3/2} \left\langle \left(\frac{h_1 - h_0}{r} \right)^{1/2} \frac{1}{h_1 - h_0} \int_{h_0}^{h_1} \varphi(h_z/r)^{3/2}\, dh_z \right\rangle \frac{1}{(2S)^{1/2}} \tag{5.54a}$$

The angle brackets in Eq. (5.54a) indicate the average. The term $[(h_1 - h_0)/r]^{1/2}$ stems from Eq. (5.53). The remainder in the angle brackets of Eq. (5.54a) is the average over $\varphi^{3/2}$. Now h_z/r, h_1/r, and h_0/r are replaced by q, q_1, and q_0, respectively:

$$\tau_p = \frac{C_1}{bL_{min}} F_{\varepsilon 0}^{3/2} \left\langle \frac{1}{(q_1 - q_0)^{1/2}} \int_{q_0}^{q_1} \varphi(q)^{3/2}\, dq \right\rangle \frac{1}{(2S)^{1/2}} \tag{5.54b}$$

Gerold and Haberkorn postulated that the actual CRSS is obtained by choosing q_0 and q_1 such that τ_p is maximized. The strategy to find the maximum is the following one. In the beginning of the search, q_0 as well as q_1 are close to $1/\sqrt{2}$. Then q_0 and q_1 are, respectively, decremented and incremented until τ_p reaches its maximum. This leads to

$$q_0 = h_0/r = 0.27$$

$$q_1 = h_1/r = 1.14$$

$$\text{Max}\left\{\left\langle\frac{1}{(q_1 - q_0)^{1/2}} \int_{q_0}^{q_1} \varphi(q)^{3/2}\, dq\right\rangle\right\} = 0.67$$

The final result for the CRSS is

$$\tau_p = \frac{8C_1 \cdot 0.67}{(\pi\omega_q)^{1/2}} (\mu_s|\varepsilon|)^{3/2}(rfb)^{1/2}/(2S)^{1/2} \tag{5.54c}$$

This is equivalent to Eq (5.51a), but α_ε [Eqs. (5.52)] is smaller by the factor 0.67:

$$\alpha_\varepsilon = \frac{8C_1}{(\pi\omega_q)^{1/2}}\, 0.67 \tag{5.54d}$$

Evidently the range $h_0 \leqslant h_z \leqslant h_1$ is only a function of r, and it is independent of f.

Actually Gerold and Haberkorn's equation differed from Eqs. (5.54) because the authors disregarded the statistical factor ω_q [Eq. (3.3b)], they replaced it by unity. Moreover, they studied a spatially uniform arrangement of particles. Therefore C_1 equaled unity in their result. Thus in their paper [5.63] α_ε was 3.0. Subsequently Gerold and Pham [5.76] used ω_q of the distribution function g_0 of particle radii [Eq. (3.6)]: $\omega_q = 2/3$. Thus the latter authors obtained $\alpha_\varepsilon = 3.7$. All these values are meant for $C_1 = 1.0$. α_ε of Eq. (5.54d) is listed in Table 5.3.

Though the above derivations have been meant for particles of uniform size, also an entry for g_{WLS} [Eq. (3.7)] is listed in Table 5.3. g_{WLS} has only been allowed for in the parameter ω_q. Actually Eqs. (5.54) involve also ω_n defined in Eq. (3.8).

5.4.2.1.3 Brown and Ham's Model. Brown and Ham's [5.16] model is similar to that of Gerold and Haberkorn, but Brown and Ham applied a quadratic averaging procedure. They based their derivation on a generalization of Eq. (4.57) with $k = 2.0$. The particles whose centers lie between h_z and $h_z + dh_z$ produce the squared CRSS contribution $d\tau_p^2(h_z, r, \varepsilon)$. Their number per unit area is $n_v dh_z$, where n_v is the number of particles per unit volume. The total squared CRSS τ_p^2 is obtained by an integration:

$$\tau_p^2 = \int_{-h_1}^{h_1} d\tau_p^2 = 2\int_0^{h_1} d\tau_p^2 \tag{5.55}$$

The limit h_1 will be determined later. $d\tau_p^2$ follows from Friedel's equation (4.26a), and again the factor C_1 is added. $n_v dh_z$ is inserted for $1/L_{\min}^2$:

$$\tau_p^2 = 2\int_0^{h_1} \frac{C_1^2}{b^2}\,[F'_{\varepsilon 0}(h_z, r, \varepsilon)]^3 n_v\, dh_z\, \frac{1}{2S} \tag{5.56a}$$

$F'_{\varepsilon 0}$ and n_v are given in Eqs. (5.44) and (3.2), respectively.

$$\tau_p^2 = \frac{2C_1^2}{b^2}\,\frac{f}{\frac{4\pi}{3}r^3}\,F_{\varepsilon 0}^3 \int_0^{h_1} \varphi^3(h_z/r)\, dh_z\, \frac{1}{2S}$$

and with Eq. (5.43) one obtains

$$\tau_p^2 = \frac{2C_1^2}{b^2}\,\frac{f}{\frac{4\pi}{3}r^3}\,64(\mu_s|\varepsilon|)^3 b^3 r^3 r \int_0^{q_1} \varphi^3(q)\; dq\, \frac{1}{2S} \tag{5.56b}$$

with

$$q = h/r$$

The integral increases monotonically with q_1; its limiting value for q_1 approaching infinity is 0.53. Therefore having q_1 approach infinity corresponds to Gerold and Haberkorn's strategy to find the actual CRSS by maximizing it. Moreover, $q_1 \to \infty$ is an obvious choice. It yields

$$\tau_p = C_1 \left(\frac{2\cdot 64\cdot 0.53}{4\pi/3}\right)^{1/2} (\mu_s|\varepsilon|)^{3/2}(rfb)^{1/2}/(2S)^{1/2} \tag{5.56c}$$

Evidently this equation is similar to Eqs. (5.51a), (5.52), and (5.54c); here α_ε is given by

$$\alpha_\varepsilon = 8C_1 \left(\frac{1.06}{4\pi/3}\right)^{1/2} \tag{5.56d}$$

For $C_1 = 1.0$, α_ε equals 4.0. This too is listed in Table 5.3. As in the case of Gerold and Haberkorn's model, the range over which the CRSS is averaged is independent of f.

5.4.2.1.4 Jansson and Melander's Model. Jansson and Melander [5.86] applied Hanson and Morris' [5.59] circle rolling technique, which has been described in Section 4.2.1.4.1. The result is again of the form of Eqs. (5.52) with α_ε equal to 3.7. Because the circle rolling technique allows for the random arrangement of the obstacles in the glide plane, this result is listed in Table 5.3 under $C_1 = 0.93$ and g_0.

Except for the minor differences in α_ε (Table 5.3), Eqs. (5.51a), (5.54c), (5.56c), and Jansson and Melander's relation are identical.

5.4.2.1.5 Gleiter's Model. Gleiter's [5.75] model differs from those presented above in two aspects:

1. Gleiter did not apply the straight-line approximation (Section 4.1.1) to obtain the maximum interaction force between an edge dislocation and a particle that has a lattice mismatch, but he used relations analogous to Eq. (5.46).
2. The range $h_1 - h_0$ over which τ_p was averaged was related to the interparticle spacing.

The result is given here in a form similar to that of Eqs. (5.52):

$$\tau_p = 11.8(\mu_s|\varepsilon|)^{3/2}(rb)^{1/2} f^{5/6}/(2S)^{1/2} \tag{5.57}$$

The main difference between this equation and those of Sections 5.4.2.1.1–5.4.2.1.4 concerns the exponent of f: It is 5/6 instead of 1/2. Gleiter's exponent is definitely at variance with the experimental results shown in Figures 4.14 and 5.19. Gerold [5.87] has pointed out that this discrepancy is caused by the point 2 above. Gerold and Haberkorn [5.63] related the range $h_1 - h_0$ of h_z over which the CRSS is averaged only to r; h_1 and h_0 were independent of f. This led to the correct variation of τ_p with f. Because Gleiter and Mott and Nabarro [5.72] averaged over the interparticle spacing, which of course is a function of f, they obtained wrong exponents of f.

5.4.2.1.6 Fuchs and Rönnpagel's Computer Simulations. Fuchs and Rönnpagel [5.83] simulated the motion of an undissociated edge dislocation through a three-dimensional arrangement of particles that have a lattice mismatch. The procedure has been detailed in Section 4.2.2.3. The authors inserted $\tau_{\text{obst}\,\varepsilon}$ of Eq. (5.41a) directly into Eq. (4.73), and they did not perform the straight-line approximation described in Section 4.1.1. The variation of the dislocation line tension S with the angle θ_d was assumed to be that of Eq. (2.11c). For $r < 3$ nm, the results obtained for undissociated edge dislocations in the copper–cobalt system agreed within 10% with the experimental data.

The motion of *dissociated* dislocations through an arrangement of misfitting particles has also been simulated. The CRSS was found to differ only by about

10% from that of undissociated dislocations. The latter surprising result will be discussed further in Section 5.4.2.1.8.

5.4.2.1.7 The Peak-Aged State. As during an Ostwald ripening treatment the average particle radius r increases at virtually constant volume fraction f, the function $\tau_p(r)$ reaches a maximum. This peak-aged state may be characterized by two entirely different situations:

1. The particles are still sheared and Eq. (4.14) holds. The particle spacing L_c in the critical moment is close to or even equal to its minimum value $L_{\min}$ [relation (4.15)]. In the latter case Eq. (4.16a) holds.
2. The Orowan process (Chapter 6) starts to operate.

Both situations have been discussed in Section 4.2.1.3.2 in connection with Figure 4.15. In case 1 the edge dislocations are so strongly bent that the maximum particle–dislocation interaction force is not given by Eq. (5.43), but it has to be derived from Eq. (5.46).

5.4.2.1.8 Comparison with Experimental Data. The CRSS of many precipitation-hardened alloys is believed to be controlled by the precipitates' lattice mismatch. Three prominent examples are (the base metal is listed first) aluminum–zinc [5.26, 5.62, 5.63, 5.88, 5.89], copper–iron [5.52, 5.53], and copper–cobalt [5.22–5.24, 5.63, 5.76, 5.83, 5.90–5.95]. The CRSS τ_p of all three systems has been found to be in qualitative or even quantitative agreement with equations of the form of Eqs. (5.52): τ_p is an approximately linear function of $(rf)^{1/2}$. The aluminum–zinc system had an offset. Haberkorn and Gerold [5.62, 5.63] gave some possible explanations. One of them was that the stress tensor σ_{obst} is not related directly to the particle radius, but there are boundary effects that reduce the effective radius. The copper–cobalt system has been studied most extensively. Its main advantage is that the spherical, coherent cobalt-rich particles are superparamagnetic. Thus their average radius r and their volume fraction f can be derived from magnetic measurements. The procedures have been described in Section 3.3.5. At 823 K, the atomic fraction of cobalt is 0.003 in the matrix and 0.93 in the particles [5.24]. Because the matrix is nearly pure copper, its CRSS τ_s is low. This renders less important the question: How do τ_s and the precipitates' contribution τ_p add up to the total, actually measured CRSS τ_t? Here Eq. (5.58), which is equivalent to Eq. (4.57), is used:

$$\tau_t^k = \tau_s^k + \tau_p^k \tag{5.58}$$

Büttner et al. [5.24] established the exponent k experimentally: $k = 1.25$.

In all derivations presented in Sections 5.4.2.1.1–5.4.2.1.7 a single dislocation has been considered. The possibility that many approximately parallel dislocations glide in the same plane had been entirely disregarded. The following dislocations press the first one against the obstacles and help it to overcome them. Nembach et al. [5.95] tensile tested thin copper–cobalt foils inside of a transmission electron microscope and observed the motion of dislocations under full load. Due to frequent cross-slip, each of them was seen to glide in a plane of its own. This may justify the disregard of the mutual dislocation interactions.

According to Figure 5.19, $\tau_p/\sqrt{f}$ of copper–cobalt alloys varies approximately linearly with $\sqrt{r}$, provided that r does not exceed about 7 nm; that is, τ_p is an approximately linear function of $(rf)^{1/2}$, in agreement with the theoretical derivations given in Sections 5.4.2.1.1–5.4.2.1.4 for underaged specimens. The dislocation's line tension S depends on the choice of the ratio of the cutoff radii R_o and R_i. In this section, $L_{\min}/b$ is inserted for this ratio. $K_S(\theta_d = 90°)$ equals 1.8 GPa at 523 K [5.24]. Because $F_{\varepsilon 0}/(2S)$ of all data is larger than 0.5, dislocation bowing cannot be expected to be only slight [relations (4.9)]. Actually the dissociation of edge dislocations into Shockley partials reduces $F_{\varepsilon 0}$ to $F_{\varepsilon \,\mathrm{dis}\, 0}$; these two forces are given by Eqs. (5.43) and (5.49), respectively. The ensuing effects will be discussed in Section 5.4.2.2. For r close to 7 nm, $\tau_p/\sqrt{f}$ reaches its maximum. For still larger particle radii, $\partial(\tau_p/\sqrt{f})/\partial\sqrt{r}$ is negative; that is, the specimens are in the overaged state (Sections 1.4 and 4.2.1.3.2). The peak-aged state has been discussed in Section 5.4.2.1.7.

A more detailed comparison between theory and experiment is presented in Figure 5.20a: There the ratio $[\tau_{p\,\mathrm{exp}}/\tau_{p0}]$ of copper–cobalt alloys is plotted versus $[F_{\varepsilon 0}/(2S)]$. τ_{p0} is defined in Eq. (5.52b), which has been meant for underaged specimens. r and f cover the following ranges: 2.0 nm $\leqslant r \leqslant$ 8.3 nm, $0.01 \leqslant f \leqslant 0.024$. The dissociation of the dislocation is only allowed for in the calculation of S: Bacon [5.96] has shown that the dissociation raises the line tension of edge dislocations in copper by the factor 1.18. $\tau_{p\,\mathrm{exp}}$ is the experimental result derived from Eq. (5.58). The shear modulus μ_F defined in Eq. (5.9a) is inserted into Eqs. (3.32) and (5.52b). μ_F and ε equal 28 GPa and 0.0162, respectively, at 523 K. In the case of agreement between theory and experiment the ratio $\tau_{p\,\mathrm{exp}}/\tau_{p0}$ is constant and equal to α_ε of Eq. (5.52a), provided that $[F_{\varepsilon 0}/(2S)]$ is smaller than 0.5. This is the condition for slight bowing of the dislocation [relations (4.9)]. Evidently none of the data presented in Figure 5.20a is below the limit. In spite of this, the 12 data points lying to the outermost left are discussed. In the original publication [5.23], 10^4 had been inserted for the ratio of the cutoff radii. This raised S above the present value and lowered the ratio $F_{\varepsilon 0}/(2S)$. The average over the theoretical results listed for α_ε in lines 2–4 of Table 5.3 is $3.63 \pm 2.6\%$; this result applies to $C_1 = 0.93$ and to particles of uniform size—that is, to g_0 of Eq. (3.6). The average over the ratios $\tau_{p\,\mathrm{exp}}/\tau_{p0}$ of the 12 outermost left data in Figure 5.20a is $1.29 \pm 1.6\%$. The quoted percentages indicate the standard deviations of the averages. Evidently the theoretical result exceeds the experimental one by the factor 2.8.

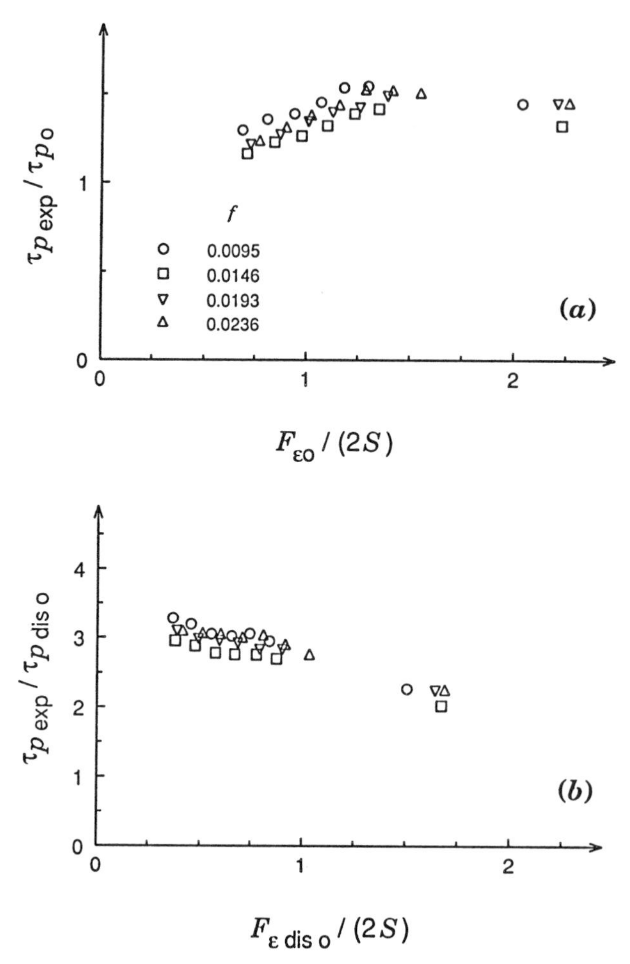

Fig. 5.20. Comparison between experimental data ($\tau_{p\,\mathrm{exp}}$) and theoretical results. System: copper–cobalt. The data have been taken at 523 K [5.23]. (*a*) The dissociation of dislocations is disregarded in τ_{p0} [Eq. (5.52b)]. (*b*) The dissociation is allowed for in $\tau_{p\,\mathrm{dis}\,0}$ [Eq. (5.59)].

Because the normalized range η_0 [Eq. (4.58g)] of all 12 data is between 0.06 and 0.11, Schwarz and Labusch's correction term $[C_1(1 + C_2\eta_0)]$ of the CRSS [Eq. (4.62)] is always between 0.96 and 0.99 and can be disregarded. The particle radius r has been inserted for the range w_y involved in η_0. So far the cobalt-rich particles have been assumed to interact with edge dislocations only via the lattice mismatch. In Section 5.2.2, it has been shown that the particles' modulus mismatch may enhance the interaction force noticeably, but only if the dislocations are undissociated. This enhancement, however, would make the discrepancy between theory and experiment even worse. Varying the outer cutoff radius R_o or the exponent k [Eq. (5.58)] within reasonable limits does not help either.

Nembach [5.22] gave the dissociation of the edge dislocations into Shockley partial dislocations as reason for the above-described disagreement. Because r of the outermost left 12 data in Figure 5.20a is below 3 nm and the width of the stacking fault ribbon of edge dislocations in copper is 3.4 nm [5.48], it is evident that the dissociation must not be disregarded. In Section 5.4.2.1.6 it has been mentioned that Fuchs and Rönnpagel's [5.83] computer simulations yielded satisfactory agreement with the experimental data, no matter whether the dissociation had been allowed for or not. These authors investigated the effect of the dissociation only for rather large particles. The reason why there was only a minor effect of the dissociation is the following: An undissociated *edge* dislocation bows symmetrically around the misfitting particle. The bent parts have some *screw* component and can therefore cut into the particle more easily than a straight *edge* dislocation. The two Shockley partials of a dissociated edge dislocation have *screw* components of opposite signs. Both partials try to increase their screw components by bending. Therefore they attempt to approach the misfitting particle from different sides—one from the left, the other one from the right. But because the partials are coupled by the stacking fault ribbon they are not free to bend to opposite sides. Thus they lose the chance to enhance their screw components and to reduce the interaction force by bending. The equations presented in Sections 5.4.2.1.1–5.4.2.1.4 were all based on the straight-line approximation introduced in Section 4.1.1. Its application implied the preclusion of the above described effects of dislocation bending. Evidently the results of the simulations justify the straight-line approximation for dissociated dislocations.

5.4.2.2 Dissociated Dislocations. Only underaged fcc materials will be considered. Equation (5.49) has been meant for this crystal structure.

5.4.2.2.1 Relations for the CRSS. In Sections 5.4.2.1.1–5.4.2.1.4 the dissociation of the edge dislocations has been disregarded. The most simple way to allow for it is to multiply the right-hand sides of Eqs. (5.51a), (5.52), (5.54c), and (5.56c) by $[\beta_1(r/b)^{\beta_2}]^{3/2}$. The coefficients β_1 and β_2 have been defined in Eq. (5.49). This procedure is equivalent to replacing $F_{\varepsilon 0}$ [Eq. (5.43)] by $F_{\varepsilon\,\mathrm{dis}\,0}$ [Eq. (5.49)] in the final equations. This is by no means a trivial operation. Because the equations for the CRSS were obtained by intricate averaging procedures, it is not correct to merely exchange the maximum interaction force in the final equations. Nembach [5.22], however, justified this approach for Ham and Brown's equation (5.56c), and it is reasonable to assume that this can be done for the other equations too. Consequently Eq. (5.52b) is generalized to

$$\tau_{p\,\mathrm{dis}\,0}(r, f, \varepsilon, \beta_1, \beta_2) = (\mu_s|\varepsilon|)^{3/2}(rfb)^{1/2}[\beta_1(r/b)^{\beta_2}]^{3/2}/(2S)^{1/2} \qquad (5.59)$$

This equation holds for dissociated as well as for undissociated edge dislocations. In the latter case, β_1 equals unity and β_2 vanishes. Equations (5.51a), (5.54c), and (5.56c) are similarly modified.

5.4.2.2.2 Comparison with Experimental Data. The data on copper–cobalt considered in Section 5.4.2.1.8 are now discussed in the light of Eq. (5.59). The CRSS τ_p presses the two partials against the cobalt-rich particles and thus reduces the width of the stacking fault ribbon in copper to about $12b$ [5.22]. Hence for $8b \leqslant r \leqslant 20b$, the parameters β_1 and β_2 equal 0.328 and 0.239, respectively (Table 5.2). In Figure 5.20b, $\tau_{p\,\mathrm{exp}}/\tau_{p\,\mathrm{dis}\,0}$ is plotted versus $F_{\varepsilon\,\mathrm{dis}\,0}/(2S)$. $\tau_{p\,\mathrm{dis}\,0}$ [Eq. (5.59)], $F_{\varepsilon\,\mathrm{dis}\,0}$ [Eq. (5.49)], and S refer to dissociated edge dislocations. There are 12 data that meet the condition that $F_{\varepsilon\,\mathrm{dis}\,0}/(2S)$ is smaller than 0.6. The standard limit for slight bowing of dislocations given in relations (4.9) is 0.5, and it is raised here to 0.6. Averaging $\tau_{p\,\mathrm{exp}}/\tau_{p\,\mathrm{dis}\,0}$ over the 12 outermost left data in Figure 5.20b yields $3.03 \pm 1.3\%$, the standard deviation of the average is quoted. This average must equal α_ε defined in Eqs. (5.52). In Section 5.4.2.1.8 the theoretical result $\alpha_\varepsilon = 3.63 \pm 2.6\%$ has been quoted; it applies to $C_1 = 0.93$ and g_0. The agreement between theory (3.63) and experiment (3.03) is quite satisfactory. It has been achieved by allowing for the dissociation of the edge dislocations. This brings the normalized range η_0 [Eq. (4.58g)] of the 12 data into the range 0.08–0.16. Hence the Schwarz–Labusch [5.10] correction factor $C_1(1 + C_2\eta_0)$ of the CRSS [Eq. (4.62)] amounts to 0.97–1.01; evidently it can be disregarded.

5.4.3 Dynamic Dislocation Effects in Copper–Cobalt Alloys

Fusenig and Nembach [5.94] measured the CRSS of copper–cobalt single crystals at temperatures T_D between 12 K and 500 K. The subscript D of T_D stands for "deformation." The average particle radius was 2.9 nm, and the volume fraction ranged from 0.0 to 0.0223. The results are presented in Figure 5.21. Evidently there is no monotonic decrease of the CRSS τ_t of the particle-strengthened specimens as T_D is raised, but there is a maximum at about 170 K. It is not caused by temperature dependencies of ε, μ_s, and S appearing in Eqs. (5.51a), (5.52), (5.54c), (5.56c), and (5.59). The low CRSSs found for $T_D < 170$ K have been interpreted as manifestations of dynamic dislocation effects. Because the matrix of these alloys is nearly pure copper (Section 5.4.2.1.8), damping of the dislocation motion is mainly due to phonons. Far below the Debye temperature their density and thus the normalized drag coefficient γ [Eq. (4.58k)] are low. Between 50 K and 300 K, γ increases approximately from 0.08 to 0.7. $F_{\varepsilon\,\mathrm{dis}\,0}/(2S)$ is always around 0.5. Because the normalized range η_0 [Eq. (4.58g)] of all specimens is close to 0.1, the normalized CRSS $\tau_{pu}^{\otimes}$ is given by the lowest drawn-out curve in Figure 4.32. The curve indicates a strong increase of $\tau_{pu}^{\otimes}$ if γ is raised from 0.15 to 2.0 (Section 4.2.2.2.2). This is just the range into which γ of the copper–cobalt alloys is estimated to fall for $50\,\mathrm{K} \leqslant T_D \leqslant 300\,\mathrm{K}$. At low temperatures there is

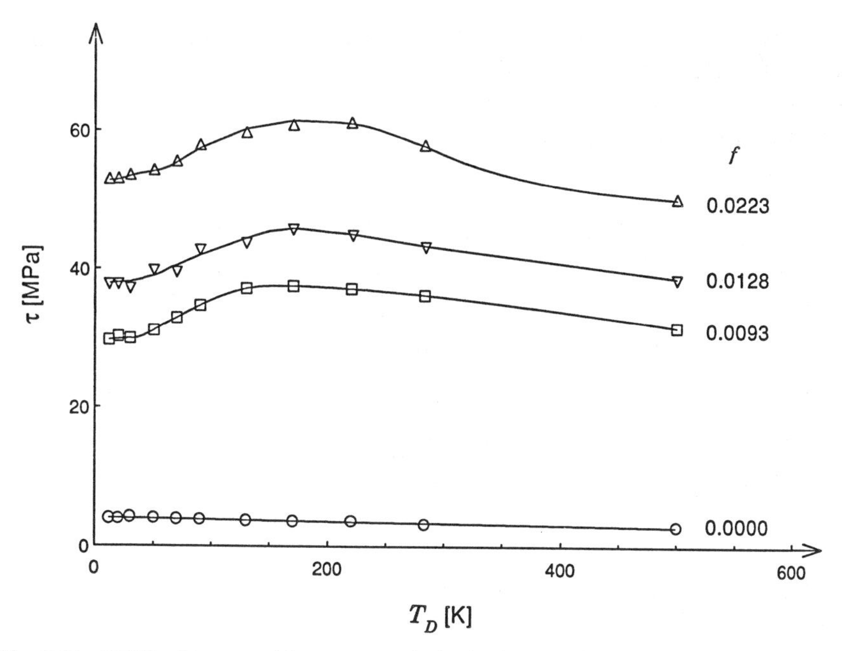

Fig. 5.21. CRSS of copper-rich copper–cobalt single crystals versus the temperature T_D, at which the specimens have been deformed. $r \approx 2.9$ nm and f is indicated. (After Ref. 5.94.)

hardly any damping of the dislocation motion and there are dynamic effects, which lead to low CRSSs. As T_D is raised, phonon damping increases and suppresses these effects; consequently the CRSS increases. At 12 K the dynamic effects are quite strong. τ_t of the specimens with $f = 0.0093$ is 37.6 MPa at 170 K and 29.7 MPa at 12 K. Evidently the dynamic effects reduce τ_t by 21% as T_D is lowered from 170 K to 12 K. The difference in CRSS is 7.9 MPa; this is nearly twice the CRSS of the solid solution matrix at 12 K.

5.5 ORDER STRENGTHENING

Order strengthening has already been touched on in Sections 4.1 and 4.2.2.3: It is illustrated in Figures 4.4–4.6 and 4.35–4.37. The coherent particles and the matrix have similar or even the same crystal structure, but the particles are long-range-ordered, whereas the matrix is disordered or short-range-ordered. The resulting interaction with dislocations is of the energy-storing type (Section 4.1). Reviews on order strengthening have been published by Nembach and Neite [5.5] and by Ardell [5.55, 5.97]. The technically most important and therefore most extensively studied examples are (see Section 5.5.1) (i) nickel-base superalloys, in which $L1_2$-long-range-ordered γ'-particles are embedded in the nickel-rich fcc matrix and (ii) isomorphous aluminum-base aluminum–lithium alloys. Some other systems for which order strengthening is believed to be relevant are the following ones; the approximate compositions of the long-range-ordered precipitates and their crystallographic structures are also given:

- Iron–titanium–silicon: Fe_2TiSi, $L2_1$ [5.98]
- Iron–nickel–aluminum–titanium: Ni(Al, Ti), B2 [5.99]
- Nickel–molybdenum: Ni_4Mo, $D1_a$ [5.100, 5.101]
- Copper–titanium: Cu_4Ti, $D1_a$ [5.102]
- Cobalt–nickel–chromium–niobium–iron: Ni_3Nb, DO_{22} [5.103]
- Iron–chromium–nickel–aluminum: NiAl, B2 [5.104]
- Maraging steels: various intermetallic phases [5.105–5.109]
- Magnesium oxide–iron oxide: magnesia ferrite, spinel [5.28]

Most of the particles are intermetallic phases. Some of their superlattices are so complex that their perfect dislocations consist of more than two matrix dislocations [5.102]. In the following, only $L1_2$-ordered particles embedded in fcc matrices will be considered. In these materials, dislocations glide in pairs.

The main constituents of maraging steels are as follows (atomic fractions are given): iron, up to 0.3 nickel, very little carbon, and additions of around 0.01 of various metals—for example, molybdenum, aluminum, and titanium. The latter constituents form intermetallic phases that are rich in iron and/or nickel. Their effects on the CRSS of maraging steels are similar to those of γ'-particles in superalloys.

5.5.1 Constitution and Crystallography of the $L1_2$-Phase

The elementary cell of the $L1_2$-structure is basically fcc. It consists of two sublattices:

A-sites: the faces of the elementary cell
B-sites: the corners of the elementary cell

Atoms occupying A-sites are called A-atoms, and the definition of B-atoms is analogous. Thus the stoichiometric composition is A_3B. Figure 5.22 shows three successive (111)-planes of the $L1_2$-structure. All 12 nearest neighbors of the B-atoms are A-atoms. The Burgers vector of a perfect dislocation in a fully ordered $L1_2$-material is of the type $a\langle 101\rangle$, where a is the lattice constant. Such a perfect dislocation dissociates into two identical partials:

$$a[10\bar{1}] = (a/2)[10\bar{1}] + (a/2)[10\bar{1}] \quad (5.60a)$$

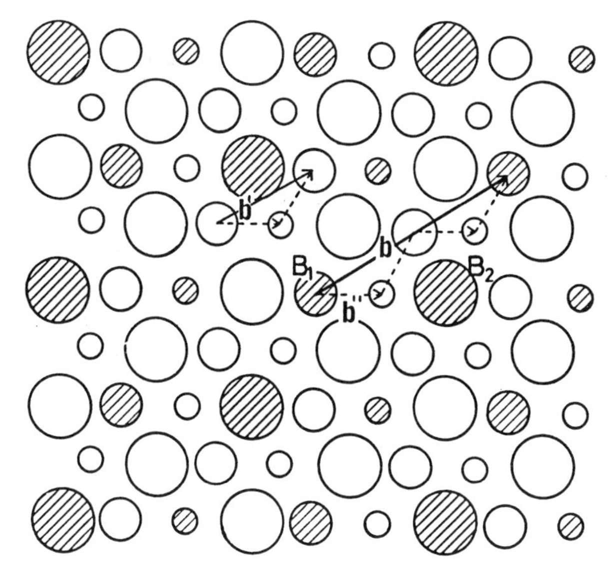

Fig. 5.22. A-atoms (open circles) and B-atoms (hatched circles) form three successive (111)-planes E_i, $1 \leqslant i \leqslant 3$, of the $L1_2$-ordered intermetallic compound A_3B. The radii of the circles decrease as i increases. The dislocations glide between E_1 and E_2. $\mathbf{b} = a[10\bar{1}]$, $\mathbf{b}' = (a/2)[10\bar{1}]$, and the dashed arrows mark Burgers vectors of the type $\mathbf{b}'' = (a/6)\langle 211 \rangle$.

These partials are perfect dislocations in the disordered fcc matrix. Each of the two $(a/2)[10\bar{1}]$-partials may dissociate further into two Shockley partials:

$$a[10\bar{1}] = (a/6)[11\bar{2}] + (a/6)[2\bar{1}\bar{1}] + (a/6)[11\bar{2}] + (a/6)[2\bar{1}\bar{1}] \quad (5.60b)$$

The $(a/6)[11\bar{2}]$-partial P1 is the leading one of a procession of four partials; the last one, P4, has the Burgers vector $(a/6)[2\bar{1}\bar{1}]$. Already P1 disturbs the long-range order: P1 makes the B-atom B_1 a nearest neighbor of another B-atom, namely, B_2. There is a complex stacking fault between P1 and P2, and the same holds for P3 and P4 [5.110]. Between P2 and P3 there is an antiphase boundary; its specific energy will be referred to as γ. γ of the $L1_2$-ordered γ'-precipitates in nickel-base superalloys is about $0.2\ \text{J/m}^2$ [5.5, 5.97, 5.111, 5.112]. P4 restores the $L1_2$-order. Let s_{ij} be the separation of Pi from Pj. s_{ij} is governed by the shear modulus and by the specific energy of the complex stacking fault (s_{12}, s_{34}), respectively, of the antiphase boundary (s_{23}). s_{23} is much larger than s_{12} and s_{34}. If $L1_2$-long-range-ordered precipitates are embedded in a disordered fcc matrix, s_{23} may exceed the interparticle spacing whereas s_{12} and s_{34} are negligible. Therefore the dissociation into Shockley partials according to Eq. (5.60b) will not be considered any further and only Eq. (5.60a) will be used. The leading $(a/2)[10\bar{1}]$-dislocation will be referred to as D1, and the trailing one will be called D2. Thus one may write symbolically: D1 = P1 + P2 and D2 = P3 + P4. Figure 4.6 shows a procession of D1–D2 pairs.

Examples of materials that are strengthened by $L1_2$-ordered precipitates are the following alloy systems; the approximate precipitate compositions are indicated:

- Binary nickel-base nickel–aluminum alloys: Ni_3Al [5.5, 5.6, 5.97, 5.113], (Section 3.2.1)
- Nickel-base superalloys: $Ni_3(Al, Ti)$ [5.5, 5.97, 5.110, 5.114, 5.115]
- Binary aluminum-base aluminum–lithium alloys: Al_3Li [5.7, 5.116–5.122]

Nickel-base superalloys and aluminum–lithium alloys are of great technical interest, whereas binary nickel–aluminum alloys serve only as model systems. The matrix of nickel-base superalloys is referred to as the γ-phase, and the $L1_2$-ordered precipitates are called the γ'-phase. It is a rather unfortunate nomenclature that the letter "γ" stands for the specific energy of the antiphase boundary and for the γ-matrix. Commercial superalloys contain many elements. Their preference for the two phases is as follows. Constituents given in parentheses show no definite preference for either phase: Their partitioning varies from alloy to alloy [5.123–5.134].

γ'-precipitates:	nickel, aluminum, titanium, niobium, vanadium, (tantalum)
γ-matrix:	chromium, cobalt, iron, molybdenum, (tungsten)

The constituents of the γ'-precipitates partition to the two sublattices A and B of the $L1_2$-structure as follows [5.5, 5.128, 5.129, 5.134–5.138]:

A-sites:	nickel, cobalt
B-sites:	aluminum, titanium, tungsten, tantalum, niobium, molybdenum

Here also those elements are listed that actually prefer the γ-matrix to the γ'-phase. Chromium has hardly any preference for A- or B-sites.

As an example for the compositions of commercial nickel-base superalloys, that of NIMONIC PE16 is given here. This material is produced by Inco Alloys International, Hereford, England. The main constituents of the overall alloy/γ'-particles are (atomic fractions are given) [5.127]: nickel 0.42/0.72, iron 0.33/0.03, chromium 0.18/0.01, molybdenum 0.02/—, aluminum 0.024/0.10, titanium 0.014/0.13, carbon 0.003/—. The low concentrations of molybdenum and carbon in the γ'-particles are not accurately known. The γ'-volume fraction f decreases as the temperature T_H of the heat treatment is raised. f is at most 0.1 [5.5]. The composition of the matrix varies with T_H and hence with f. The composition of the γ'-particles is independent of T_H. Dislocation processes in NIMONIC PE16 are discussed in Sections 1.2, 4.1, 5.5.2, 5.5.4, 6.2.2, and 7.2. Unless their radius exceeds about 100 nm, the $L1_2$-ordered γ'-precipitates in nickel-base superalloys are spherical.

5.5.2 Dislocation Configurations in Materials Strengthened by Long-Range-Ordered Particles

Nembach et al. [5.139] tensile-tested thin single-crystal foils of the commercial nickel-base superalloy NIMONIC PE16 (Section 5.5.1) inside of a transmission electron microscope and observed the configurations of the dislocations under full load. A micrograph of an underaged (Section 1.4) specimen has been presented in Figure 4.6. The leading $(a/2)\langle 110\rangle$-dislocation D1 is strongly scalloped, whereas the trailing one D2 is nearly straight or just follows the overall curvature of D1. Though it was not possible to image the long-range-ordered γ'-particles and the dislocations simultaneously, the authors concluded on the basis of micrographs like that shown in Figure 4.6 that in the critical moment when D1 breaks free from the γ'-particles, D2 may be considered as straight. In this case the ratio d_2/L_2 follows from Eq. (3.4):

$$d_2/L_2 = f \tag{5.61a}$$

where d_i is the mean length of the dislocation Di lying inside of a γ'-particle, and L_i is the mean spacing of γ'-particles along Di. d_i and L_i are meant for the critical configuration, and they are indicated in Figure 5.23. Pretorius and Rönnpagel's [5.140] computer simulations, which have been described in Section 4.2.2.3 (Figures 4.35a and 4.37), led to the conclusion that the antiphase boundaries pull D2 forward so that it lies outside of all γ'-precipitates. This means that Eq. (5.61a) should be replaced by

$$d_2/L_2 = 0 \tag{5.61b}$$

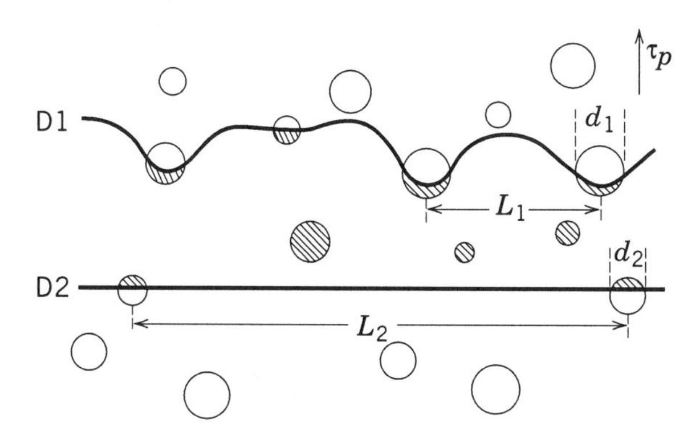

Fig. 5.23. Critical configurations of the paired dislocations D1 and D2 in a material strengthened by long-range-ordered γ'-particles, shown schematically. d_i = mean length of Di lying inside of a γ'-particle, L_i = mean spacing of γ'-particles along Di. Antiphase boundaries are shown hatched.

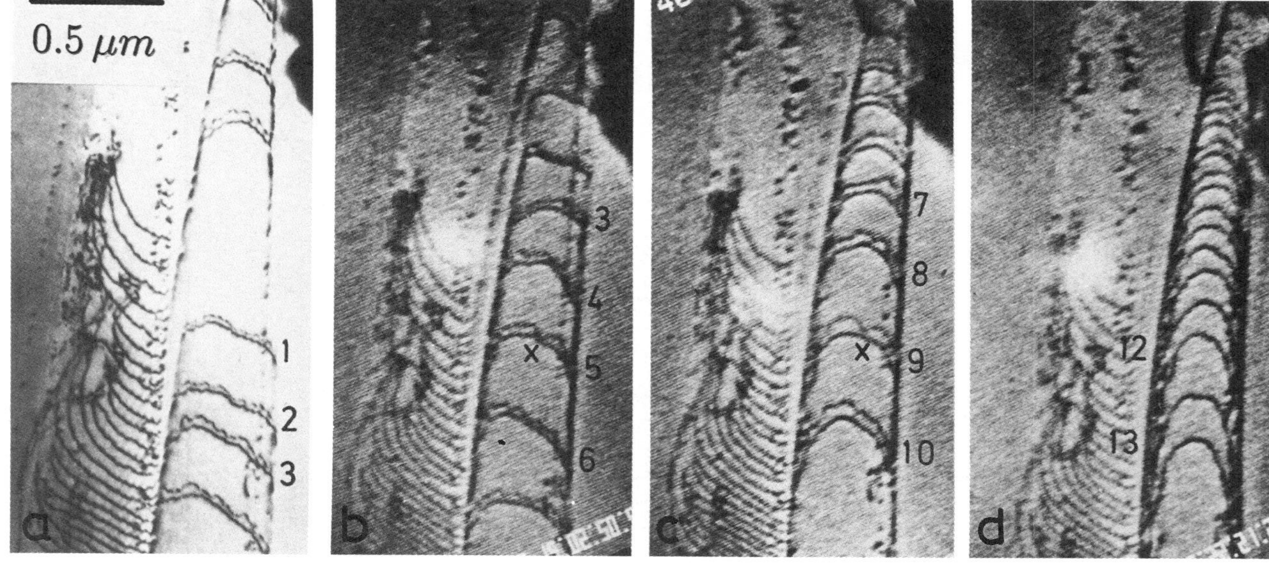

Fig. 5.24. Sequence of transmission electron micrographs of dislocations gliding in a single-crystal thin foil of the commercial nickel-base superalloy NIMONIC PE16. $f = 0.09$, $r = 4$ nm. All micrographs have been taken under full load. The pairs are numbered. Pairs 5 and 9 are held up at the same spot marked by $\times$. The pairs visible in (*a*) ahead of pair 1 glide in a different, though parallel, plane. (After Ref. 5.139.)

Since it is open to discussion which of these two equations describes the actual configuration of D2 better, the ratio d_2/L_2 is written as

$$d_2/L_2 = \alpha_{\gamma'} f \tag{5.61c}$$

with $0 \leqslant \alpha_{\gamma'} \leqslant 1.0$. $\alpha_{\gamma'}$ has to be derived from experimental data. The subscript γ' of $\alpha_{\gamma'}$ reminds one of "γ'-strengthening."

The evolution of the configurations of D1 and D2 as the strain is increased is shown in Figure 5.24. Two parallel glide planes have been activated. The average radius r of the γ'-precipitates equals 4 nm and their volume fraction is 0.09. At the point marked by $\times$ in Figures 5.24b and 5.24c there is a strong obstacle, which stops all pairs. The mean radius r_r of the γ'-particle–glide-plane intersection equals $\omega_r r$. The parameters r_r and ω_r have been defined in Section 3.1. Thus r_r equals 3.3 nm, $\omega_r = 0.82$ of g_{WLS} [Eq. (3.7)] has been inserted. The mean number n_p of pairs needed to cut the γ'-particles in two is given by

$$n_p = \frac{2r_r}{2b} = \frac{r_r}{b}$$

Here and throughout the discussion of order strengthening, b is the length of the Burgers vector of D1 or D2—that is, $b = a/\sqrt{2}$. b of NIMONIC PE16 equals 0.25 nm. This yields $n_p = 13$. As can be seen in Figure 5.24d, there is no more pairing noticeable for "pair" 13. γ'-particles cut in two are shown in Figure 1.4b.

5.5.3 Modeling of the CRSS

Though many pairs of dislocations may glide in the same plane, only a single pair will be considered in the analytical derivations presented below; that is, the elastic interaction between pairs will be disregarded. The analogous neglections have been made in Sections 5.1–5.4, where other particle-strengthening mechanisms have been analyzed. Pretorius and Rönnpagel [5.140], however, allowed for the mutual interaction of pairs in their computer simulations. This will be detailed in Section 5.5.4.1. Castagné et al. [5.141] and Shimanuki and Doi [5.142] treated this interaction analytically.

For the sake of clarity the long-range-ordered precipitates will always be referred to as γ'-precipitates and the matrix will be referred to as γ-matrix; that is, the designation of the phases is that of nickel-base alloys, even though the models are applicable to other order-strengthened systems. The resolved shear stress $\tau_{\text{obst}\,\gamma}$ due to a γ'-particle follows from Eq. (4.71):

Inside of the γ'-particle: $\tau_{\text{obst}\,\gamma} = \pm\gamma/b$
Outside of the γ'-particle: $\tau_{\text{obst}\,\gamma} = 0$

γ is the specific energy of the antiphase boundaries. γ appears also as a subscript in $\tau_{\text{obst}\,\gamma}$ and in F_γ defined below. The plus/minus signs in the above equation apply to D2/D1, respectively. Figure 4.5a shows $\tau_{\mathrm{obst}\,\gamma}$ of D1. The straight-line approximation [Eq. (4.2)] yields the following for the force profile $F_\gamma(y, -h_z)$ of a spherical γ'-particle of radius ρ centered at $(0, 0, h_z)$: $\rho > (h_z^2 + y^2)^{1/2}$; that is, part of the dislocation lies inside of the γ'-particle:

$$F_\gamma(y, -h_z) = \pm 2(\rho^2 - h_z^2 - y^2)^{1/2}\gamma$$

$\rho \leqslant (h_z^2 + y^2)^{1/2}$; that is, the entire dislocation lies outside of the γ'-particle:

$$F_\gamma(y, -h_z) = 0$$

$F_\gamma(y, -h_z = -\rho/\sqrt{2})$ is shown in Figure 4.5c. The absolute maximum of $|F_\gamma(y, -h_z)|$ equals $2\rho\gamma$.

The critical configuration of a dislocation pair shearing coherent, spherical long-range-ordered precipitates is sketched in Figure 5.23. In the critical moment the following balances of forces are established:

For the leading dislocation D1: $\quad \tau_p bL_1 + \tau_{\mathrm{mut}} bL_1 - \gamma d_1 = 0 \quad$ (5.62a)

For the trailing dislocation D2: $\quad \tau_p bL_2 - \tau_{\mathrm{mut}} bL_2 + \gamma d_2 = 0 \quad$ (5.62b)

These two equations follow from Eqs. (4.14) and (4.73). τ_p is the CRSS and L_i is the average spacing of γ'-particles along D*i*. τ_{mut} represents the repulsive elastic interaction between D1 and D2. The average forces exerted by a γ'-particle on D1 and D2 are $-\gamma d_1$ and γd_2, respectively; d_i is the average length of D*i* lying inside of a γ'-particle. The two forces $\tau_p bL_1$ and $\tau_{\mathrm{mut}} bL_1$ drive D1 forward. The corresponding forces $\tau_p bL_2$ and $-\tau_{\mathrm{mut}} bL_2$ acting on D2 nearly cancel. Therefore D1 bows out strongly between the γ'-particles, whereas D2 is nearly straight or follows the overall curvature of D1. This is clearly visible in Figure 4.6, and it is indicated in Figure 5.23. Eliminating τ_{mut} from Eqs. (5.62) yields

$$\tau_p = \frac{\gamma}{2b}\left\{\frac{d_1}{L_1} - \frac{d_2}{L_2}\right\} \tag{5.63}$$

The forces γd_1 and γd_2 have been calculated on the basis of the straight-line approximation introduced in Section 4.1.1. d_1 is a function of the individual radii ρ of the sheared γ'-particles and of the coordinates h_z of their centers (Figure 4.2). In contrast to Sections 5.3 and 5.4.2, where elaborate averaging procedures over h_z were required, here $2r_r$ is inserted for d_1. r_r is the average radius of the intersection between a γ'-particle and the glide plane (Section 3.1). The ratio d_2/L_2 is given in Eq. (5.61c). The separation of D2 from D1 depends on r, γ, and the effective shear modulus that governs τ_{mut}. There are two unknown parameters in Eq. (5.63): L_1 and $\alpha_{\gamma'} = d_2/(L_2 f)$. Because L_1 is the critical length L_c of D1, L_1 has to stay within the limits L_{min} and L_{max} [relation

(4.15)]. L_{min} and L_{max} have been introduced in Section 3.1. In the underaged state, L_1 is larger than L_{min}, whereas in the peak-aged state, L_1 equals L_{min}. This has been detailed in Section 4.2.1.3.2. The various models for underaged order strengthened materials differ primarily in what the authors inserted for L_1. Three different models are presented below. Though in their original papers some authors assumed that $\alpha_{\gamma'}$ vanishes whereas others inserted unity for it, the general term $\alpha_{\gamma'}$ is kept throughout. w_x and w_y are the ranges of the γ'-particle–D1 interaction force in the x- and y-direction, respectively. The latter is the forward direction of the dislocation glide. With the only exception of Section 5.5.3.1.3, w_x will be disregarded. Unity will be inserted for all combinations of the statistical parameters ω_n [Eq. (3.5b)].

Equations (5.62) and (5.63) imply that the maximum interaction force between a γ'-particle and D1 is entirely due to the long-range order of the γ'-particle. The view that contributions of lattice or modulus mismatches can often be disregarded is supported by Figures 4.5d and 5.8. In Figure 4.5d the constrained lattice mismatch is about 10 times as large as in actual nickel-base superalloys or aluminum–lithium alloys. Commercial order-strengthened materials are designed to have a very small lattice mismatch: It is normally below 0.002 [5.5]. It cannot be ruled out, however, that in some order-strengthened alloy systems a lattice or a modulus mismatch modifies the particle–dislocation interaction forces slightly [5.3, 5.65]. In the analyses of the experimental CRSS data obtained for several of the systems listed in Section 5.5, the authors discussed the relevance of other interaction mechanisms.

Gleiter and Hornbogen [5.143, 5.144] have published a complete treatment of order strengthening. These authors studied the configurations of D1 and D2 in great detail and related the CRSS to the radius and the volume fraction of the γ'-particles.

5.5.3.1 Underaged State

5.5.3.1.1 $L_1 = L_{cF}$, $w_x = w_y = 0$. Brown and Ham [5.16] and Raynor and Silcock [5.145] inserted the Friedel length L_{cF} of Eq. (4.27a) for L_1 appearing in Eq. (5.63). This implies that the ranges w_x and w_y are disregarded. The result for the CRSS τ_p is

$$\tau_p = \frac{\gamma}{2b}\left\{C_1\left(\frac{4\omega_r^3}{\pi\omega_q}\right)^{1/2}\left[\frac{\gamma r f}{S}\right]^{1/2} - \alpha_{\gamma'} f\right\} \tag{5.64a}$$

where γ, r, and f are the specific antiphase boundary energy, the average radius, and the volume fraction of the γ'-precipitates, respectively. Here the statistical factors C_1 (Section 4.2.2.2.1, Table 4.1), ω_r [Eq. (3.5a)] and ω_q [Eq. (3.5d)] have been introduced. ω_r and ω_q depend on the distribution function g of the particle radii. In his original derivation, Friedel assumed a uniform arrangement of the particles in the glide plane. This leads to $C_1 = 1.0$. If they are distributed at random, C_1 is about 10% smaller (Sections 4.2.1.4.1 and 4.2.2). Schwarz and Labusch [5.10] found for the latter case $C_1 = 0.94$ (Table 4.1). S is the line tension of D1. Equations similar in structure to Eq. (5.64a) have been derived by Ham [5.146] and by Guyot [5.147]. Gleiter and Hornbogen's [5.143, 5.144] relation was also of this structure.

For comparisons with similar equations and with experimental data, Eq. (5.64a) is rewritten as

$$\frac{2b\tau_p}{f} = A_1\gamma^{3/2}[r/(fS)]^{1/2} + A_2\gamma \tag{5.65}$$

A_1 and A_2 of Eq. (5.64a) are given by

$$A_1 = C_1\left(\frac{4\omega_r^3}{\pi\omega_q}\right)^{1/2} \tag{5.64b}$$

$$A_2 = -\alpha_{\gamma'} \tag{5.64c}$$

For $g = g_{\mathrm{WLS}}$ [Eq. (3.7)] and $C_1 = 0.94$, A_1 of Eq. (5.64b) equals 0.91. In the derivation of the Friedel length it had been assumed that the dislocations are only slightly bent (Section 4.2.1.3.1)—that is, that the obstacles are weak. This limitation is carried over to Eqs. (5.64).

5.5.3.1.2 $L_l = L_{cL}$, $w_x = 0$, $w_y > 0$. Haasen and Labusch [5.148] and Nembach and Neite [5.5] inserted Schwarz and Labusch's length L_{cL} [Eq. (4.67)] for L_1. Thus the range w_y of the interaction force between a γ'-particle and D1 was allowed for. w_y was assumed to be proportional to r_r; this view is supported by Figure 4.27b:

$$w_y = \xi r_r = \xi\omega_r r \tag{5.66}$$

r_r and ω_r have been defined in Section 3.1. Thus one obtains the following for the CRSS τ_p:

$$\tau_p = \frac{\gamma}{2b}\left\{C_1\left(\frac{4\omega_r^3}{\pi\omega_q}\right)^{1/2}\left[\frac{\gamma r f}{S}\right]^{1/2} + \left[\frac{2C_1C_2\xi\omega_r^2}{\pi\omega_q} - \alpha_{\gamma'}\right]f\right\} \tag{5.67a}$$

The parameters C_1 and C_2 have been listed in Table 4.1. Those of energy-storing obstacles have to be chosen. Transforming this equation to the form of Eq. (5.65) yields

$$A_1 = C_1\left(\frac{4\omega_r^3}{\pi\omega_q}\right)^{1/2} \tag{5.67b}$$

and

$$A_2 = \frac{2C_1C_2\xi\omega_r^2}{\pi\omega_q} - \alpha_{\gamma'} \qquad (5.67c)$$

With $C_1 = 0.94$, $C_2 = 0.82$, $\omega_r = 0.82$, and $\omega_q = 0.75$, this leads to

$$A_2 = 0.44\xi - \alpha_{\gamma'} \qquad (5.67d)$$

The values inserted for ω_r and ω_q are those of g_{WLS} defined in Eq. (3.7).

Evidently taking w_y of D1 into consideration affects only A_2, but not A_1. If ξ is disregarded, Eqs. (5.64a) and (5.67a) are identical. In view of the uncertainties concerning $\alpha_{\gamma'}$ [Eqs. (5.61)], Reppich's [5.25] suggestion to allow for w_y of D2 is not followed here. It must be remembered that Schwarz and Labusch's [5.10] computer simulations had been based on the assumption that f is not too large: According to supposition 9 listed at the beginning of Section 4.2.2.2, f should be smaller than about 0.2. This limitation is carried over to Eqs. (5.67).

Nembach and Neite [5.5] inserted the effective dislocation line tension $S_{\text{eff }1}$ [Eq. (4.50a)] for S. This allowed for the relatively strong bowing of the dislocations (Figures 4.6 and 5.24) and guaranteed self-consistency of the evaluations of the data presented in Section 5.5.4.1: L_1 calculated on the basis of Eq. (4.67) stayed within the expected range: $L_{\min} \leqslant L_1 \leqslant L_{\max}$ (Section 3.1). It must, however, be stressed that L_{cF} (Section 5.5.3.1.1) as well as L_{cL} (Section 5.5.3.1.2) should only be applied if the dislocations are only slightly bent. Those shown in Figure 4.6 are outside of the allowed range of bowing (Section 4.1.2).

5.5.3.1.3 High Volume Fractions. Because the γ'-volume fraction of technical superalloys exceeds 0.2, Ardell et al. [5.149] derived a relation for the CRSS which is supposed to be applicable even if f is rather high. The authors started out with Ham's [5.146] model of the CRSS (Section 4.2.1.4.2) but eliminated the supposition that f should be low. The range w_x parallel to the dislocation is thus implicitly allowed for. Friedel's supposition of the "steady state" (supposition 4 in Section 4.2.1.3.1) was maintained. The result was

$$\tau_p = \frac{\gamma}{2b}\{u - \alpha_{\gamma'}f\} \qquad (5.68a)$$

with

$$u = \frac{-B + (B^2/3 + 4B)^{1/2}}{2(1 - B/6)} \qquad (5.68b)$$

and

$$B = \left[\frac{4\omega_r^3}{\pi\omega_q}\right] \frac{\gamma r f}{S(\theta_d = 90°)} \tag{5.68c}$$

In the original paper the numerical factor was $3\pi^2/32$ instead of $4\omega_r^3/(\pi\omega_q)$. In most cases the difference will be negligible. If all particles have the same radius, the difference vanishes. In the limit that B is very small, this is, for example, the case if f is very small, u reduces to $\sqrt{B}$, and Eq. (5.68a) is identical with Eq. (5.64a) with $C_1 = 1.0$.

In advanced versions of Eqs. (5.68), Ardell [5.150] allowed for the variation of the line tension with the bowing angle. The problems encountered in such an endeavor have been touched on in Section 4.2.1.4.3. Because the result was what the author himself called "not particularly transparent," it is presented here only in the limit when u is very small:

$$\tau_p = \frac{\gamma}{2b}\left\{\left[\frac{4\omega_r^2}{\pi\omega_q}\right]^{1/2} [2fy(1 + 36y^2)]^{1/2} - \alpha_{\gamma'} f\right\} \tag{5.69a}$$

with

$$y = \frac{\gamma r_r}{2S(\theta_d = 90°)} \tag{5.69b}$$

where y is a measure of the angle through which the dislocation D1 is bent.

If $\alpha_{\gamma'}$ is positive AND (logic AND) r is very small, Eqs. (5.64), (5.67), (5.68), and (5.69) yield negative values for the CRSS τ_p. This unreasonable result is caused by the approximations made in the derivations of these equations. Moreover, very small γ'-particles are no real obstacles to the glide of dislocations.

5.5.3.2 *Peak-Aged State.* Here it is assumed that in the peak-aged state L_1 equals its lower limit L_{min}. The alternative view that this state marks the transition to the operation of the Orowan process (Section 4.2.1.3.2) is disregarded. Substituting L_{min} [Eq. (3.3c)] for L_1 in Eq. (5.63) yields

$$\tau_p = \frac{\gamma}{2b}\left\{\frac{2\omega_r}{(\pi\omega_q)^{1/2}} f^{1/2} - \alpha_{\gamma'} f\right\} \tag{5.70}$$

The CRSS of peak-aged order-strengthened materials is evidently independent of the radius r, and τ_p depends only on f. If one wishes to allow for the randomness of the distribution of the γ'-particles in the glide plane, one may multiply the term involving $f^{1/2}$ in Eq. (5.70) by a factor around 0.9 (Section

4.2.1.4.1). Equation (5.61c) relates the parameter $\alpha_{\gamma'}$ to the configuration of the trailing dislocation D2. Though $\alpha_{\gamma'}$ of underaged specimens may differ from $\alpha_{\gamma'}$ of peak-aged ones, the same designation $\alpha_{\gamma'}$ is used in Eqs. (5.64a), (5.67a), (5.68a), (5.69a), and (5.70).

Hüther and Reppich [5.151] analyzed the case of large average γ'-particle radii. If they exceed the separation of D2 from D1, in the critical moment D2 just touches the periphery of those γ'-particles in which D1 lies. This situation is similar to the one that may be encountered in materials that are strengthened by particles that have a stacking fault energy mismatch [Section 5.3, Eq. (5.36a)].

5.5.4 Comparison with Experimental Data

5.5.4.1 Underaged Specimens. Most of the experimental data on order strengthening concern binary nickel-base nickel–aluminum alloys [5.5, 5.6, 5.97, 5.113, 5.152], binary aluminum base aluminum–lithium alloys [5.7, 5.117–5.122], and multicomponent nickel-base superalloys [5.5, 5.97, 5.110]. In all of these systems the matrices are fcc and the coherent precipitates have the $L1_2$-superlattice. Unless they are very large, they are spherical. In aluminum–lithium alloys and in superalloys the lattice mismatch is very small: below about 0.002. In nickel-base alloys the particles consist of the stable γ'-phase and in aluminum–lithium alloys of the metastable δ'-phase. Normally the average radius r of the particles and their volume fraction f are determined by TEM.

One of the most extensively studied order-strengthened materials is the commercial nickel-base superalloy NIMONIC PE16 [5.5, 5.97] produced by Inco Alloys International, Hereford, England. Its overall composition and that of the γ'-precipitates have been discussed in Section 5.5.1. f is governed by the temperature T_H at which the Ostwald ripening treatment is carried out and r by its duration t. The subscript H of T_H stands for "heat" treatment. Results for the function $r(t)$ have been presented in Figure 3.5. f ranges from 0.029 at $T_H = 1119$ K to 0.10 at $T_H = 949$ K. Since f decreases and the growth rate of the γ'-particles increases as T_H is raised (Section 3.2.1.2), specimens with high f AND (logic AND) large r cannot be prepared within reasonable times.

In Figure 5.25 the total CRSS τ_t of NIMONIC PE16 single crystals is plotted versus $\sqrt{r}$. The peak-aged state is reached for f equal to 0.089 and 0.061, and the overaged state is reached only for 0.061. From such τ_t data, the γ'-particles' contribution τ_p to τ_t is calculated with the aid of Eq. (5.58); the exponent k of underaged as well as of peak-aged specimens equals 1.23 [5.5]. Its derivation will be described in Section 7.2.1. The strategy is to assign such a value to k that τ_p merely reflects the weak temperature dependence of the dislocation line tension S, even though τ_t and the CRSS τ_s of the γ-matrix vary strongly with the deformation temperature T_D; T_D ranged from 90 K to 500 K. The subscript D of T_D refers to "deformation." The ratio τ_s/τ_t is a function of r, f, and T_D. The maximum ratio of underaged specimens exceeds 0.5.

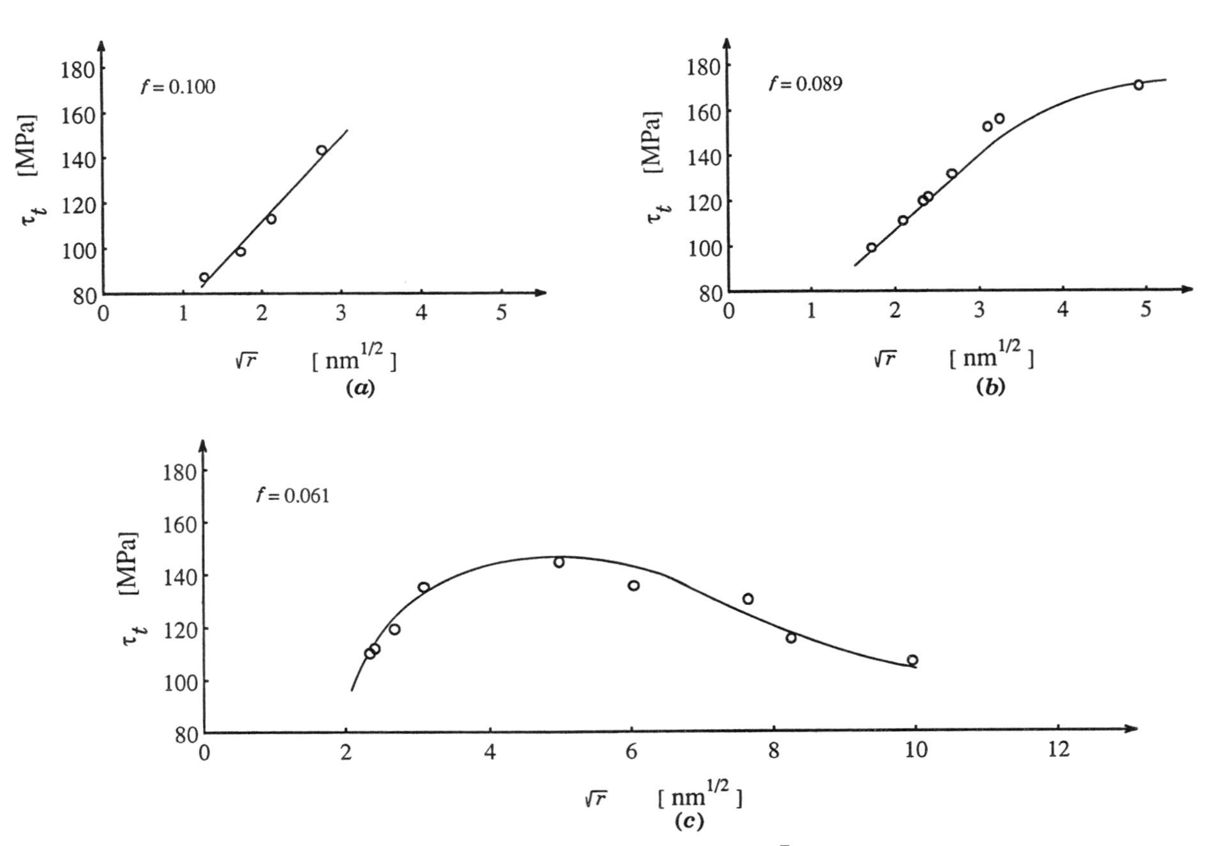

Fig. 5.25. Total CRSS τ_t of NIMONIC PE16 single crystals versus $\sqrt{r}$. The data have been taken at ambient temperature. f is indicated in the upper left. (After Ref. 5.5.)

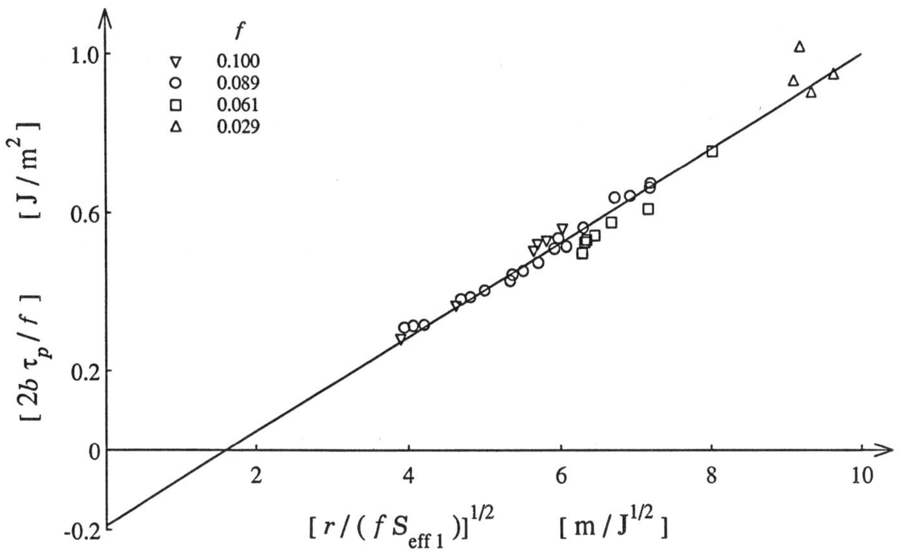

Fig. 5.26. According to Eqs. (5.65) and (5.67): $2b\tau_p/f$ is plotted versus $[r/(fS_{\text{eff }1})]^{1/2}$ for underaged NIMONIC PE16 single crystals. The data have been taken at temperatures between 90 K and 500 K. The straight line represents Eq. (5.67a) fitted to the data. (After Ref. 5.5.)

Figure 5.26 shows the variation of $2b\tau_p/f$ of underaged NIMONIC PE16 single crystals with $[r/(fS_{\text{eff }1})]^{1/2}$. These terms have been chosen according to Eqs. (5.67). The ranges of r and f are $3\text{ nm} \leqslant r \leqslant 11\text{ nm}$ and $0.029 \leqslant f \leqslant 0.10$. The straight line represents Eqs. (5.67) fitted to the data. Its slope yields the specific antiphase boundary energy γ and its ordinate intercept the parameter A_2. The following parameters have been inserted: $S = S_{\text{eff }1}$ of Eq. (4.50a) = geometric mean, $R_o/R_i = (L_{cL} - 2r_r)/b$, where L_{cL} is the Labusch length of Eq. (4.67), $C_1 = 0.94$ (Table 4.1), $C_2 = 0.82$ (Table 4.1), and ω_r and ω_q of g_{WLS} (Table 3.2). At ambient temperature the line tension parameters $K_S(\theta_d = 0°)$ and $K_S(\theta_d = 90°)$ of NIMONIC PE16 are 15.7 GPa and 3.5 GPa, respectively [5.49]. The length of the Burgers vector in the γ-matrix is 0.25 nm. The results are: $\gamma = 0.258\text{ J/m}^2$ and $\gamma A_2 = -0.19\text{ J/m}^2$ [5.5]. Evidently Eqs. (5.67) describe the data well. Because the term γA_2 is obtained by an extrapolation, it is somewhat less accurate.

The value 0.258 J/m^2 of the specific antiphase boundary energy falls within the expected range [5.111, 5.112]. Because the ordinate intercept is negative, $\alpha_{\gamma'}$ is positive [Eqs. (5.67c) and (5.67d)]. This supports the view expressed by Eq. (5.61a) that D2 does not avoid all γ'-particles. Unless additional information on $\alpha_{\gamma'}$ is available, the parameter ξ defined in Eq. (5.66) cannot be derived from measurements of τ_p as a function of f and r. The term A_2 [Eqs. (5.67c) and (5.67d)], which is experimentally accessible, involves a combination of ξ and $\alpha_{\gamma'}$. Inserting unity for $\alpha_{\gamma'}$—that is, assuming Eq. (5.61a) to hold—yields

$\xi = 0.6$. This is quite reasonable. With this value for ξ, the normalized range η_0 [Eq. (4.58g)] of all data shown in Figure 5.26 is between 0.08 and 0.19 and the Labusch correction factor $[C_1(1 + C_2\eta_0)]$ [Eq. (4.62)] of the CRSS falls in the range 1.00 and 1.08. The effect of η_0 on A_2 [Eqs. (5.67)] amounts to about 35%.

The exponent k equals 1.23; k, which governs the superposition of solid solution and γ'-particle hardening [Eq. (5.58)], needs some further consideration. If $k = 1.0$ instead of $k = 1.23$ is used in Eq. (5.58), the data in a plot like that presented in Figure 5.26 exhibit a definite dependence on the deformation temperature T_D. Consequently, $k = 1.0$ has to be rejected. If only the size and the volume fraction of the γ'-precipitates are varied, but not T_D, any deficiencies of k are far less apparent.

For $S = S_{\text{eff }1}$, $R_o/R_i = (L_{cF} - 2r_r)/b$, and $C_1 = 1.0$ Eqs. (5.64) yield $\gamma = 0.247\ \text{J/m}^2$ and $\alpha_{\gamma'} = +0.69$. This result for γ differs by 4% from that derived from Eqs. (5.67). The reasons are the differences in R_o and C_1.

Because the derivation of the ordinate intercept $[A_2\gamma]$ of Eq. (5.65) involves the extrapolation of $[r/(fS)]^{1/2}$ to zero, $A_2\gamma$ is rather sensitive to the choice of the outer cutoff radius R_o. This has been demonstrated by Nembach [5.153] by an example: Assigning constant values to R_o yielded positive ordinate intercepts, whereas relating R_o to $L_{\min}$ or to $L_{\max}$ led to negative intercepts. For $R_o = L_{cF}$, $|A_2\gamma|$ was very small. Evidently one must be very careful in drawing conclusions from A_2. γ itself is far less sensitive to R_o [5.5, 5.153].

In Figure 5.27 the same experimental data are presented as in Figure 5.26; but with reference to Eqs. (5.68), $2b\tau_p/f$ is plotted versus u/f. $S = S(\theta_d = 90°)$ with $R_o = L_{\min}$ and $R_i = b$ has been inserted. The results are $\gamma = 0.202\ \text{J/m}^2$ and $\alpha_{\gamma'} = 0.74$. The relative standard deviations of the data from the fitted straight line are 50% higher than in Figure 5.26, which is based on Eqs. (5.67). Because $S(\theta_d = 90°)$ is smaller than $S_{\text{eff }1}$, Eqs. (5.68) lead to a lower value for γ than Eqs. (5.67) do. Equations (5.67) and (5.68), which allow for w_y and w_x (roughly speaking), respectively, represent the data about equally well. Both sets of equations involve two adjustable parameters:

Eqs. (5.67): γ and A_2; A_2 is a combination of ξ and $\alpha_{\gamma'}$
Eqs. (5.68): γ and $\alpha_{\gamma'}$

Both sets of equations lead to positive values of $\alpha_{\gamma'}$. Equations (5.68) yield $\alpha_{\gamma'} = +0.74$.

Results obtained for binary aluminum–lithium alloys strengthened by $L1_2$-long-range-ordered δ'-precipitates have been compiled in Figure 5.28 [5.122]. f ranges from 0.03 to 0.16. Again Eq. (5.67a) has been fitted to the data, and it is represented by the straight line. The results of the fit are as follows: $\gamma = 0.11\ \text{J/m}^2$ and $A_2 = [+0.3]$. Such a positive ordinate intercept can only be explained on the basis of Eqs. (5.67), which involve the parameter ξ characteristic of the range w_y. In Figure 5.28, the outer cutoff radius R_o equals $L_{\min}$. Setting R_o equal to L_{cF} or $1000b$ leads to positive values of A_2.

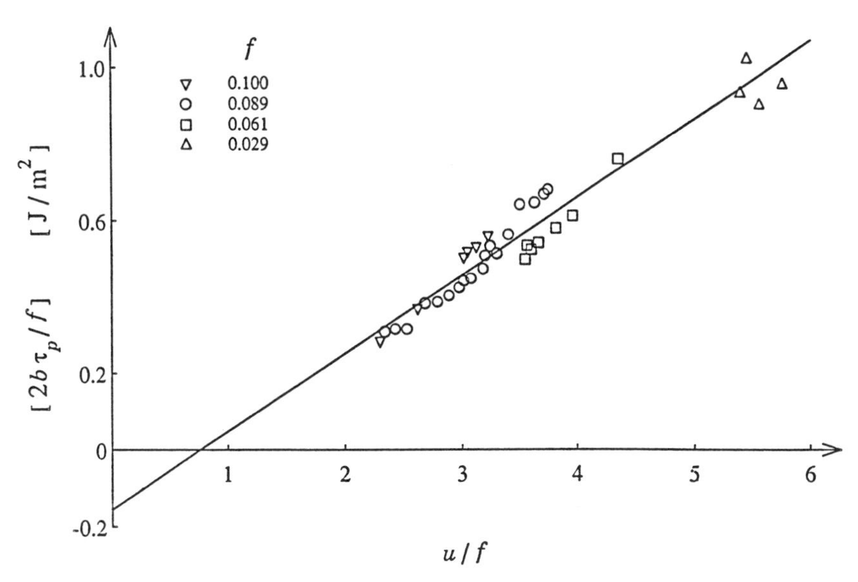

Fig. 5.27. According to Eqs. (5.68): $2b\tau_p/f$ versus u/f. The data are the same as in Figure 5.26. The straight line represents Eqs. (5.68) fitted to the data. (After Ref. 5.5.)

Thus it can be concluded that Eqs. (5.67) describe order strengthening well provided that f is below about 0.2.

Pretorius and Rönnpagel [5.140] simulated the glide of dislocation pairs in NIMONIC PE16 in a computer. The procedures involved have been described in Section 4.2.2.3. Some of the resulting configurations of the dislocations D1 and D2 have been shown in Figures 4.35–4.37. They strikingly resemble those observed in the transmission electron microscope (Figures 4.35c and 4.36c). Such simulations yield detailed information on how D1 attacks the γ'-particles and on how D2 lies relative to them in the critical moment of D1.

Pretorius and Rönnpagel aimed at achieving quantitative agreement between experimental τ_p data and those derived from the simulations. They started out with the experimental result 0.26 J/m^2 for the specific antiphase boundary energy γ. $\mu_F = 46$ GPa has been used as μ_s in Eq. (4.72b). τ_p simulated for a single dislocation pair turned out to be much higher than the experimental result. By increasing the width of the glide plane in the x-direction, the probability of soft spots, which give rise to unzipping (Sections 4.2.1.3.1 and 4.2.2.1), was raised and thus τ_p lowered. A further reduction of τ_p was achieved by having many pairs glide simultaneously in the same plane. This corresponds to the real situations shown in Figures 4.6 and 5.24. This finally led to an approximate agreement between experiments and simulations. This is, however, rather surprising. The analytical models presented in Section 5.5.3.1 and the simulations relate the same experimental CRSS data to the same

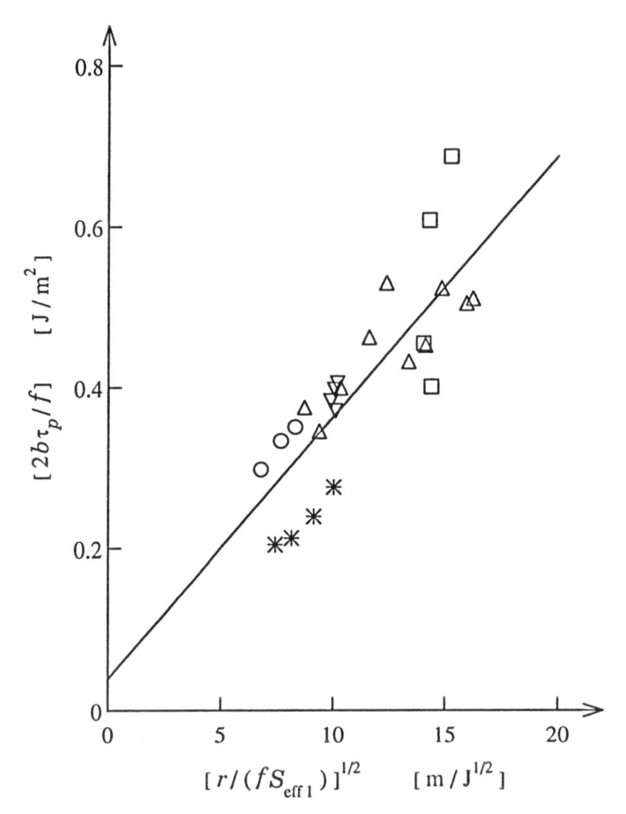

Fig. 5.28. According to Eqs. (5.67): $2b\tau_p/f$ is plotted versus $[r/(fS_{\mathrm{eff}\ 1})]^{1/2}$ for underaged aluminum-rich aluminum–lithium alloys. □ [5.7], ○ [5.117], ▽ [5.118], * [5.121], △ [5.122]. The straight line represents Eq. (5.67a) fitted to the data. (After Ref. 5.122.)

value of γ, but only the simulations allow for many pairs gliding in the same plane. This lowers the CRSS significantly. So far it remains unknown why in spite of this important difference both approaches led to the same results. Possible explanations are (i) the assumptions on which Eqs. (5.67) have been based (e.g., slight bending of the dislocations) and (ii) differences in the line tensions, especially in the outer cutoff radius.

5.5.4.2 Peak-Aged Specimens. Nembach and Neite [5.5] investigated order-strengthening of underaged and peak-aged single crystals of the commercial nickel-base superalloy NIMONIC PE16 and of binary nickel-rich nickel–aluminum alloys. The specific antiphase boundary energy γ of *underaged* specimens of both systems was found to be $0.260 \pm 0.003\ \mathrm{J/m^2}$. In Figure 5.29 experimental τ_p data of *peak-aged* specimens of both systems are plotted versus the γ'-volume fraction f. The curve in Figure 5.29 represents Eq. (5.70) with

$\gamma = 0.26\,\text{J/m}^2$, $\alpha_{\gamma'} = 1.0$, $\omega_r = 0.82$, and $\omega_q = 0.75$. The latter two statistical parameters have been defined in Eqs. (3.5a) and (3.5d), here the parameters of the distribution function g_{WLS} of Eq. (3.7) have been inserted. Evidently the curve in Figure 5.29 represents the experimental data satisfactorily. This justifies the assumption under which Eq. (5.70) had been derived: The γ'-particles are sheared and their spacing L_1 along D1 equals $L_{\min}$ in the critical moment. Thus the CRSS of peak-aged specimens is not governed by the onset of the operation of the Orowan process (Chapter 6). Glazer and Morris [5.154] and Ardell and Huang [5.155] analyzed the transition from shearing of long-range-ordered particles to circumventing them. Their aim was to obtain γ. In some instances, TEM studies of superalloys revealed Orowan loops next to dislocation pairs [5.156, 5.157]. This does, however, not prove that the onset of the operation of the Orowan process actually controls the macroscopic CRSS.

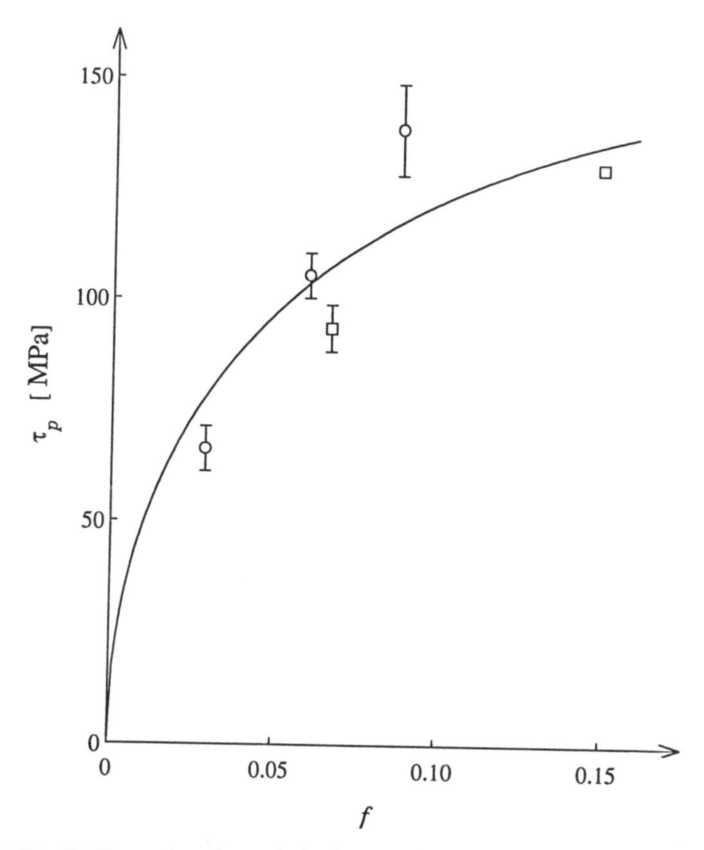

Fig. 5.29. CRSS τ_p of peak-aged single crystals versus the γ'-volume fraction f. The curve represents Eq. (5.70) with $\gamma = 0.26\,\text{J/m}^2$ and $\alpha_{\gamma'} = 1.0$. ○ NIMONIC PE16, $r = 25$ nm, the data have been averaged over the range $90\,\text{K} \leqslant T_D \leqslant 500\,\text{K}$; □ binary nickel-rich nickel–aluminum alloys, $r = 7.5$ nm, $T_D = 293$ K. (After Ref. 5.5.)

Schänzer and Nembach [5.158] investigated a series of γ'-strengthened nickel-base superalloys that were based on the commercial one NIMONIC 105 produced by Inco Alloys International. The γ'-volume fraction f ranged from 0.07 to 0.5, and there were underaged as well as peak-aged specimens. In both cases the CRSS had been found to increase much more strongly with f than predicted by Eqs. (5.67) and (5.70). Equations (5.67) are not supposed to be applied to materials with f beyond about 0.2. If f exceeds this limit, Eqs. (5.67) are expected to overestimate, but not to underestimate, τ_p. Equations (5.68), which had been meant for high f, did not represent the data better than Eqs. (5.67) did. Therefore, Schänzer and Nembach gave the tentative interpretation that the antiphase boundary energy increases with f. This has been linked to chemical inhomogeneities of the γ'-precipitates.

5.6 SUMMARY

In Chapter 4, strengthening of metals and alloys by shearable coherent particles has been analyzed from a general point of view. In Sections 5.1–5.5, five different mechanisms of particle strengthening have been analyzed in the light of the models presented in Chapter 4: chemical (Γ), modulus mismatch ($\Delta\mu$), stacking fault energy mismatch ($\Delta\gamma$), lattice mismatch (ε), and order (γ) strengthening. The parameters that govern the respective particle–dislocation interaction are given here in parentheses.

The resulting relations for the CRSS τ_p of underaged specimens are of the form

$$\tau_p = \tau_p(\Delta, r, f, S) \tag{5.71}$$

The particles are assumed to be spherical; their mean radius is r and their volume fraction f. Δ in Eq. (5.71) stands for the above parameters Γ, $\Delta\mu$, $\Delta\gamma$, ε, or γ. S is the dislocation line tension. Most relations have been based on the supposition that dislocations bow out only slightly. This allows one to (i) perform the straight-line approximation (Section 4.1.1) in the calculation of the interaction force between a particle and a dislocation and (ii) consider S as constant. If τ_p is derived from Friedel's equations (4.26), the result is often of the form

$$\tau_p = A\Delta^{3/2}\left(\frac{rf}{2S}\right)^{1/2} \tag{5.72}$$

where A is constant. Equations (5.52) are of this structure and Eq. (5.65) involves such a term.

If the peak-aged state is *not* governed by the onset of the Orowan process, Eqs. (4.16) relate τ_p to r, f, and Δ. Equation (5.70), which is applicable to peak-aged order-strengthened materials, has been derived from Eq. (4.16a).

The most difficult problem is to determine the relevant interaction mechanism between particles and dislocations—that is, to determine Δ appearing in Eq. (5.71). If this problem can be solved, the models describe the functional dependence of τ_p on r and f quite satisfactorily; the quantitative agreement between theory and experiment is within a factor of two or better. In Section 6.3 it will be shown that the same error margins apply to the Orowan process. Thus particle strengthening is probably the best understood sector of the plastic deformation of metals and alloys.

REFERENCES

5.1 A. Pineau and F. Baudier, *Scripta Metall.*, **3**, 757, 1969.

5.2 B. Reppich, *Acta Metall.*, **23**, 1055, 1975.

5.3 A. Melander and P. Å. Persson, *Metal Science*, **12**, 391, 1978.

5.4 E. Nembach, *phys. stat. sol. (a)*, **78**, 571, 1983.

5.5 E. Nembach and G. Neite, *Prog. Mater. Sci.*, **29**, 177, 1985.

5.6 V. A. Phillips, *Philos. Mag.*, **16**, 103, 1967.

5.7 Y. Miura, A. Matsui, M. Furukawa, and M. Nemoto, *Proceedings of the 3rd International Al–Li Conference, Oxford* (edited by C. Baker, P. J. Gregson, S. J. Harris, and C. J. Peel), p. 427, The Institute of Metals, London, 1986.

5.8 P. M. Kelly, *Int. Met. Rev.*, **18**, 31, 1973.

5.9 J. Friedel, *Les Dislocations*, Gauthier-Villars, Paris, 1956.

5.10 R. B. Schwarz and R. Labusch, *J. Appl. Phys.*, **49**, 5174, 1978.

5.11 D. A. Porter and K. E. Easterling, *Phase Transformations in Metals and Alloys*, Van Nostrand Reinhold, New York, 1981.

5.12 R. Wagner and R. Kampmann, in *Materials Science and Technology*, Vol. 5 (edited by R. W. Cahn, P. Haasen, and E. J. Kramer), p. 213, VCH Verlagsgesellschaft mbH, Weinheim, 1991.

5.13 A. Kelly and M. E. Fine, *Acta Metall.*, **5**, 365, 1957.

5.14 V. Gerold and K. Hartmann, *Proc. Int. Conference on the Strength of Metals and Alloys, Trans. JIM Suppl.*, **9**, 509, 1968.

5.15 F. Rabe and P. Haasen, *Proceedings of the 8th International Conference on the Strength of Metals and Alloys*, Vol. 1 (edited by P. O. Kettunen, T. K. Lepistö, and M. E. Lehtonen), p. 561, Pergamon Press, Oxford, 1988.

5.16 L. M. Brown and R. K. Ham, in *Strengthening Methods in Crystals* (edited by A. Kelly and R. B. Nicholson), p. 9, Applied Science Publishers, London, 1971.

5.17 R. L. Fleischer, in *Electron Microscopy and Strength of Crystals* (edited by G. Thomas and J. Washburn), p. 973, Interscience Publishers/John Wiley & Sons, New York, 1963.

5.18 J. P. Hirth and J. Lothe, *Theory of Dislocations*, 2nd edition, John Wiley & Sons, New York, 1982.

5.19 G. J. Shiflet, Y. W. Lee, H. I. Aaronson, and K. C. Russell, *Scripta Metall.*, **15**, 719, 1981.

5.20 M. Breu, W. Gust, and B. Predel, *Z. Metallkde.*, **82**, 279, 1991.

5.21 R. Hattenhauer and P. Haasen, *Philos. Mag. A*, **68**, 1195, 1993.

5.22 E. Nembach, *Scripta Metall.*, **18**, 105, 1984.

5.23 E. Nembach and M. Martin, *Acta Metall.*, **28**, 1069, 1980.

5.24 N. Büttner, K.-D. Fusenig, and E. Nembach, *Acta Metall.*, **35**, 845, 1987.

5.25 B. Reppich, in *Materials Science and Technology*, Vol. 6 (edited by R. W. Cahn, P. Haasen, and E. J. Kramer), p. 311, VCH Verlagsgesellschaft mbH, Weinheim, 1993.

5.26 V. Gerold, in *Dislocations in Solids*, Vol. 4 (edited by F. R. N. Nabarro), p. 219, North-Holland, Amsterdam, 1979.

5.27 S. D. Harkness and J. J. Hren, *Met. Trans.*, **1**, 43, 1970.

5.28 H. Knoch and B. Reppich, *Acta Metall.*, **23**, 1061, 1975.

5.29 C. Teodosiu, *Elastic Models of Crystal Defects*, Springer-Verlag, Berlin, 1982.

5.30 C. N. Reid, *Deformation Geometry for Materials Scientists*, Pergamon Press, Oxford, 1973.

5.31 R. W. Weeks, S. R. Pati, M. F. Ashby, and P. Barrand, *Acta Metall.*, **17**, 1403, 1969.

5.32 G. Knowles and P. M. Kelly, *BSC/ISI Conference Scarborough*, p. 9, The Iron and Steel Institute, London, 1971.

5.33 A. Melander and P. Å. Persson, *Acta Metall.*, **26**, 267, 1978.

5.34 R. Peierls, *Proc. Phys. Soc. London*, **52**, 34, 1940.

5.35 F. R. N. Nabarro, *Proc. Phys. Soc. London*, **59**, 256, 1947.

5.36 J. D. Eshelby, *Philos. Mag.*, **40**, 903, 1949.

5.37 R. Bullough and V. K. Tewary, in *Dislocations in Solids*, Vol. 2 (edited by F. R. N. Nabarro), p. 1, North-Holland, Amsterdam, 1979.

5.38 K. C. Russel and L. M. Brown, *Acta Metall.*, **20**, 969, 1972.

5.39 R. Siems, P. Delavignette, and S. Amelinckx, *phys. stat. sol.*, **2**, 636, 1962.

5.40 U. Dehlinger and E. Kröner, *Z. Metallkde.*, **51**, 457, 1960.

5.41 M. Comninou and J. Dundurs, *J. Appl. Phys.*, **43**, 2461, 1972.

5.42 S. D. Gavazza and D. M. Barnett, *Int. J. Eng. Sci.*, **12**, 1025, 1974.

5.43 A. Reuss, *Z. Angew. Math. Mech.*, **9**, 49, 1929.

5.44 W. Voigt, *Lehrbuch der Kristallphysik*, p. 962, reprinted by Johnson Reprint Corporation, New York, 1966.

5.45 G. Simmons and H. Wang, *Single Crystal Elastic Constants and Calculated Aggregate Properties: A Handbook*, 2nd edition, MIT Press, Cambridge, Massachusetts, 1971.

5.46 E. S. Fisher and D. Dever, *Trans. Met. Soc. AIME*, **239**, 48, 1967.

5.47 W. C. Overton, Jr. and J. Gaffney, *Phys. Rev.*, **98**, 969, 1955.

5.48 D. J. H. Cockayne, M. L. Jenkins, and I. L. F. Ray, *Philos. Mag.*, **24**, 1383, 1971.

5.49 H. Pottebohm, G. Neite, and E. Nembach, *Mater. Sci. Eng.*, **60**, 189, 1983.

5.50 F. Wallow, G. Neite, W. Schröer, and E. Nembach, *phys. stat. sol. (a)*, **99**, 483, 1987.

5.51 G. Mima, S. Hori, and S. Saji, *Technol. Rep. Osaka Univ.*, **19**, 401, 1969.

5.52 K. Matsuura, M. Kitamura, and K. Watanabe, *Trans. JIM*, **19**, 53, 1978.

5.53 H. Wendt and R. Wagner, *Acta Metall.*, **30**, 1561, 1982.

5.54 I. A. Ibrahim and A. J. Ardell, *Mater. Sci. Eng.*, **36**, 139, 1978.

5.55 A. J. Ardell, *Met. Trans. A.*, **16A**, 2131, 1985.

5.56 H. Gleiter, *Acta Metall.*, **15**, 1213, 1967.

5.57 H. Gleiter, *Acta Metall.*, **15**, 1223, 1967.

5.58 G. I. Taylor, *J. Inst. Met.*, **62**, 307, 1938.

5.59 K. Hanson and J. W. Morris, Jr., *J. Appl. Phys.*, **46**, 983, 1975.

5.60 D. Gerlich and E. S. Fisher, *J. Phys. Chem. Sol.*, **30**, 1197, 1969.

5.61 K.-H. Dünkeloh, G. Kralik, and V. Gerold, *Z. Metallkde*, **65**, 773, 1974.

5.62 H. Haberkorn and V. Gerold, *phys. stat. sol.*, **15**, 167, 1966.

5.63 V. Gerold and H. Haberkorn, *phys. stat. sol.*, **16**, 675, 1966.

5.64 P. S. Dobson, P. J. Goodhew, and R. E. Smallman, *Philos. Mag.*, **16**, 9, 1967.

5.65 U. Messerschmidt and M. Bartsch, *Mater. Sci. Eng.*, **A164**, 332, 1993.

5.66 P. B. Hirsch and A. Kelly, *Philos. Mag.*, **12**, 881, 1965.

5.67 N. F. Mott, in *Imperfections in Nearly Perfect Crystals* (edited by W. Shockley, J. H. Hollomon, R. Maurer, and F. Seitz), p. 173, John Wiley & Sons, New York, Chapman & Hall, London, 1952.

5.68 F. Ernst and P. Haasen, *phys. stat. sol.* (*a*), **104**, 403, 1987.

5.69 T. B. Massalski, editor, *Binary Alloy Phase Diagrams*, Vol. 1, p. 3, American Society for Metals, Metals Park, Ohio, 1986.

5.70 E. Nembach, *Scripta Metall.*, **20**, 763, 1986.

5.71 R. Hattenhauer, M. Bartsch, U. Messerschmidt, P. Haasen, and P.-J. Wilbrandt, *Philos. Mag. A*, **70**, 447, 1994.

5.72 N. F. Mott and F. R. N. Nabarro, *Proc. Phys. Soc. London*, **52**, 86, 1940.

5.73 J. D. Eshelby, in *Solid State Physics*, Vol. 3 (edited by F. Seitz and D. Turnbull), p. 79, Academic Press, New York, 1956.

5.74 H. Wiedersich, *Proceedings International Conference on the Strength of Metals and Alloys, Trans. JIM, Suppl.*, **9**, 34, 1968.

5.75 H. Gleiter, *Z. Angew. Phys.*, **23**, 108, 1967.

5.76 V. Gerold and H.-M. Pham, *Scripta Metall.*, **13**, 895, 1979.

5.77 V. Gerold and H.-M. Pham, *Z. Metallkde.*, **71**, 286, 1980.

5.78 K. Fuchs, *Proc. R. Soc. London A*, **157**, 444, 1936.

5.79 E. Kröner, *Acta Metall.*, **2**, 302, 1954.

5.80 J. D. Eshelby, *Proc. R. Soc., Series A, London*, **241**, 376, 1957.

5.81 D. M. Barnett, *Scripta Metall.*, **5**, 261, 1971.

5.82 D. M. Barnett, J. K. Lee, H. I. Aaronson, and K. C. Russell, *Scripta Metall.*, **8**, 1447, 1974.

5.83 A. Fuchs and D. Rönnpagel, *Mater. Sci. Eng.*, **A164**, 340, 1993.

5.84 C. Wagner, *Z. Elektrochem.*, **65**, 581, 1961.

5.85 I. M. Lifshitz and V. V. Slyozov, *J. Phys. Chem. Solids*, **19**, 35, 1961.

5.86 B. Jansson and A. Melander, *Scripta Metall.*, **12**, 497, 1978.

5.87 V. Gerold, *Acta Metall.*, **16**, 823, 1968.

5.88 J. Dash and M. E. Fine, *Acta Metall.*, **9**, 149, 1961.

5.89 H. Haberkorn, K. Hartmann, and V. Gerold, *Z. Metallkde.*, **62**, 200, 1971.

5.90 J. D. Livingston, *Trans. Met. AIME*, **215**, 566, 1959.

5.91 M. Witt and V. Gerold, *Z. Metallkde.*, **60**, 482, 1969.

5.92 M. Witt and V. Gerold, *Scripta Metall.*, **3**, 371, 1969.

5.93 K. E. Amin, V. Gerold, and G. Kralik, *J. Mater. Sci.*, **10**, 1519, 1975.

5.94 K.-D. Fusenig and E. Nembach, *Acta Metall. Mater.*, **41**, 3181, 1993.

5.95 E. Nembach, K. Suzuki, M. Ichihara, and S. Takeuchi, *Mater. Sci. Eng. A*, **101**, 109, 1988.

5.96 D. J. Bacon, *Philos. Mag. A*, **38**, 333, 1978.

5.97 A. J. Ardell, in *Intermetallic Compounds*, Vol. 2 (edited by J. H. Westbrook and R. L. Fleischer), p. 257, John Wiley & Sons, New York, 1994.

5.98 D. H. Jack and R. W. K. Honeycombe, *Acta Metall.*, **20**, 787, 1972.

5.99 I. O. Smith and M. G. White, *Met. Trans. A*, **7A**, 293, 1976.

5.100 J. W. Goodrum and B. G. Lefevre, *Met. Trans. A*, **8A**, 939, 1977.

5.101 P. R. Okamoto and G. Thomas, *Acta Metall.*, **19**, 825, 1971.

5.102 J. Greggi and W. A. Soffa, *Proceedings of the 5th International Conference on the Strength of Metals and Alloys*, Vol. 1 (edited by P. Haasen, V. Gerold, and G. Kostorz), p. 651, Pergamon Press, Toronto, 1979.

5.103 M. C. Chaturvedi and D. W. Chung, *Met. Trans. A*, **12A**, 77, 1981.

5.104 R. Taillard and A. Pineau, *Mater. Sci. Eng.*, **56**, 219, 1982.

5.105 L. K. Singhal and J. W. Martin, *Acta Metall.*, **16**, 947, 1968.

5.106 S. Floreen, *Met. Rev.*, **13**, 115, 1968.

5.107 R. B. Nicholson, *BSC/ISI Conference Scarborough*, p. 1, The Iron and Steel Institute, London, 1971.

5.108 J. W. Christian, in *Strengthening Methods in Crystals* (edited by A. Kelly and R. B. Nicholson), p. 261, Applied Science Publishers, London, 1971.

5.109 R. B. Nicholson, in *Strengthening Methods in Crystals* (edited by A. Kelly and R. B. Nicholson), p. 535, Applied Science Publishers, London, 1971.

5.110 D. P. Pope and S. S. Ezz, *Int. Met. Rev.*, **29**, 136, 1984.

5.111 N. Baluc, H. P. Karnthaler, and M. J. Mills, *Philos. Mag. A*, **64**, 137, 1991.

5.112 H. F. Yu, I. P. Jones, and R. E. Smallman, *Philos. Mag. A*, **70**, 951, 1994.

5.113 V. Munjal and A. J. Ardell, *Acta Metall.*, **23**, 513, 1975.

5.114 W. Betteridge and J. Heslop, editors, *The Nimonic Alloys*, 2nd edition, Edward Arnold, London, 1974.

5.115 C. T. Sims, N. S. Stoloff, and W. C. Hagel, editors, *Superalloys II*, John Wiley & Sons, New York, 1987.

5.116 E. A. Starke, Jr., in *Alloying* (edited by J. L. Walter, M. R. Jackson, and C. T. Sims), p. 165, American Society for Metals, Metals Park, Ohio, 1988.

5.117 P. Sainfort and P. Guyot, *Proceedings of the 7th International Conference on the Strength of Metals and Alloys*, Vol. 1 (edited by H. J. McQueen, J.-P. Bailon, J. I. Dickson, J. J. Jonas, and M. G. Akben), p. 441, Pergamon Press, Oxford, 1985.

5.118 M. Furukawa, Y. Miura, and M. Nemoto, *Trans. Jpn. Inst. Met.*, **26**, 230, 1985.

5.119 J. C. Huang and A. J. Ardell, *Mater. Sci. Eng.*, **A104**, 149, 1988.

5.120 E. J. Lavernia, T. S. Srivatsan, and F. A. Mohamed, *J. Mater. Sci.*, **25**, 1137, 1990.

5.121 V. Gerold, H.-J. Gudladt, and J. Lendvai, *phys. stat. sol.* (*a*), **131**, 509, 1992.

5.122 C. Schlesier and E. Nembach, *Acta Metall. Mater.*, **43**, 3983, 1995.

5.123 O. H. Kriege and J. M. Baris, *Trans. ASM*, **62**, 195, 1969.

5.124 E. H. van der Molen, J. M. Oblak, and O. H. Kriege, *Met. Trans.*, **2**, 1627, 1971.

5.125 R. L. Dreshfield and J. F. Wallace, *Met. Trans.*, **5**, 71, 1974.

5.126 K. M. Delargy and G. D. W. Smith, *Met. Trans. A*, **14A**, 1771, 1983.

5.127 W. Mangen, E. Nembach, and H. Schäfer, *Mater. Sci. Eng.*, **70**, 205, 1985.

5.128 D. Blavette and A. Bostel, *Acta Metall.*, **32**, 811, 1984.

5.129 D. Blavette, A. Bostel, and J. M. Sarrau, *Met. Trans. A*, **16A**, 1703, 1985.

5.130 J. P. Collier, P. W. Keefe, and J. K. Tien, *Met. Trans. A*, **17A**, 651, 1986.

5.131 R. Schmidt and M. Feller-Kniepmeier, *Met. Trans. A*, **23A**, 745, 1992.

5.132 R. Schmidt and M. Feller-Kniepmeier, *Scripta Metall. Mater.*, **26**, 1919, 1992.

5.133 C. C. Jia, K. Ishida, and T. Nishizawa, *Met. Mater. Trans. A*, **25A**, 473, 1994.

5.134 S. Duval, S. Chambreland, P. Caron, and D. Blavette, *Acta Metall. Mater.*, **42**, 185, 1994.

5.135 A. V. Karg, D. E. Fornwalt, and O. H. Kriege, *J. Inst. Met.*, **99**, 301, 1971.

5.136 S. Ochiai, Y. Oya, and T. Suzuki, *Acta Metall.*, **32**, 289, 1984.

5.137 M. Enomoto and H. Harada, *Met. Trans. A*, **20A**, 649, 1989.

5.138 A. Marty, M. Bessiere, F. Bley, Y. Calvayrac, and S. Lefebvre, *Acta Metall. Mater.*, **38**, 345, 1990.

5.139 E. Nembrach, K. Suzuki, M. Ichihara, and S. Takeuchi, *Philos. Mag. A*, **51**, 607, 1985.

5.140 T. Pretorius and D. Rönnpagel, *Proceedings of the 10th International Conference on the Strength of Materials* (edited by H. Oikawa, K. Maruyama, S. Takeuchi, and M. Yamaguchi), p. 689, JIM, 1994.

5.141 J.-L. Castagné, A. Pineau, and M. Sindzingre, *C. R. Acad. Sci. Paris, Ser. C*, **263**, 1465, 1966.

5.142 Y. Shimanuki and H. Doi, *Trans. JIM*, **16**, 123, 1975.

5.143 H. Gleiter and E. Hornbogen, *phys. stat. sol.*, **12**, 235, 1965.

5.144 H. Gleiter and E. Hornbogen, *Mater. Sci. Eng.*, **2**, 285, 1967/8.

5.145 D. Raynor and J. M. Silcock, *Met. Sci. J.*, **4**, 121, 1970.

5.146 R. K. Ham, *Trans. JIM Suppl.*, **9**, 52, 1968.

5.147 P. Guyot, *Philos. Mag.*, **24**, 987, 1971.

5.148 P. Haasen and R. Labusch, *Proceedings of the 5th International Conference on the Strength of Metals and Alloys*, Vol. 1 (edited by P. Haasen, V. Gerold, and G. Kostorz), p. 639, Pergamon Press, Toronto, 1979.

5.149 A. J. Ardell, V. Munjal, and D. J. Chellman, *Met. Trans. A*, **7A**, 1263, 1976.

5.150 A. J. Ardell, *Met. Sci.*, **14**, 221, 1980.

5.151 W. Hüther and B. Reppich, *Z. Metallkde.*, **69**, 628, 978.

5.152 V. A. Phillips, *Acta Metall.*, **14**, 1533, 1966.
5.153 E. Nembach, *Scripta Metall.*, **16**, 1261, 1982.
5.154 J. Glazer and J. W. Morris, Jr., *Philos. Mag. A*, **56**, 507, 1987.
5.155 A. J. Ardell and J. C. Huang, *Philos. Mag. Lett.*, **58**, 189, 1988.
5.156 V. Martens and E. Nembach, *Acta Met.*, **23**, 149, 1975.
5.157 E. Nembach, S. Schänzer, and K. Trinckauf, *Philos. Mag. A*, **66**, 729, 1992.
5.158 S. Schänzer and E. Nembach, *Acta Metall. Mater.*, **40**, 803, 1992.

CHAPTER 6

The Critical Resolved Shear Stress of Particle-Hardened Materials. Circumventing of Particles

If the particles are incoherent, dislocations circumvent or bypass them. The relevant process was suggested by Orowan [6.1] in 1947; subsequently it was named after him. It is sketched in Figure 6.1: The external shear stress τ_{ext} increases from left to right. τ_{ext} causes the dislocation to bow out between the particles. The segments $d\mathbf{s}_1$ and $d\mathbf{s}_2$ in Figure 6.1c have opposite signs and thus attract each other. If τ_{ext} is sufficiently high, $d\mathbf{s}_1$ contacts $d\mathbf{s}_2$ and both of them annihilate; the same holds for the segments $d\mathbf{s}_3$ and $d\mathbf{s}_4$. The final state is shown in Figure 6.1d: The dislocation has left a loop around each particle and is now free to continue its glide to the right. These loops lead to high work-hardening rates unless they can escape from the glide plane by cross-slip [6.2–6.6]. Strengthening by incoherent oxide particles is often referred to as dispersion strengthening.

The Orowan process operates not only if the particles are incoherent, but also (1) if the particles are coherent, but dislocations cannot glide in them—that is, if they are not ductile—or (2) if the particles are coherent and ductile, but they are so widely spaced that the stress required to shear them exceeds that to bypass them. Case 2 has been discussed in Section 4.2.1.3.2 in connection with Figure 4.15. Such a material is called *overaged* (Section 1.4). Orowan's estimate of the CRSS τ_p was given in Eq. (4.28):

$$\tau_p = \frac{2S}{bL_p}$$

or with the line tension S of Eq. (4.49c) one obtains

$$\tau_p = \frac{\mu_s b}{L_p} \tag{6.1}$$

where μ_s is the shear modulus of the matrix, b the length of the Burgers vector, and L_p the *free* space between particles. L_p is indicated in Figure 6.1a. Actually

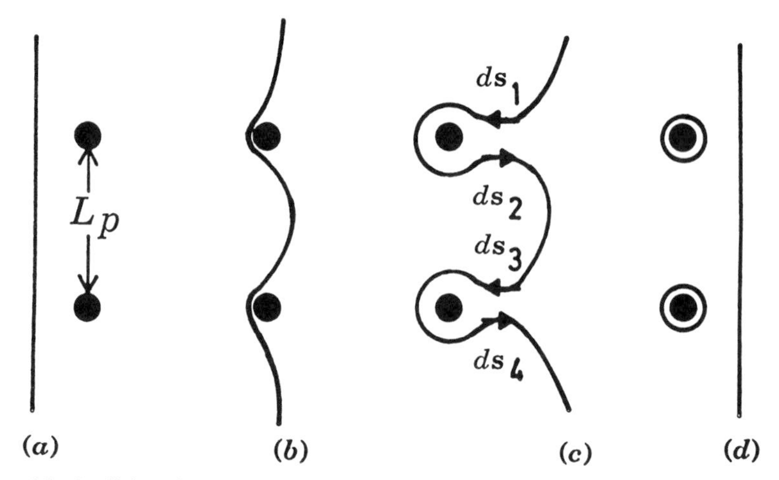

Fig. 6.1. A dislocation bypasses impenetrable particles, shown schematically. The external stress increases from left to right.

Orowan's result for τ_p was twice that of Eq. (6.1) because he inserted $[\mu_s b^2]$ instead of $[\mu_s b^2/2]$ for S [6.7].

Subsequent refinements of Eq. (6.1) primarily concern the following points; the sections in which they will be dealt with are given in parentheses:

1. Brown's [6.8] concept of the dislocation self-stress (Section 6.1.1)
2. The randomness of the particle arrangement (Section 6.1.2)
3. The elastic anisotropy (Section 6.1.1)

Points 1 and 3 have already been touched upon in Section 4.2.1.4.3.

6.1 MODELING OF THE CRSS

6.1.1 Critical Orowan Stress of Collinear, Equidistant Particles

Bacon et al. [6.9] calculated the CRSS τ_p of the Orowan process on the basis of Brown's [6.8] concept of the dislocation self-stress. Thus the self-interactions between different parts of a bent dislocation are fully allowed for; for example, the attractions between the segments ds_2 and ds_1 and between the segments ds_2 and ds_3 (Figure 6.1c) are taken into consideration. The attractions $ds_2 \leftrightarrow ds_1$ and $ds_2 \leftrightarrow ds_3$ are governed by the particles' diameter d_p measured in the glide plane and by the interparticle spacing L_p (Figure 6.1a), respectively. Bacon et al. calculated τ_p numerically for pure screw and for pure edge dislocations. They made the following assumptions: The particles lie on a straight line, they are equidistantly spaced (Figure 4.20b), and the material is elastically isotropic. The authors presented their numerical results in the following form, where θ_{d0} is the angle between the dislocation and its Burgers vector before bowing out:

Screw dislocations:

$$\tau_p(\theta_{d0} = 0^\circ) = \frac{\mu_s b}{2\pi L_p} \frac{1}{(1-\nu)} [\ln(D/R_i) + B] \tag{6.2a}$$

Edge dislocations:

$$\tau_p(\theta_{d0} = 90^\circ) = \frac{\mu_s b}{2\pi L_p} [\ln(D/R_i) + B] \tag{6.2b}$$

where R_i is the dislocations' inner cutoff radius and ν is Poisson's ratio, which was assumed to equal 1/3. The numerical constant B is about 0.65 [6.9, 6.10]. D is a function of L_p and d_p:

$$\frac{1}{D} = \frac{1}{L_p} + \frac{1}{d_p} \tag{6.3}$$

Usually L_p is much larger than d_p, and D approaches d_p. In this case the dislocation segment $d\mathbf{s}_2$ in Figure 6.1c mainly interacts with the segment $d\mathbf{s}_1$. In the other limit that L_p is much smaller than d_p, D is close to L_p. Evidently Eq. (6.3) yields the expected limiting values for D. No allowance has been made for the dislocations' core energy. If one wishes to take it into account, one may adjust R_i correspondingly. This has been demonstrated in Section 2.2. For future reference it is convenient to introduce the dislocation energy parameter $K_E(\theta_d)$ (Section 2.2) into Eqs. (6.2a) and (6.2b):

Screw dislocations:

$$\tau_p(\theta_{d0} = 0^\circ) = \frac{2K_E(\theta_d = 90^\circ)b}{L_p} [\ln(D/R_i) + B] \tag{6.2c}$$

Edge dislocations:

$$\tau_p(\theta_{d0} = 90^\circ) = \frac{2K_E(\theta_d = 0^\circ)b}{L_p} [\ln(D/R_i) + B] \tag{6.2d}$$

It is stressed that the critical Orowan stress is governed by the dislocation line *energy* and not by the line *tension*. This will be elaborated on below. Evidently the Orowan stress of screw dislocations exceeds that of edge dislocations by about 50%.

Scattergood and Bacon [6.10] extended the above-mentioned calculations in two respects: They allowed for the elastic anisotropy of the matrix and they also treated dislocations of mixed character. First the effects of the elastic anisotropy are reported. Figure 6.2 shows numerical results for the critical shear stress τ_p of edge and of screw dislocations. τ_p is given in units of $\mu_k b/L_p$. The shear modulus μ_K has been defined in Eq. (5.9c), and below it will be discussed further. The critical configurations of copper and nickel are shown in Figure 4.22. The authors tried to maintain Eqs. (6.2) in spite of the elastic anisotropy by redefining μ_s, ν, and K_E. Equations (6.2c) and (6.2d) hold for screw and edge dislocations if K_E is calculated under full allowance for the elastic anisotropy. This is stressed here by the term "aniso" in the argument of K_E:

Screw dislocations:

$$\tau_p(\theta_{d0} = 0°) = \frac{2K_E(\theta_d = 90°,\ \text{aniso})b}{L_p}\,[\ln(D/R_i) + B] \tag{6.4a}$$

Edge dislocations:

$$\tau_p(\theta_{d0} = 90°) = \frac{2K_E(\theta_d = 0°,\ \text{aniso})b}{L_p}\,[\ln(D/R_i) + B] \tag{6.4b}$$

K_E has been listed by Bacon et al. [6.11, 6.12].

Scattergood and Bacon [6.10] introduced the anisotropic shear modulus μ_K and the anisotropic Poisson's ratio ν_K:

$$\mu_K = 4\pi K_E(\theta_d = 0°,\ \text{aniso}) \tag{6.5a}$$

$$\nu_K = 1 - K_E(\theta_d = 0°,\ \text{aniso})/K_E(\theta_d = 90°,\ \text{aniso}) \tag{6.5b}$$

Inserting μ_K and ν_K into Eqs. (6.4a) and (6.4b) leads to equations that are analogous to Eqs. (6.2a) and (6.2b):

Screw dislocations:

$$\tau_p(\theta_{d0} = 0°) = \frac{\mu_K b}{2\pi L_p}\frac{1}{(1 - \nu_K)}\,[\ln(D/R_i) + B] \tag{6.4c}$$

Edge dislocations:

$$\tau_p(\theta_{d0} = 90°) = \frac{\mu_K b}{2\pi L_p}\,[\ln(D/R_i) + B] \tag{6.4d}$$

The straight lines in Figure 6.2 represent Eqs. (6.4c) and (6.4d). The agreement with the numerical data is excellent. At ambient temperature, μ_K and ν_K of copper equal 42 GPa and 0.43, respectively [6.10]. The two other shear moduli μ_F [Eq. (5.9a)] and μ_R [Eq. (5.9b)] introduced in Section 5.2.1 are 31 GPa and 43 GPa, respectively. Because τ_p of edge dislocations depends only on μ_K, but

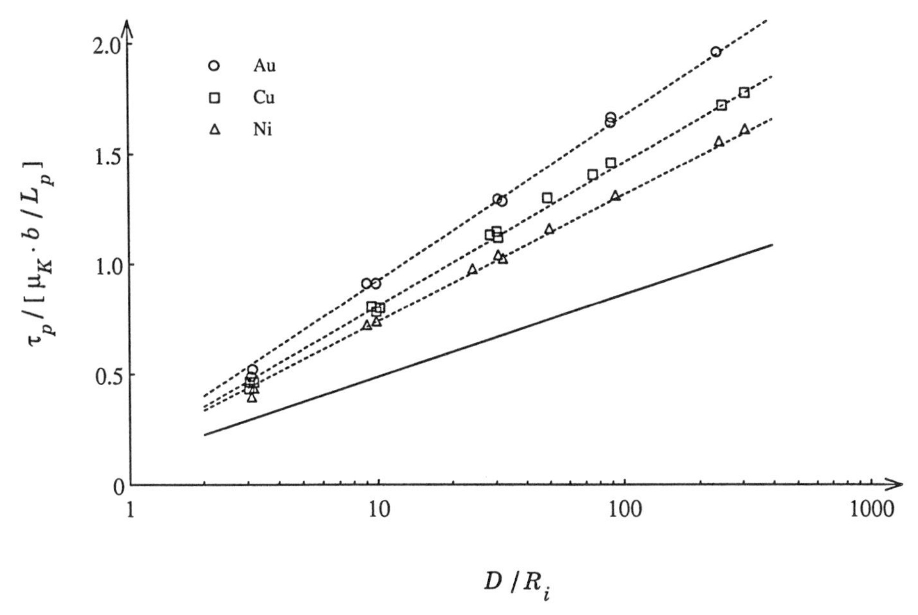

Fig. 6.2. Numerical results for the critical Orowan shear stress τ_p of gold, copper, and nickel are plotted versus D/R_i. τ_p is given in units of $\mu_K b/L_p$. D and μ_K are defined in Eqs. (6.3) and (6.5a), respectively. The abscissa is logarithmically divided. The dashed straight lines represent Eq. (6.4c) (screw dislocations). The drawn-out line represents Eq. (6.4d) (edge dislocations) for all three metals. The elastic anisotropy is allowed for. (After Ref. 6.10.)

not on ν_K, the straight lines for the three metals gold, copper, and nickel coincide in Figure 6.2.

Equations (6.4a) and (6.4b) lend themselves to the following interpretation [6.4, 6.10, 6.13]. It is given here for the general case of a dislocation of mixed character. Unless the dislocation has exact screw or exact edge character, its bow-out is not normal to its base line; this has been shown in Figures 4.21 and 4.23. The critical configuration just before the bow-out expands indefinitely is sketched in Figure 6.3. The directions of the dislocation next to the particles are characterized by the tangent angles ψ_1 and ψ_2. If the bow-out is virtually advanced by δ, the stress τ_p does the work $(\delta \sin\psi \cdot \tau_p b L_p)$. ψ is the arithmetic mean of ψ_1 and $(180° - \psi_2)$. In general the angles ψ_1 and ψ_2 do not add up to 180°. The work done by τ_p is stored in the two new dislocation segments of length δ at B and C; that is, τ_p must be high enough to pull out dipoles. The energy E_l per unit length of each segment is written according to Eq. (2.2):

$$E_l(\theta_d) = K_E(\theta_d) b^2 \ln(R_o/R_i)$$

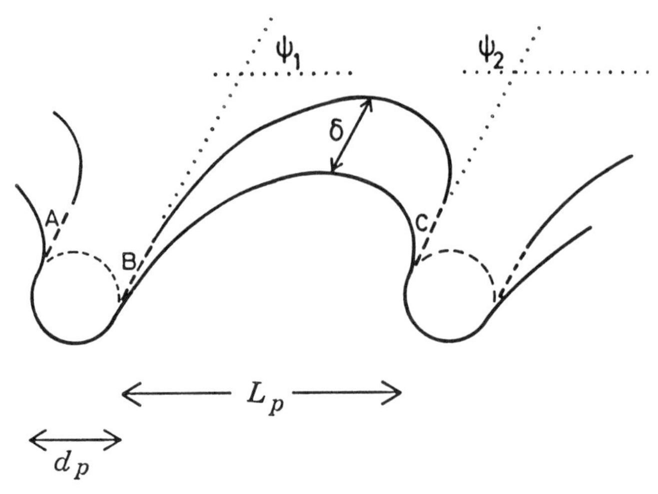

Fig. 6.3. A dislocation of mixed character that has been bent to its critical configuration is virtually advanced by δ. (After Refs. 6.4 and 6.10.)

θ_d equals the difference between ψ and θ_{d0}:

$$\theta_d = \psi - \theta_{d0} \tag{6.6}$$

The energy balance yields

$$\delta \sin\psi \cdot \tau_p b L_p = 2\delta E_l(\theta_d)$$

$$\tau_p(\theta_{d0}) = \frac{2E_l(\psi - \theta_{d0}, \text{aniso})}{bL_p \sin\psi} = \frac{2K_E(\psi - \theta_{d0},\ \text{aniso})b \ln(R_o/R_i)}{L_p \sin\psi} \tag{6.7}$$

"aniso" in the arguments of E_l and of K_E indicates that the elastic anisotropy of the material is taken into consideration. For screw and edge dislocations, the angles ψ and θ_{d0} are

Screw dislocation: $\psi = 90°,\quad \theta_{d0} = 0°$
Edge dislocation: $\psi = 90°,\quad \theta_{d0} = 90°$

Thus if $\ln(R_o/R_i)$ is replaced by $\ln(D/R_i) + B$, Eqs. (6.4a) and (6.4b) are recovered. Scattergood and Bacon called the above derivation "an approximation in the spirit of the line-tension approximation."

Scattergood and Bacon [6.10] calculated the critical Orowan stress τ_p of dislocations of mixed character numerically. Their results for gold and copper are reproduced in Figure 6.4. L_p and d_p equal $1000R_i$ and $100R_i$, respectively. Figure 6.4a shows the critical tangent angle ψ as a function of θ_{d0}. τ_p, in units

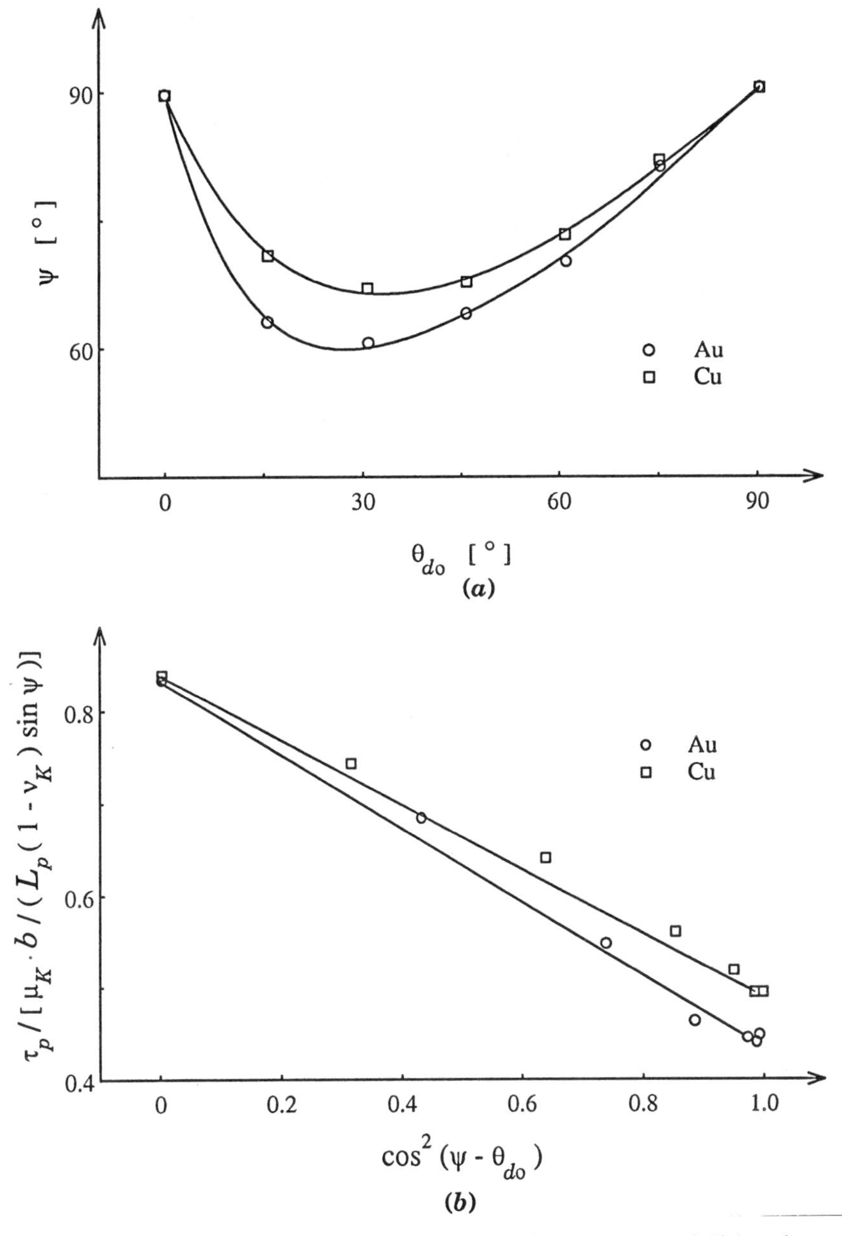

Fig. 6.4. Numerical results for the critical Orowan shear stress τ_p of dislocations of mixed character in gold and copper. The elastic anisotropy is allowed for. Free interparticle spacing $L_p = 1000R_i$, particle diameter $d_p = 100R_i$. (*a*) Critical tangent angle ψ versus θ_{d0} = angle between the dislocation and its Burgers vector before bowing out. (*b*) τ_p in units of $\mu_K b/[L_p(1 - \nu_K) \sin\psi]$ versus $\cos^2(\psi - \theta_{d0})$. The straight lines represent Eq. (6.8). (After Ref. 6.10.)

of $[\mu_K b/\{L_p(1-\nu_K)\sin\psi\}]$, is plotted versus $\cos^2(\psi-\theta_{d0})$ in Figure 6.4b. In order to obtain a compact representation of the numerical data presented in Figure 6.4b, the authors assumed that the variation of $K_E(\theta_d, \text{aniso})$ [Eq. (6.7)] with θ_d is that given by Eq. (2.3b), which has been meant for elastically isotropic materials. $K_E(\theta_d = 0°, \text{aniso})$ and $K_E(\theta_d = 90°, \text{aniso})$ are calculated anisotropically and Eq. (2.3b) is used as an interpolation scheme for $K_E(\theta_d, \text{aniso})$ with $0° < \theta_d < 90°$. Thus after inserting $[\ln(D/R_i) + B]$ for $\ln(R_o/R_i)$ in Eq. (6.7), it assumes the following form:

$$\tau_p(\theta_{d0}) = \frac{\mu_K b}{2\pi(1-\nu_K)} \frac{1-\nu_K\cos^2(\psi-\theta_{d0})}{L_p\sin\psi}[\ln(D/R_i)+B] \tag{6.8}$$

μ_K and ν_K have been defined in Eqs. (6.5). The straight lines in Figure 6.4b represent Eq. (6.8) with $B = 0.65$; ψ in the arguments of $\cos^2(\psi-\theta_{d0})$ and of $\sin\psi$ has to be read from Figure 6.4a. The agreement with the numerical data is satisfactory.

Scattergood and Bacon also studied the extent to which deWit and Koehler's [6.14] dislocation line tension scheme is applicable. Scattergood and Bacon considered the outer cutoff radius R_o in the line tension as an adjustable parameter. They derived it from the following postulate concerning the area A_c swept out by the dislocation before it reaches its critical configuration: A_c calculated on the basis of the line tension model must coincide with A_c obtained from Brown's [6.8] self-stress model under full allowance for the elastic anisotropy. The relevance of A_c will become apparent at the beginning of the following section. Good overall agreement was achieved if R_o was set equal to $L_p/5$. This holds for $d_p < L_p/4$. For L_p approaching d_p, R_o must approach L_p.

6.1.2 Critical Orowan Stress of a Random Array of Particles

In order to obtain the critical shear stress of real particle arrays the equations presented in the preceding section have to be developed further in two respects:

1. A suitable average over τ_p of screw and of edge dislocations has to be taken.
2. The effective particle spacing L_p of a random array of particles has to be derived.

Point 1 has been dealt with by Hirsch and Humphreys [6.5]. In the case of weak shearable obstacles, that type of dislocation governs the CRSS that yields the highest CRSS (Section 4.2). If the shearable particles are so strong that the dislocations are strongly bent, averaging procedures are required (Sections 4.2.1.4.3 and 5.5.4.1). Hirsch and Humphreys pointed out that a dislocation loop that expands between two impenetrable particles must be able to reach its critical configuration before it touches an additional particle. The critical configuration is very close to a semiellipse; this holds even if the material is elastically anisotropic [6.12, 6.15]. The axes are l_0 and l_{90}: l_0 is parallel to the Burgers vector and l_{90} is normal to it. For the sake of clarity, only edge and screw dislocations are analyzed here. Thus the area A_c swept out by the half-loop in the critical configuration is given by

$$A_c = \frac{\pi}{2}\frac{l_0 l_{90}}{4} \tag{6.9}$$

Hirsch and Humphreys' postulate implies

$$A_c \leqslant \frac{\pi}{2}\frac{(L_{\min} - 2r_r)^2}{4} \tag{6.10}$$

$L_{\min}$ is the square lattice spacing and r_r is the average radius of the particle–glide-plane intersection (Section 3.1). Inserting Eq. (6.9) into relation (6.10) and taking its upper limit yields

$$l_0 l_{90} = (L_{\min} - 2r_r)^2 \tag{6.11a}$$

The ratio l_0/l_{90} of the axes is very close to $1/(1 - \nu_K)$, even if the dislocation's self-stress and the elastic anisotropy are allowed for [6.12, 6.15]. The anisotropic Poisson's ratio ν_K has been defined in Eq. (6.5b). Eliminating l_{90} from Eq. (6.11a) leads to

$$l_0^2(1 - \nu_K) = (L_{\min} - 2r_r)^2 \tag{6.11b}$$

and with Eq. (6.5b) one obtains

$$l_0^2[K_E(\theta_d = 0°, \text{aniso})/K_E(\theta_d = 90°, \text{aniso})] = (L_{\min} - 2r_r)^2$$
$$l_0 = (L_{\min} - 2r_r)[K_E(\theta_d = 90° \text{ aniso})/K_E(\theta_d = 0°, \text{aniso})]^{1/2} \tag{6.11c}$$

Inserting l_0 for the particle spacing L_p in Eq. (6.4a) yields the critical Orowan stress τ_p of a screw dislocation:

$$\tau_p = \frac{2[K_E(\theta_d = 0°,\ \text{aniso}) \cdot K_E(\theta_d = 90°,\ \text{aniso})]^{1/2} b}{L_{\min} - 2r_r}[\ln(D/R_i) + B] \tag{6.12a}$$

Because exactly the same result is obtained for edge dislocations, θ_{d0} is not used as argument of τ_p. Both types of dislocations yield the same critical Orowan shear stress τ_p because relations (6.9) and (6.10) are assumed to hold. Evidently τ_p is proportional to the geometric mean of the energy parameters K_E [Eq. (2.2)] of edge and screw dislocations. Below, K_{Eg} will be written for this mean:

$$K_{Eg} = [K_E(\theta_d = 0°, \text{aniso}) \cdot K_E(\theta_d = 90°, \text{aniso})]^{1/2} \tag{6.12b}$$

The subscript g of K_{Eg} indicates "geometric" mean. There is some similarity between K_{Eg} and the effective dislocation line tension $S_{\mathrm{eff}\,1}$, which has been defined in Eq. (4.50a) and applied in Section 5.5.4.1.

Bacon et al. [6.9], who derived Eqs. (6.2) assuming elastic isotropy, have suggested a procedure to allow for the randomness of the distribution of the impenetrable particles in the glide plane. The authors adapted the derivations presented in Chapter 4 for shearable obstacles to the present impenetrable ones. The reasoning is detailed here for edge dislocations, whose critical shear stress τ_p is given by Eq. (6.2b). Besides the array of impenetrable obstacles, the authors considered another array of penetrable (i.e., shearable) obstacles. Both arrays are supposed to yield the same CRSS τ_p and to have the same volume fraction f and the same average radius r of particles. From these conditions the maximum interaction force F_0 between the shearable particles and edge dislocations can be calculated. F_0 and the critical angle $\phi_c/2$, which is equivalent to F_0, have been defined in Eqs. (4.7a) and (4.8), respectively. Equation (4.14) relates F_0 to τ_p and the critical length L_c; here L_c equals L_p:

$$\tau_p b L_p = F_0$$

Inserting for τ_p from Eq. (6.2b) yields

$$F_0 = \frac{\mu_s b}{2\pi L_p}\,[\ln(D/R_i) + B]\cdot bL_p$$

Equation (4.8a) relates F_0 to the dislocation line tension S and to $\phi_c/2$. Because elastic isotropy was assumed, Bacon et al. [6.9] approximated S by $[(\mu_s b^2/4\pi)\ln(L_p/R_i)]$. This yields the following for F_0:

$$F_0 = 2\cdot\frac{\mu_s b^2}{4\pi}\ln(L_p/R_i)\cdot\cos(\phi_c/2)$$

Thus one sees

$$\cos(\phi_c/2) = \frac{\ln(D/R_i) + B}{\ln(L_p/R_i)} \tag{6.13}$$

In this way, even impenetrable obstacles can be assigned a critical breaking angle $\phi_c/2$. This is, however, merely formal.

In many systems that contain impenetrable particles, $\phi_c/2$ is far larger than the expected 0°. $\cos(\phi_c/2)$ of the experimental data shown in Figure 6.5 equals 0.78 on an average, and this corresponds to $\phi_c/2 = 39°$. More experimental data for $\phi_c/2$ can be calculated from $P \approx [\cos(\phi_c/2)]^{1/2}$ [Eq. (6.15)] listed in Table 6.1. The reason for these surprisingly large $\phi_c/2$-values is that the dislocation's self-interaction across the particles—for example, the attraction between the segments $d\mathbf{s}_1$ and $d\mathbf{s}_2$ in Figure 6.1c—lowers τ_p drastically. Bacon et al. [6.9] proceeded to insert $\cos(\phi_c/2)$ of Eq. (6.13) into Eq. (4.26b) (Friedel's equation), which had been meant for penetrable obstacles; again the above approximation $S = (\mu_s b^2/4\pi)\ln(L_p/R_i)$ was used:

$$\tau_p = \frac{2\mu_s b^2}{4\pi b L_{\min}} \ln(L_p/R_i) \left[\frac{\ln(D/R_i) + B}{\ln(L_p/R_i)}\right]^{3/2} \tag{6.14a}$$

In view of Eq. (6.12a), $L_{\min}$ is replaced by $L_p = (L_{\min} - 2r_r)$. Thus the finite extension of the particles (Section 4.2.1.4.2) is taken into account. Now the following two substitutions are made:

$$\mu_s = 4\pi[K_E(\theta_d = 0°, \text{aniso}) \cdot K_E(\theta_d = 90°, \text{aniso})]^{1/2} = 4\pi K_{Eg}$$

and

$$B \approx 0.65 \approx \ln 2$$

r_r is the average radius of the particle–glide-plane intersection (Section 3.1). Finally the factor $Y \approx 0.9$ is introduced. It allows for the randomness of the particle arrangement in the glide plane. Y has been discussed at length in Section 4.2.1.4.1. Y has an analogous function as C_1 in Schwarz and Labusch's equation (4.62). Thus one obtains the following as final result for the Orowan stress:

$$\tau_p = Y \frac{2K_{Eg} b}{L_{\min} - 2r_r} \frac{\{\ln(2D/R_i)\}^{3/2}}{\{\ln[(L_{\min} - 2r_r)/R_i]\}^{1/2}} \tag{6.14b}$$

D is given in Eq. (6.3), and d_p equals $2r_r$. Evidently the above procedure to allow for the random arrangement of the impenetrable particles introduces the factors Y and P [Eq. (6.15)]; Eqs. (6.12a) and (6.14b) differ by them:

$$P = \{\ln(2D/R_i)/\ln[(L_{\min} - 2r_r)/R_i]\}^{1/2} \approx \{\cos(\phi_c/2)\}^{1/2} \tag{6.15}$$

The average P-value of the experimental data shown in Figure 6.5 is 0.89.

Inserting for $L_{\min}$ and r_r from Eqs. (3.3c) and (3.5a), respectively, one gets the following instead of Eq. (6.14b):

$$\tau_p = Y \frac{2K_{Eg} b}{r[(\pi\omega_q/f)^{1/2} - 2\omega_r]} \frac{\{\ln(2D/R_i)\}^{3/2}}{\{\ln[r[(\pi\omega_q/f)^{1/2} - 2\omega_r]/R_i]\}^{1/2}} \tag{6.14c}$$

In most cases $L_{\min}$ is much larger than r_r and Eq. (6.14c) can be slightly simplified:

$$\tau_p = Y \frac{2K_{Eg} b}{r[(\pi\omega_q/f)^{1/2} - 2\omega_r]} \frac{\{\ln(4\omega_r r/R_i)\}^{3/2}}{\{\ln[r(\pi\omega_q/f)^{1/2}/R_i]\}^{1/2}} \tag{6.14d}$$

This function $\tau_p(r, f = \text{const.})$ is shown in Figure 4.15.

Kocks [6.16] reported an equation similar to Eq. (6.14a), but with $2D$ instead of D. Thus Kocks' version of Eq. (6.14d) reads

$$\tau_p = Y \frac{2K_{Eg}b}{r[(\pi\omega_q/f)^{1/2} - 2\omega_r]} \frac{\{\ln(8\omega_r r/R_i)\}^{3/2}}{\{\ln[r(\pi\omega_q/f)^{1/2}/R_i]\}^{1/2}} \tag{6.16}$$

Because it is open to discussion whether one should insert b or $2b$ for R_i in Eqs. (6.14d) and (6.16), the difference between these two equations is of hardly any consequence. In Table 6.1 a numerical example is presented for overaged NIMONIC PE16: The results derived from these two equations differ by $\pm 8\%$.

The derivation of Friedel's equation (4.26b) involved the supposition that dislocations bow out only slightly (Section 4.2.1.3.1); hence relation (4.9b) requires that $\phi_c/2$ calculated from Eq. (6.13) should be larger than 60°. As can be calculated from P [Eq. (6.15)] listed in Table 6.1, $\phi_c/2$ of actual systems in which the Orowan process operates is likely to be below this limit. But since $P \approx \{\cos(\phi_c/2)\}^{1/2}$ is close to unity (Table 6.1) and leads only to a minor correction of Eqs. (6.12), the violation of the "slight bowing condition" has no serious consequences.

For the sake of conformity, Eq. (6.12a) is repeated here with two alterations: Y is added and B is replaced by ln 2:

$$\tau_p = Y \frac{2K_{Eg}b}{L_{\min} - 2r_r} [\ln(2D/R_i)] \tag{6.17}$$

In the discussion of experimental data in Section 6.2, it will become apparent that the differences in τ_p calculated from Eqs. (6.14c), (6.14d), (6.16), and (6.17) are small. The Orowan stress is roughly proportional to $1/L_{\min}$ and hence to $\sqrt{f}/r$. This is the classical example of overaging.

6.2 COMPARISON WITH EXPERIMENTAL DATA

Rühle [6.17–6.19] has listed materials that are strengthened by incoherent oxide particles; the emphasis of his overviews is on the technical aspects of dispersion strengthening. Two classes of materials will be discussed below in the light of the models and equations presented in Section 6.1:

1. *Dispersion-strengthened materials:* Copper strengthened by incoherent silicon oxide particles [6.4, 6.20] and copper strengthened by incoherent beryllium oxide particles [6.21].
2. *Overaged materials strengthened by coherent precipitates:* Copper strengthened by coherent cobalt-rich precipitates [6.22] and the nickel-base superalloy NIMONIC PE16 (Section 5.5) strengthened by coherent γ'-precipitates [6.23].

In order to obtain the particles' contribution τ_p to the total measured CRSS, the solid solution matrix' CRSS τ_s has to be subtracted. This will be done on the basis of Eq. (5.58). Because τ_s of the three copper-base alloy systems amounts to only a few megapascals, the choice of the exponent k in Eq. (5.58) does not really matter and unity will be inserted for k of these systems. k of overaged NIMONIC PE16 single crystals has been established experimentally (Section 7.2.1) [6.23]: $k = 1.45$. The strong effects of choosing alternative value for k of NIMONIC PE16 are demonstrated in Table 6.1.

Though the CRSSs of the four alloy systems listed above have been measured at various temperatures between 77 K and 500 K, with the exception of Figure 6.6 only the room temperature data will be analyzed with reference to Eqs. (6.14c), (6.14d), (6.16), and (6.17). The agreement between theory ($\tau_{p\,\mathrm{theo}}$) and experiment ($\tau_{p\,\mathrm{exp}}$) will be judged on the basis of the ratio $\tau_{p\,\mathrm{exp}}/(\tau_{p\,\mathrm{theo}}/Y)$. In the case of perfect agreement this ratio equals $Y \approx 0.9$ (Section 6.1.2). The geometric mean K_{Eg} [Eq. (6.12b)] of the dislocation line energy parameters is needed in all evaluations. According to its definition, K_{Eg} has to be calculated anisotropically. At ambient temperature K_{Eg} equals 4.50 GPa for the copper alloys [6.22] and 6.68 GPa for NIMONIC PE16 [6.23]. The dissociation of the dislocations into Shockley partial dislocations is disregarded. Bacon [6.24] found that this dissociation lowers K_{Eg} of copper by about 7%. The statistical parameters ω_r and ω_q depend on the distribution function g of the particle radii. The following parameters ω_r and ω_q will be used below: for the incoherent oxide particles those of g_0 [Eq. (3.6)] and for the coherent precipitates those of g_{WLS} [Eq. (3.7)]. The length b of the Burgers vector will be inserted for R_i.

6.2.1 Dispersion-Strengthened Copper

Ebeling and Ashby [6.4, 6.20] and Jones [6.21] produced fine spherical silicon oxide particles, respectively fine fairly equiaxed beryllium oxide particles in copper by internal oxidation. This process has been described in Section 3.2.3. The mean radii r of the particles and their volume fractions f spanned the following ranges:

Silicon oxide:	$29\,\mathrm{nm} \leqslant r \leqslant 92\,\mathrm{nm}$,	$0.0033 \leqslant f \leqslant 0.01$
Beryllium oxide:	$3.6\,\mathrm{nm} \leqslant r \leqslant 11.8\,\mathrm{nm}$,	$0.0082 \leqslant f \leqslant 0.041$

In contrast to Ebeling and Ashby, who used the geometric mean particle radius in the analysis of their CRSS data, the arithmetic mean will be inserted below.

The ratio $\tau_{p\,\mathrm{exp}}/(\tau_{p\,\mathrm{theo}}/Y)$ is plotted versus $\tau_{p\,\mathrm{exp}}$ in Figure 6.5. $\tau_{p\,\mathrm{theo}}$ has been calculated from Eq. (6.14c). The average ratio equals $0.675 \pm 3.8\%$ for the silicon oxide particles, $0.447 \pm 4.3\%$ for the beryllium oxide particles, and $0.546 \pm 5.2\%$ for both alloy systems evaluated together. This ratio is expected to be equal to $Y \approx 0.9$ (Section 6.1.2). Evidently there is a discrepancy by nearly the factor 2. Alternative evaluations are listed in Table 6.1; their respective results agree within $\pm 22\%$. This holds even if the exponent k of Eq.

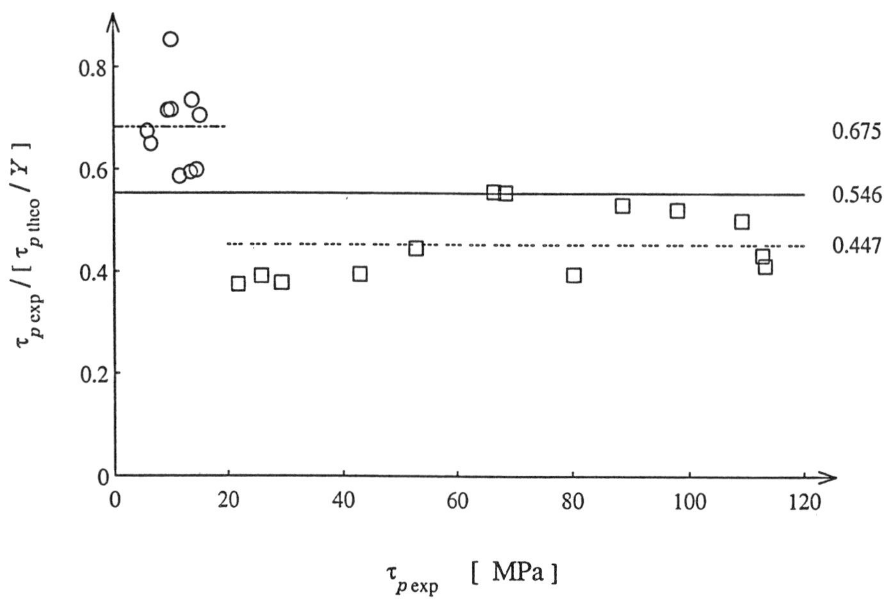

Fig. 6.5. The ratios $[\tau_{p\,exp}/(\tau_{p\,theo}/Y)]$ of copper containing incoherent silicon oxide particles (○ —--) [6.4, 6.20] and of copper containing incoherent beryllium oxide particles (□---) [6.21] are plotted versus $\tau_{p\,exp}$. $\tau_{p\,exp}$ is the experimental result for the CRSS, and $\tau_{p\,theo}$ is the theoretical result derived from Eq. (6.14c); $Y \approx 0.9$. The horizontal lines and the numbers on the right-hand side indicate the respective averages; —— overall average. The data have been taken at ambient temperature.

(5.58) is raised to 2.0. The ratio of the copper alloys containing beryllium oxide particles amounts only to about 66% of the ratio of those alloys that contain silicon oxide particles. The reason for this difference remains unknown; possible explanations are stresses associated with the oxide particles and/or minor differences in their shape.

According to the above evaluations, the standard deviation of the ratio $\tau_{p\,exp}/(\tau_{p\,theo}/Y)$ of Ebeling and Ashby's data amounts to 3.8%; the corresponding percentage of Jones' data is 4.3%. $\tau_{p\,theo}$ has been calculated from Eq. (6.14c). Hence it is concluded that this equation represents the *functional* dependence of τ_p on r and f satisfactorily. There is, however, no *quantitative* agreement between theory and experiment. It has been shown in the preceding paragraph that $\tau_{p\,theo}$ is too high, roughly by the factor two.

6.2.2 Overaged Materials Strengthened by Coherent Precipitates

Büttner et al. [6.22] studied *overaged* copper–cobalt alloys; *underaged* ones have been discussed in Sections 5.4.2.1.8 and 5.4.2.2.2. The spherical precipitates were rich in cobalt. There were two volume fractions (0.0076 and 0.13) and one average particle radius (29.4 nm). The results for the ratio $\tau_{p\,exp}/$

Table 6.1. Comparison Between Theoretical CRSS Data $\tau_{p\,\text{theo}}$ and Experimental Data $\tau_{p\,\text{exp}}$[a]

	Equation Number	k	$\tau_{p\,\text{exp}}/(\tau_{p\,\text{theo}}/Y)$	P
Cu—SiO_2 [6.4, 6.20]	6.17	1.00	0.596 ± 4.1%	0.883
Cu—BeO [6.21]	6.17	1.00	0.397 ± 4.4%	0.887
Cu—SiO_2 and Cu—BeO	6.17	1.00	0.484 ± 5.3%	0.885
Cu—SiO_2	6.14c	1.00	0.675 ± 3.8%	
Cu—BeO	6.14c	1.00	0.447 ± 4.3%	
Cu—SiO_2 and Cu—BeO	6.14c	1.00	0.546 ± 5.2%	
Cu—SiO_2	6.16	1.00	0.569 ± 4.0%	
Cu—BeO	6.16	1.00	0.345 ± 4.2%	
Cu—SiO_2 and Cu—BeO	6.16	1.00	0.442 ± 6.1%	
Cu—SiO_2	6.14c	2.00	0.805 ± 4.2%	
Cu—BeO	6.14c	2.00	0.465 ± 3.9%	
Cu—SiO_2 and Cu—BeO	6.14c	2.00	0.612 ± 6.5%	
Cu—Co [6.22]	6.17	1.00	0.93	0.889
$f = 0.0076, 0.0130$	6.14c	1.00	1.05	
$r = 29.4$ nm	6.14c	2.00	1.20	
NIMONIC PE16 [6.23]	6.17	1.45	1.48	0.931
$f = 0.029$, $r = 69$ nm	6.14c	1.45	1.59	
	6.14d	1.45	1.54	
	6.16	1.45	1.33	
	6.14c	1.00	0.98	
	6.14c	2.00	2.01	

[a]The parameters k, Y, and P are defined in Eqs. (5.58), (6.14b), and (6.15), respectively.

$(\tau_{p\,\text{theo}}/Y)$ are listed in Table 6.1: 1.05 for Eq. (6.14c) and 0.93 for Eq. (6.17). The agreement with $Y \approx 0.9$ is quite satisfactory, but it is probably fortuitous, because the cobalt-rich precipitates have a lattice mismatch (Section 5.4.2.1.8), which is expected to raise the CRSS. This may, however, be compensated at least in part by some particle shearing.

Nembach and Neite [6.23] measured the CRSS of overaged single crystals of the nickel-base superalloy NIMONIC PE16, which is strengthened by spherical γ'-precipitates. The results for underaged and for peak-aged specimens have been reported in Sections 5.5.4.1 and 5.5.4.2, respectively. f and r of the overaged single crystals covered the following ranges: $0.029 \leqslant f \leqslant 0.061$ and $69\text{ nm} \leqslant r \leqslant 125\text{ nm}$. In Figure 6.6, the ratio $\tau_{p\,\text{exp}}/(\tau_{p\,\text{theo}}/Y)$ is plotted versus the average radius r of the γ'-precipitates. $\tau_{p\,\text{theo}}$ has been calculated on the basis of Eq. (6.14d). This had been done in order to stay as close as possible to the original publication, in which Eq. (6.16) had been used. Evidently there is a definite increase of the ratio with r and a decrease with f. The average ratio is 1.71 instead of $Y \approx 0.9$. The effects of choosing other equations or setting the

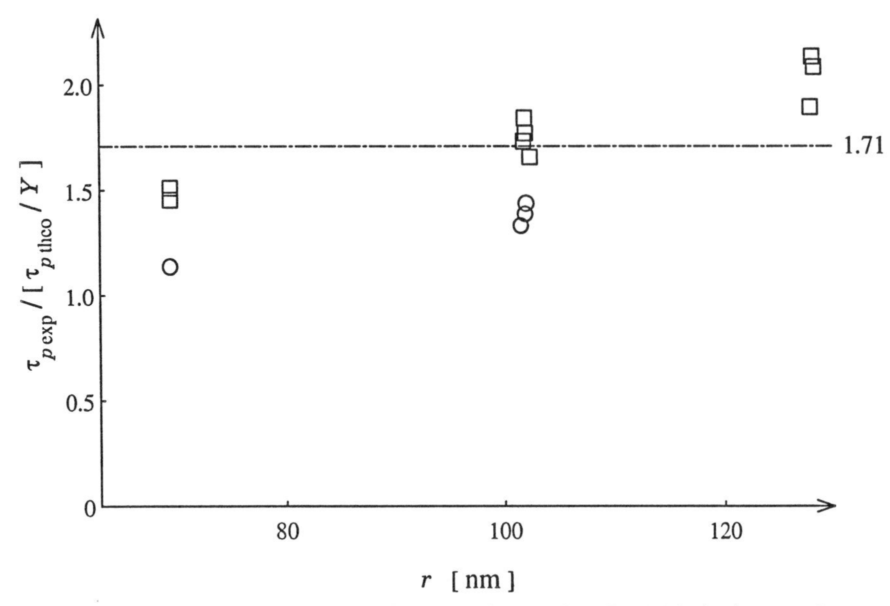

Fig. 6.6. The ratios $\tau_{p\,\mathrm{exp}}/(\tau_{p\,\mathrm{theo}}/Y)$ of overaged NIMONIC PE16 single crystals are plotted versus the average radius r of the γ'-particles; ○ $f = 0.061$, □ $f = 0.029$. The data have been taken at temperatures between 90 K and 500 K. $\tau_{p\,\mathrm{exp}}$ is the experimental result for the CRSS, $\tau_{p\,\mathrm{theo}}$ has been calculated with Eq. (6.14d), and $Y \approx 0.9$. The horizontal line indicates the average ratio. (After Ref. 6.23.)

exponent k of Eq. (5.58) equal to 1.0 or 2.0 instead of 1.45 [6.23] are demonstrated in Table 6.1 for $f = 0.029$ and $r = 69$ nm. All evaluations lead to an increase of the ratio $[\tau_{p\,\mathrm{exp}}/(\tau_{p\,\mathrm{theo}}/Y)]$ with r. Because the CRSS τ_s of the matrix amounts to more than 50% of the total CRSS τ_t, it is of exceeding importance to insert the right value for the exponent k in Eq. (5.58). $k = 1.45$ has been established experimentally [6.23]. The ratios listed in Table 6.1 for Eqs. (6.14c), (6.14d), (6.16), and (6.17) for $k = 1.45$ and the ratios plotted in Figure 6.6 are around 1.5; they exceed $Y \approx 0.9$ by about 60%. Because the γ'-precipitates are coherent, some of them will be sheared. The probability of shearing will decrease as r increases. This may explain the increase of the ratio $\tau_{p\,\mathrm{exp}}/(\tau_{p\,\mathrm{theo}}/Y)$ with r (Figure 6.6), but the high values of this ratio remain unaccounted for.

6.3 SUMMARY

Equations (6.14c), (6.14d), (6.16), and (6.17) relate the resolved shear stress τ_p needed for the operation of the Orowan process to the radius r and the volume fraction f of the particles. In the case of copper strengthened by incoherent oxide particles, Eq. (6.14c) represents the functional variation of τ_p with r and f satisfactorily. For overaged NIMONIC PE16 single crystals hardened by coherent γ'-precipitates, the quoted equations are inferior descriptions of the functional variation of τ_p with r and f. For all analyzed systems, the ratio $\tau_{p\,\exp}/(\tau_{p\,\text{theo}}/Y)$ has been found to range from about 0.4 to about 2.0. This ratio is supposed to equal $Y \approx 0.9$. $\tau_{p\,\exp}$ and $\tau_{p\,\text{theo}}$ are the experimental and theoretical results, respectively, for τ_p. Thus in the case of the Orowan process, theory and experiment agree within a factor of about 2.0. This is the same margin as that found for shearing of particles (Section 5.6). Because in the latter case there is often some doubt about the dominant particle–dislocation interaction mechanism and rather involved averaging procedures are necessary, one might have expected a better agreement between theory and experiment if the particles are bypassed than if they are sheared. But in the case of bypassing, the experimental conditions are often somewhat less well defined. It is very difficult to obtain spatially homogeneous oxide particle distributions by internal oxidation. If the particles are coherent, their stress field may contribute to the CRSS and there may be some particle shearing.

REFERENCES

6.1 E. Orowan, *Symposium on Internal Stresses in Metals and Alloys*, p. 451, Institute of Metals, London, 1948.

6.2 H. Gleiter, *Acta Metall.*, **15**, 1213, 1967.

6.3 H. Gleiter, *Acta Metall.*, **15**, 1223, 1967.

6.4 M. F. Ashby, in *Physics of Strength and Plasticity* (edited by A. S. Argon), p. 113, MIT Press, Cambridge, Massachusetts, 1969.

6.5 P. B. Hirsch and F. J. Humphreys, in *Physics of Strength and Plasticity* (edited by A. S. Argon), p. 189, MIT Press, Cambridge, Massachusetts, 1969.

6.6 M. F. Ashby, in *Oxide Dispersion Strengthening* (edited by G. S. Ansell, T. D. Cooper, and F. V. Lenel), Second Bolton Landing Conference, p. 143, Gordon and Breach, New York, 1966.

6.7 F. R. N. Nabarro, *Symposium on Internal Stresses in Metals and Alloys*, p. 237, Institute of Metals, London, 1948.

6.8 L. M. Brown, *Philos. Mag.*, **10**, 441, 1964.

6.9 D. J. Bacon, U. F. Kocks, and R. O. Scattergood, *Philos. Mag.*, **28**, 1241, 1973.

6.10 R. O. Scattergood and D. J. Bacon, *Philos. Mag.*, **31**, 179, 1975.

6.11 D. J. Bacon and R. O. Scattergood, *J. Phys. F: Metal Phys.*, **4**, 2126, 1974.

6.12 D. J. Bacon, D. M. Barnett, and R. O. Scattergood, *Prog. Mater. Sci.*, **23**, 51, 1979.

6.13 M. F. Ashby, *Acta Metall.*, **14**, 679, 1966.

6.14 G. deWit and J. S. Koehler, *Phys. Rev.*, **116**, 1113, 1959.

6.15 R. O. Scattergood and D. J. Bacon, *phys. stat. sol. (a)*, **25**, 395, 1974.

6.16 U. F. Kocks, *Mater. Sci. Eng.*, **27**, 291, 1977.

6.17 M. Rühle, *Z. Metallkde.*, **71**, 1, 1980.

6.18 M. Rühle, *Z. Metallkde.*, **71**, 65, 1980.

6.19 M. Rühle, *Metall.*, **36**, 1280, 1982.
6.20 R. Ebeling and M. F. Ashby, *Philos. Mag.*, **13**, 805, 1966.
6.21 R. L. Jones, *Acta Metall.*, **17**, 229, 1969.
6.22 N. Büttner, K.-D. Fusenig, and E. Nembach, *Acta Metall.*, **35**, 845, 1987.
6.23 E. Nembach and G. Neite, *Prog. Mater. Sci.*, **29**, 177, 1985.
6.24 D. J. Bacon, *Philos. Mag. A*, **38**, 333, 1978.

CHAPTER 7

Multiple Strengthening and Synergisms

In most of the models of particle strengthening discussed in Chapters 4 and 5 it has been assumed that all particles present in the specimen produce the same resolved particle–dislocation interaction shear stress τ_{obst} (Section 2.5) or that a suitable average shear stress can be defined. The subscript "obst" of τ_{obst} indicates a possible generalization: Not only particles, but also other types of stationary obstacles to the glide of dislocations, may be considered—for example, solute atoms.

There are at least two reasons why there will always be a range of stresses τ_{obst}: The particle size has some distribution and the glide plane does not intersect all particles at the same "geographical latitude." In the case of stacking fault (Section 5.3) and lattice mismatch (Section 5.4.2) strengthening, elaborate averaging procedures have been developed. Moreover, often the particle–dislocation interaction *force* depends on the character of the dislocation. Many materials simultaneously contain different types of obstacles to the glide of dislocations, and it is not possible to represent these different obstacles by an average one. The most frequently encountered combination is that of solid solution with particle strengthening. Precipitation hardening is virtually always accompanied by at least some solid solution hardening of the matrix because some of the atoms that had been added to the material to produce the precipitates stay solved in the matrix. In the analyses of experimental data on particle strengthening presented in Chapters 5 and 6, this superposition has been taken into consideration.

Two examples of multiple strengthened technical materials are as follows:

1. Alloy MA 6000 (Section 3.2.2) [7.1] is strengthened by various solute atoms, by coherent $L1_2$-long-range-ordered γ'-precipitates (Section 5.5.1), and by incoherent oxide particles (Section 6.2.1).
2. Commercial aluminum-base alloys contain solute atoms and different types of precipitates [7.2–7.5].

If different types of solutes are simultaneously solved in the matrix, they are represented by an average one.

The above-described situations of multiple strengthening have to be distinguished from those where there is only one type of particle present in a specimen but each particle interacts with dislocations via various mechanisms: A particle may, for example, be long-range-ordered (Sections 4.1 and 5.5) and it may have a modulus (Section 5.2) mismatch. This has been discussed in Sections 4.1 and 4.2.

7.1 MODELING OF MULTIPLE STRENGTHENING

Let there be N distinct types of obstacles O_i, $1 \leqslant i \leqslant N$, in the specimen; their respective numbers per unit area of the glide plane is n_{si}, $1 \leqslant i \leqslant N$. The different types of obstacles are supposed to be intermixed: the positions of O_i- and O_j-obstacles, $1 \leqslant i, j \leqslant N$, $i \neq j$, are uncorrelated; no part of the specimen is supposed to be dominated by either type.

If there are only O_i-obstacles, $1 \leqslant i \leqslant N$—that is, $n_{si} > 0$ and $n_{sj} = 0$ for $1 \leqslant j \leqslant N, j \neq i$—the CRSS is τ_i:

$$\tau_i = \tau_i(n_{si}, H_i) \tag{7.1}$$

where H_i stands for "hardening"; it represents all those parameters that, besides n_{si}, govern the respective hardening mechanism. In the case of lattice mismatch strengthening (Section 5.4), H_i symbolizes the shear modulus of the matrix, the constrained lattice mismatch of the particles, their radius, and the dislocation line tension. If many different types of obstacles are simultaneously present, the CRSS is τ_t; the subscript t of τ_t stands for "total." τ_t depends on n_{si} and H_i, $1 \leqslant i \leqslant N$:

$$\tau_t = \tau_t(n_{si}, H_i, 1 \leqslant i \leqslant N) \tag{7.2a}$$

At least in principle, it may be possible to establish the function $\tau_t(n_{si}, H_i, 1 \leqslant i \leqslant N)$, but what one really wants to know is τ_t as a function of $\tau_i, 1 \leqslant i \leqslant N$:

$$\tau_t = \tau_t(\tau_i, 1 \leqslant i \leqslant N) \tag{7.2b}$$

This function must not involve any of the individual parameters n_{si} or $H_i, 1 \leqslant i \leqslant N$.

For the sake of clarity, in most cases to be discussed below, N will be limited to 2. Two equations that are of the form postulated by Eq. (7.2b) have been suggested [7.6–7.9]:

$$\tau_t = \tau_1 + \tau_2 \tag{7.3a}$$

and

$$\tau_t^2 = \tau_1^2 + \tau_2^2 \tag{7.3b}$$

Brown and Ham's [7.8] law of mixtures is not of the structure required by Eq. (7.2b) because it explicitly involves n_{s1} and n_{s2}:

$$\tau_t = \left(\frac{n_{s1}}{n_{s1} + n_{s2}}\right)^{1/2} \tau_1 + \left(\frac{n_{s2}}{n_{s1} + n_{s2}}\right)^{1/2} \tau_2 \tag{7.4}$$

Equations (7.3) are special cases of Eqs. (4.57) and (5.58), which are repeated here:

$$\tau_t^k = \tau_1^k + \tau_2^k \tag{7.5}$$

The exponent k is supposed to lie between 1.0 and 2.0. High values of k suppress the effect of the smaller one of the two stresses τ_1 and τ_2. The higher k is, the closer is τ_t to the larger one of τ_1 and τ_2. Nembach and Neite [7.10] found that neither the linear equation (7.3a) nor the Pythagorean equation (7.3b) describe the superposition of solid solution and γ'-particle hardening of the nickel-base superalloy NIMONIC PE16 (Section 5.5.4.1) satisfactorily. Therefore these authors introduced Eq. (7.5) as an ad hoc generalization of Eqs. (7.3). k of underaged and peak-aged NIMONIC PE16 single crystals was found to equal 1.23 (Section 5.5.4.1), k of overaged ones equaled 1.45 (Section 6.2). The procedures for deriving k will be detailed in Section 7.2.1. So far Eq. (7.5) has no theoretical basis. It has been successfully used to represent the results of Foreman and Makin's [7.11] computer simulations of the glide of dislocations through random arrays of point obstacles of two different strengths. This has been communicated in Section 4.2.2.1. There were, however, some deficiencies: k varied with the ratio of the two obstacle strengths. This has been illustrated in Figure 4.26. Only if k is independent of the strengths of the two classes of obstacles, Eq. (7.5) is of the form postulated by Eq. (7.2b).

Equations (7.3) have been rationalized for the following cases:

1. The linear equation (7.3a) if two conditions are fulfilled:
   (a) Both types of obstacles are sheared by dislocations.
   (b) There are few strong O_2-obstacles mixed in with many weak O_1-obstacles.
2. The Pythagorean equation (7.3b) in the two different cases 2a and 2b:
   (a) If two conditions are fulfilled:
      (i) Both types of obstacles are sheared by dislocations.
      (ii) Both types of obstacles are of approximately the same strength.
   (b) If both types of obstacles are circumvented by dislocations.

Kocks [7.7, 7.9, 7.12] gave the following rationale for the linear equation (7.3a). Let the maximum of the absolute value of the interaction force between O_i- obstacles and a dislocation be F_{0i}, $1 \leqslant i \leqslant 2$. F_0 has been defined in Eq. (4.7a). If there are only O_i-obstacles, the CRSS is τ_i and the spacing of

O_i-obstacles along the dislocation is L_{ci} in the critical moment. If both types of obstacles are present, the CRSS is τ_t and the critical spacing of O_i-obstacles is L_{cti}. Kocks made three assumptions:

(i) $F_{02} \gg F_{01}$
(ii) $L_{c2} \gg L_{c1}$
(iii) $L_{cti} \approx L_{ci}$, $1 \leqslant i \leqslant 2$

Assumptions i and ii are equivalent to the condition 1b given above for the applicability of the linear equation (7.3a). τ_i follows from Eq. (4.14):

$$\tau_i b L_{ci} = F_{0i} \tag{7.6}$$

where b is the length of the Burgers vector. If thermal activation can be disregarded, Eq. (7.6) holds also for solute atoms.

Now the combination of intermixed O_1- and O_2-obstacles is considered. In the critical moment the two O_2-obstacles O_2' and O_2'' are nearest neighbors along the dislocation. The forces that the weak O_1-obstacles lying between O_2' and O_2'' exert on the dislocation are smoothed out into a kind of back-stress $[F_{01}/(bL_{ct1})]$. Because the applied stress τ_t is reduced by this back-stress, Eq. (7.6) is replaced by [7.9, 7.12]:

$$\left(\tau_t - \frac{F_{01}}{bL_{ct1}}\right) bL_{ct2} = F_{02}$$

$$\tau_t b L_{ct2} = F_{02} + F_{01}\frac{L_{ct2}}{L_{ct1}} \tag{7.7a}$$

The latter equation lends itself to an alternative interpretation [7.7]. The force $\tau_t b L_{ct2}$, which the external stress exerts on the dislocation length L_{ct2}, is balanced by one O_2-obstacle and by many O_1-obstacles. Their number is estimated to be L_{ct2}/L_{ct1}.

Dividing Eq. (7.7a) by bL_{ct2} yields

$$\tau_t = \frac{F_{02}}{bL_{ct2}} + \frac{F_{01}}{bL_{ct1}} \tag{7.7b}$$

By virtue of the above assumption iii, L_{ci} can be inserted for L_{cti}:

$$\tau_t = \frac{F_{02}}{bL_{c2}} + \frac{F_{01}}{bL_{c1}}$$

Substituting for $F_{0i}/(bL_{ci})$ from Eq. (7.6) leads to the linear equation (7.3a):

$$\tau_t = \tau_2 + \tau_1$$

The questionable points of the above derivations are assumption iii: $L_{cti} \approx L_{ci}$, $1 \leqslant i \leqslant 2$, and the term $[F_{01}/(bL_{ct1})]$ on the left-hand side of Eq. (7.7a).

The Pythagorean equation (7.3b) is first justified for the above case 2a: Both types of obstacles are sheared and they are of nearly equal strength; that is, $F_{01} \approx F_{02} = F_0$. Friedel's equation (4.26a) is assumed to be applicable; the dependence of the dislocation line tension S on n_{si} via the outer cutoff radius is disregarded. n_{si} is the number of O_i-obstacles per unit area of the glide plane. $[n_{s1} + n_{s2}]^{-1/2}$ and $n_{si}^{-1/2}$ is written instead of the respective square lattice spacing L_{min} of the obstacles [Eq. (3.1a)]:

$$\tau_t^2 = (n_{s1} + n_{s2}) \frac{F_0^3}{2Sb^2}$$

$$\tau_t^2 = n_{s1} \frac{F_{01}^3}{2Sb^2} + n_{s2} \frac{F_{02}^3}{2Sb^2}$$

$$\tau_t^2 = \tau_1^2 + \tau_2^2$$

Thus the Pythagorean equation (7.3b) is recovered. The basis of the above reasoning is that τ_i^2 is proportional to n_{si} and that all other parameters involved are the same for both types of obstacles. Since also Eqs. (6.14), (6.16), and (6.17), which relate the critical Orowan stress to the particles' radius and volume fraction, approximately fulfill the condition that τ_i^2 is proportional to n_{si}, the Pythagorean equation (7.3b) also describes strengthening by two types of bypassed particles. This is the above case 2b.

It appears from the above derivations that before applying either of Eqs. (7.3) one must check carefully whether the assumptions under which the respective equation has been derived are actually fulfilled for the system under investigation. It has been demonstrated in Figures 4.26a and 4.26b that Eqs. (7.3) are poor representations of the results obtained by Foreman and Makin [7.11] for double strengthening.

Hanson and Morris [7.13] and Melander and coworkers [7.14, 7.15] modeled multiple strengthening on the basis of Hanson and Morris' [7.16] circle rolling procedure described in Section 4.2.1.4.1. For the sake of completeness, also the averaging techniques applied in Sections 5.3 and 5.4.2 are mentioned in the present context.

7.2 EXPERIMENTAL INVESTIGATIONS OF MULTIPLE STRENGTHENING

The densities n_{s1} and n_{s2} of the obstacles of types O_1 and O_2, respectively, are varied independently and the total CRSS τ_t is measured as a function of the individual ones τ_1 and τ_2. The two most widely studied combinations of O_1- and O_2-obstacles are:

1. Solute atoms and particles
2. Two different types of particles

The latter combination is likely to involve also some solid solution hardening. The above examples 1 and 2 of double strengthening will be discussed in Sections 7.2.1 and 7.2.2, respectively.

7.2.1 Solid Solution and Particle Strengthening

The following parameters are defined:

O_1-obstacles: solute atoms of atomic fraction c; they yield $\tau_s = \tau_1$

O_2-obstacles: particles of volume fraction f and average radius r; they produce $\tau_p = \tau_2$

τ_s is the CRSS of the matrix of the particle-hardened material. Because τ_s hardly ever vanishes, τ_p cannot be directly measured (see beginning of this chapter). Only the total CRSSs τ_t and τ_s are experimentally accessible. Hence if one wants to compare theoretical results on particle strengthening with experimental ones, one has to subtract τ_s somehow from τ_t. In Chapters 5 and 6 this has been done on the basis of Eq. (5.58), which is identical with Eq. (7.5) and equivalent to Eq. (4.57). Evidently the exponent k is required.

Nembach and Neite [7.10] studied γ'-strengthening of underaged, peak-aged, and overaged single crystals of the nickel-base superalloy NIMONIC PE16 (Sections 5.5.4 and 6.2.2). It was found that the superposition of solid solution hardening of the matrix and of hardening by the $L1_2$-long-range-ordered γ'-particles could not be represented by Eqs. (7.3). This led the quoted authors to Eq. (7.5); the exponent k was determined by an optimization procedure. It is sketched here for underaged specimens, its basis is the strong variation of τ_s with the deformation temperature T_D. τ_p, however, is nearly independent of T_D. Because the authors did not wish to presuppose the validity of any of the equations presented in Section 5.5.3.1 for order strengthening, they wrote the following for τ_{pmn} of the γ'-particle-hardened specimen m deformed at T_{Dn}:

$$\tau_{pmn} = \frac{A_m^*}{[S_{\text{eff }1}(T_{Dn})]^{1/2}} + B_m^* \tag{7.8}$$

This equation has the form of those derived in Section 5.5.3.1. $S_{\text{eff }1}$ is the mean dislocation line tension defined in Eq. (4.50a). The parameters A_m^* and B_m^* are independent of T_{Dn}, but are functions of the radius and the volume fraction of the γ'-particles. τ_{pmn} varies only very slightly with T_{Dn}: τ_{pmn} merely reflects the temperature dependence of $S_{\text{eff }1}^{1/2}$. Measurements were performed at four temperatures T_{Dn} between 90 K and 500 K. The optimum exponent $k_{\text{opt }m}$ was chosen such that the four pairs of τ_{pmn} values derived from Eqs. (7.5) (experimental τ_p data) and (7.8) (theoretical τ_p data) agreed as closely with each other as possible. $k_{\text{opt }m}$, A_m^* and B_m^* were adjustable parameters. In a final

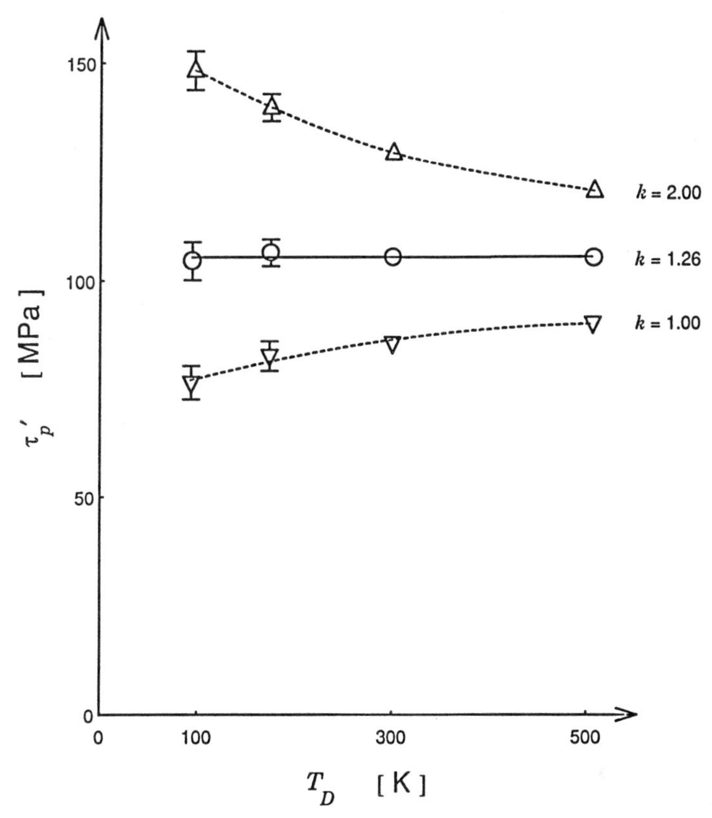

Fig. 7.1. Temperature adjusted CRSS τ_p' [Eq. (7.9)] of underaged NIMONIC PE16 single crystals versus the deformation temperatures T_D. The γ'-volume fraction is 0.10, and the average γ'-particle radius is 7.5 nm. The exponent k inserted into Eq. (7.9) is given on the right-hand side; the optimum value of k is 1.26. Error limits are indicated if they exceed the size of the symbols. (After Ref. 7.10.)

step, $k_{\mathrm{opt}\,m}$ was averaged over all underaged γ'-particle-hardened specimens. The result was $k_{\mathrm{opt}} = [1.23 \pm 0.07]$. The importance of determining k_{opt} correctly is demonstrated in Figure 7.1. There τ_p' is plotted versus T_D. Since τ_p' equals τ_p adjusted for the above-mentioned very slight variation of τ_p with T_D, τ_p' must be independent of T_D:

$$\tau_p' = [\tau_t^k(T_D) - \tau_s^k(T_D)]^{1/k} \, \frac{\tau_{p\,\mathrm{theo}}(T_D = 293\ \mathrm{K})}{\tau_{p\,\mathrm{theo}}(T_D)} \tag{7.9}$$

The subscripts m and n have been suppressed. $\tau_{p\,\mathrm{theo}}$ has been calculated from Eq. (5.67a) with the experimental results reported in Section 5.5.4.1 for A^* and

B^*. In the example given in Figure 7.1, k_{opt} equals 1.26. τ_t decreases by about 25% as T_{Dn} is raised from 90 K to 500 K. As can be seen in Figure 7.1, equating k with 1.0 or with 2.0, leads to temperature dependencies of τ_p' which are definitely outside of the experimental limits of error. If and only if the correct exponent k is used in Eq. (7.5), the data plotted in Figure 5.26 are independent of T_D. If only the size and the volume fraction of the γ'-precipitates are varied, but not T_D, any deficiences of k are far less apparent.

The optimum value k_{opt} of peak-aged and of overaged NIMONIC PE16 single crystals was determined by procedures similar to those described above for underaged specimens. The results were: 1.23 for the peak-aged state and 1.45 ± 0.16 for the overaged state. The latter exponent has been used in Table 6.1. There also the strong effects of alternative k-values on τ_p are demonstrated. Varying k from 1.00 through 1.45 to 2.00 changes τ_p of overaged specimens from 1.0 through 1.6 to 2.0, in suitably normalized units. These effects are so strong because τ_s of NIMONIC PE16 is rather high. The fact that k_{opt} varies with the state of aging is not quite compatible with the postulate expressed by Eq. (7.2b).

Büttner et al. [7.17] investigated the superposition of solid solution and particle strengthening for underaged copper-base copper–gold–cobalt alloys. The atomic fraction c_{Au} of gold ranged from 0.00 to 0.06. It stays solved in the matrix and raises τ_s, whereas most of the cobalt precipitates as cobalt-rich particles, which have a lattice mismatch. Their effect on the CRSS of binary copper–cobalt alloys has been analyzed in Sections 5.4.2.1.8 and 5.4.2.2.2. Equations (5.52) and (5.59) relate τ_p to the average radius r and the volume fraction f of the precipitates, to their constrained lattice mismatch ε, to the shear modulus μ_s of the matrix, and to the dislocation line tension S. Equations (5.52a) and (5.59) are repeated here with the dependence of μ_s, ε, and S on c_{Au} and on the deformation temperature T_D explicitly indicated:

$$\tau_p = \alpha_\varepsilon \tau_{p\,\text{dis}\,0} \tag{7.10a}$$

$$\tau_{p\,\text{dis}\,0}(c_{\text{Au}}, r, f, T_D) = \{\mu_s(c_{\text{Au}}, T_D)|\varepsilon(c_{\text{Au}}, T_D)|\}^{3/2}(rfb)^{1/2}[\beta_1(r/b)^{\beta_2}]^{3/2}/[2S(c_{\text{Au}}, T_D)]^{1/2} \tag{7.10b}$$

α_ε is a numerical constant, and it is close to 3.6 (Section 5.4.2.1.8). The subscript "dis" of $\tau_{p\,\text{dis}\,0}$ and the parameters β_1 and β_2 show that the dissociation of the edge dislocation into Shockley partial dislocations is taken into consideration. $\tau_{p\,\text{exp}}(k, c_{\text{Au}}, r, f, T_D)$, which is calculated from experimental τ_t data on the basis of Eq. (7.5), is divided by $\tau_{p\,\text{dis}\,0}(c_{\text{Au}}, r, f, T_D)$. k is adjusted such that this ratio $\tau_{p\,\text{exp}}/\tau_{p\,\text{dis}\,0}$ is independent of c_{Au}, r, f, and T_D. According to Eqs. (7.10), this ratio is supposed to equal $\alpha_\varepsilon \approx 3.6$. For $0.0 \leqslant c_{\text{Au}} \leqslant 0.032$, this optimization procedure yields the value 1.25 for k_{opt}. This is very close to the above result 1.23 for underaged NIMONIC PE16 single crystals. For c_{Au} exceeding 0.032, Eq. (7.5) is not applicable.

Evidently the *linear* equation (7.3a) is not applicable to NIMONIC PE16 or to copper–gold–cobalt alloys. The effects of concentration fluctuations in the solid solution matrix need some attention. Such fluctuations raise the maximum F_{0s} of the absolute value of the solute–dislocation interaction force [7.18, 7.19]. Thus increasing c_{Au} of the ternary copper–gold–cobalt alloys is likely to raise F_{0s}. Foreman and Makin's [7.11] computer simulations have shown a dependence of the exponent k of Eq. (7.5) on the relative strengths of the two types of obstacles (Section 4.2.2.1). This may explain why Eq. (7.5) could not be applied to copper–gold–cobalt alloys if c_{Au} covered the whole range 0.0–0.06. In the case of NIMONIC PE16 the matrix composition was nearly constant; it varied only very slightly with the γ'-volume fraction.

Ebeling and Ashby [7.20] studied the strengthening effects of incoherent silicon oxide particles in pure copper (Section 6.2.1) and in copper solid solution hardened by $c_{Au} = 0.0065$ gold. In both alloy systems the oxide particles were produced by internal oxidation at temperatures T_H between 1023 K and 1323 K. The subscript H of T_H reminds one of "heat" treatment. The authors measured the volume fraction f and the average radius r of the oxide particles only for the gold-free specimens and supposed that at given T_H and given duration t of the oxidation treatment, f and r were the same for the gold-containing and for the gold-free specimens; that is, the gold content was supposed not to affect the growth of the oxide particles. The authors concluded from their CRSS data that the linear equation (7.3a) described the superposition of τ_s due to the solved gold atoms and of τ_p due to the silicon oxide particles well.

If, however, the gold content does affect the growth of the oxide particles, $\tau_t(T_H, t, c_{Au} = 0.0065)$ must not be compared with $\tau_t(T_H, t, c_{Au} = 0.000)$, because the oxide particle distributions are not the same. An estimate of the ensuing effects on the optimum exponent k_{opt} in Eq. (7.5) is obtained as follows. Suppose that because of differences in the particle distributions, $\tau_t(T_H, t, c_{Au} = 0.0065)$ has been overestimated by merely 7%. If $[0.93 \cdot \tau_t(T_H, t, c_{Au} = 0.0065)]$ is compared with $\tau_t(T_H, t, c_{Au} = 0.000)$ for $T_H = 1123$ K and $T_H = 1223$ K, an optimization procedure similar to those described above yields $k_{opt} = 1.24$. The two reported temperatures are the only ones for which both sets of τ_t data are available. If, in addition, τ_t data of $T_H = 1273$ K and of $T_H = 1323$ K are to be evaluated, the authors' extra- and interpolation schemes have to be followed. Reducing τ_t of the four gold-containing specimens again by 7% leads to $k_{opt} = 1.34$. These evaluations shed some doubt on Ebeling and Ashby's proof of the validity of the linear equation (7.3a). The assumption that $\tau_t(T_H, t, c_{Au} = 0.0065)$ has been underestimated leads to optimum k-values that are less than unity.

7.2.2 Strengthening by Bimodal Particle Distributions

Bimodal particle distributions can be produced by exploiting the temperature dependence of the solubility limits (Figure 3.3). First the specimen is heat-treated at the relatively high temperature T_{H1}. This yields the class 1 particles of average radius r_1' and volume fraction f_1'. During a subsequent treatment at T_{H2} below T_{H1}, new class 2 particles nucleate and grow in between the old class 1 particles, whose size and volume fraction also increase with time. At the end of the second treatment there are two classes of precipitates:

Class 1: r_1, f_1
Class 2: r_2, f_2

r_i and f_i are the average radius and volume fraction, respectively, of class i particles. Normally r_1 and f_1 exceed the respective values r_2 and f_2. The particles of either class will have a distribution g of their radii. After prolonged treatments at T_{H2}, only class 1 particles survive; their function g tends to be somewhat broader than that found after isothermal treatments [7.21]. Moreover, the distribution function $g^*(\lambda/L_{\min})$ of interparticle spacings λ (Section 3.3) is likely to be affected by the second heat treatment.

The following parameters have to be distinguished, and their definitions are given below:

$r_i, f_i, \tau_{ti}, \tau_{pi}, 1 \leqslant i \leqslant 2$
$\tau_{t1,2}, \tau_{p1,2}$

Though the CRSS of the matrix is always referred to as τ_s, its variation with T_H is supposed to be taken into consideration. If, after the treatment at T_{H2}, all particles of class j are removed from the specimen, the CRSS is $\tau_{ti}, i \neq j$; the class i particles' contribution to τ_{ti} is τ_{pi}. If both classes of particles are present, the total CRSS is $\tau_{t,1,2}$; the combined contribution of both classes of particles to $\tau_{t1,2}$ is $\tau_{p1,2}$. The derivation of the experimental function $\tau_{p1,2}(\tau_{p1}, \tau_{p2})$ involves allowing somehow for τ_s.

The various just-mentioned superpositions are summarized by the following equations, the different exponents k are given appropriate subscripts:

$$\tau_{ti}^{k_{si}} = \tau_s^{k_{si}} + \tau_{pi}^{k_{si}}, \qquad 1 \leqslant i \leqslant 2 \tag{7.11a}$$

$$\tau_{t1,2}^{k_{st}} = \tau_s^{k_{st}} + \tau_{p1,2}^{k_{st}} \tag{7.11b}$$

$$\tau_{p1,2}^{k_p} = \tau_{p1}^{k_p} + \tau_{p2}^{k_p} \tag{7.11c}$$

Evidently there are four different exponents: k_{s1}, k_{s2}, k_{st}, and k_p. Often one assumes that k_{s1}, k_{s2}, and k_{st} are the same; one just writes k_s for all three of them. Equations (7.11) are special forms of Eq. (7.5). Depending on how strictly one wishes to fulfill the postulate expressed by Eq. (7.2b), one may tolerate that the exponents k_{s1}, k_{s2}, and k_{st}, which refer to solute atoms, differ from k_p referring to particles. The higher the exponent k_p is, the closer is $\tau_{p1,2}$ to the larger one of the two individual stresses τ_{p1} and τ_{p2}.

Most of the investigations of strengthening by bimodal particle distributions have been carried out for nickel-base alloys containing two classes of $L1_2$-long-range-ordered γ'-precipitates [7.10, 7.21–7.24]. Order strengthening has been discussed in Section 5.5. In the examples to be presented below, the class 1 particle distributions were slightly or highly overaged.

Reppich et al. [7.22] produced two classes of γ'-particles in polycrystals of the commercial nickel-base superalloy NIMONIC PE16. This material has already been referred to in Sections 4.1, 4.2.2.3, 5.5, and 6.2.2. Reppich et al. inserted 1.2 for k_s. The superpositions of τ_{p1} and τ_{p2} could be represented by Eqs. (7.11c) and (7.5), but the exponent k_p was found to vary with the relative strengths of the γ'-particles of the two classes. As a measure of their strengths, a parameter similar to τ_p^* defined in Eq. (4.18) has been used, even if the Orowan process operated. If the exponent k_p in Eq. (7.11c) varies with the strength of the particles, Eq. (7.11c) is not of the form of Eq. (7.2b), which is assumed not to involve individual parameters of the particles, but only τ_{pi}. Reppich et al.'s optimum k_p-values ranged from 1.0 to 1.5. Nembach and Neite [7.10] performed similar experiments on NIMONIC PE16 single crystals. $k_s = 1.23$ has been used. The representation of the experimental data by Eq. (7.11c) is much better with $k_p = 2.0$ than with $k_p = 1.0$. The optimum exponent is estimated to be close to 1.7.

Chellman et al. [7.23] continued and elaborated earlier work by Chellman and Ardell [7.24] on binary nickel-rich nickel–aluminum single crystals containing two classes of γ'-precipitates. Equation (7.11c) was found to represent the function $\tau_{p1,2}(\tau_{p1}, \tau_{p2})$ satisfactorily with k between 1.5 and 1.7. Chellman et al. subtracted τ_s linearly. Huang and Ardell [7.4] investigated strengthening of aluminum polycrystals by two types of second-phase particles. The authors described the superposition of the various contributions to the yield strength by equations analogous to Eqs. (7.11) with $k_s = 1.0$ and $k_p = 1.4$.

Thus it appears that strengthening by two classes of particles cannot be described by the linear equation (7.3a) or by the Pythagorean equation (7.3b), but Eq. (7.11c) with an exponent k_p between 1.0 and 2.0 has to be used. If the uncertainties concerning the subtraction of solid solution hardening are taken into consideration, the error limits of the optimum exponent k_p in Eq. (7.11c) are wide.

Koppenaal and Kuhlmann-Wilsdorf [7.6] strengthened copper single crystals by neutron irradiation and by work hardening. The superposition of both hardening mechanisms could be described by the Pythagorean equation (7.3b).

7.3 SYNERGISMS

The validity of equations like Eqs. (7.2b), (7.3), (7.5), and (7.11) rests on the assumption that the presence of type O_i-obstacles does not alter the strengthening effects due to the O_j-obstacles, $j \neq i$. In most cases, however, multiple strengthening leads to cross-terms and thus to synergisms. Nembach

[7.25] discussed three examples, two of which are referenced here. The obstacle combinations are:

(i) O_1: precipitates O_2: solute atoms
(ii) O_1: grain boundaries O_2: precipitates

First the O_1-obstacles are introduced into the material and τ_1 is determined; adding O_2-obstacles is found to change τ_1.

1. In the ternary copper–gold–cobalt system, cobalt-rich precipitates are the O_1-obstacles, and the gold atoms stay solved in the copper-rich matrix and act as O_2-obstacles. The precipitates' contribution τ_1 to the total CRSS is given by Eqs. (7.10). The gold atoms lower the shear modulus μ_s of the matrix and the dislocation line tension S and raise the cobalt-rich precipitates' lattice mismatch ε [7.17]. An atomic fraction of 0.06 of gold increases the maximum particle–dislocation interaction force by 41% and hence τ_1 by 71% [7.25]. This is a strong synergism.

2. Mangen and Nembach [7.26] studied the combination of grain boundary ($=O_1$-obstacles) with γ'-particle ($=O_2$-obstacles) strengthening for the nickel-base superalloy NIMONIC PE16, which has been referred to in Sections 4.1, 4.2.2.3, 5.5, 6.2.2, and 7.2. The Hall [7.27]–Petch [7.28] equation relates the yield strength σ_y of polycrystals to their grain size d:

$$\sigma_y = M\tau_p + k_{\mathrm{HP}} d^{-1/2} \tag{7.12}$$

where τ_p is the CRSS of single crystals of the same material and M is the Taylor [7.29] factor. Evidently the coefficient k_{HP} governs the effect of the grain boundaries on σ_y. Mangen and Nembach found that Eq. (7.12) described their data well, but k_{HP} was a function of the size and of the volume fraction of the γ'-precipitates. k_{HP} turned out to be proportional to $1/(L_{\min} - 2r_r)^{0.6}$, where $L_{\min}$ is the γ'-particles' square lattice spacing and r_r is the average radius of their intersection with glide planes. $L_{\min}$ and r_r have been defined in Section 3.1. Nembach [7.30] rationalized the function $k_{\mathrm{HP}}(L_{\min}, r_r)$: The γ'-precipitates limit the maximum length of grain boundary dislocation sources. Evidently this constitutes another synergism: The introduction of the γ'-precipitates ($=O_2$-obstacles) modifies the effect of grain boundaries ($=O_1$-obstacles) on the yield strength.

The latter synergism has another important implication. During an Ostwald ripening treatment (Section 3.2.1.2), which raises the average radius r of the particles but leaves their volume fraction f approximately constant, the CRSS τ_p and the yield strength σ_y peak at different radii. τ_p and σ_y reach their maxima, when $\partial\tau_p/\partial r$ and $[M\partial\tau_p/\partial r + (\partial k_{\mathrm{HP}}/\partial r)/d^{1/2}]$, respectively, vanish.

Since in underaged specimens, $\partial \tau_p / \partial r$ is positive and $\partial k_{HP} / \partial r$ negative, σ_y peaks at smaller values of r than τ_p does. The smaller the grains, the larger the difference; it may exceed 50% of the peak radius of τ_p [7.26].

REFERENCES

7.1 J. D. Whittenberger, *Met. Trans. A*, **15A**, 1753, 1984.

7.2 E. M. Dunn, A. P. Davidson, J. P. Faunce, C. G. Levi, S. Maitra, and R. Mehrabian, in *Aluminum, Properties and Physical Metallurgy* (edited by J. E. Hatch), 3rd printing, p. 25, American Society for Metals, Metals Park, Ohio, 1988.

7.3 E. A. Starke, Jr., in *Alloying* (edited by J. L. Walter, M. R. Jackson, and C. T. Sims), p. 165, American Society for Metals, Metals Park, Ohio, 1988.

7.4 J. C. Huang and A. J. Ardell, *Acta Metall.*, **36**, 2995, 1988.

7.5 T. H. Sanders, Jr. and E. A. Starke, Jr., *Proceedings of the 5th International Aluminum–Lithium Conference* (edited by T. H. Sanders, Jr. and E. A. Starke, Jr.), p. 1, MCEP, Birmingham, England, 1989.

7.6 T. J. Koppenaal and D. Kuhlmann-Wilsdorf, *Appl. Phys. Lett.*, **4**, 59, 1964.

7.7 U. F. Kocks, in *Physics of Strength and Plasticity* (edited by A. S. Argon), p. 143, MIT Press, Cambridge, Massachusetts, 1969.

7.8 L. M. Brown and R. K. Ham, in *Strengthening Methods in Crystals* (edited by A. Kelly and R. B. Nicholson), p. 9, Applied Science Publishers, London, 1971.

7.9 U. F. Kocks, A. S. Argon, and M. F. Ashby, *Prog. Mater. Sci.*, **19**, 1, 1975.

7.10 E. Nembach and G. Neite, *Prog. Mater. Sci.*, **29**, 177, 1985.

7.11 A. J. E. Foreman and M. J. Makin, *Can. J. Phys.*, **45**, 511, 1967.

7.12 U. F. Kocks, *Proceedings of the International Conference on the Strength of Metals and Alloys, Trans. JIM, Suppl.*, **9**, 1, 1968.

7.13 K. Hanson and J. W. Morris, Jr., *J. Appl. Phys.*, **46**, 2378, 1975.

7.14 A. Melander and P. Å. Persson, *Acta Metall.*, **26**, 267, 1978.

7.15 B. Jansson and A. Melander, *Scripta Metall.*, **12**, 497, 1978.

7.16 K. Hanson and J. W. Morris, Jr., *J. Appl. Phys.*, **46**, 983, 1975.

7.17 N. Büttner, K.-D. Fusenig, and E. Nembach, *Acta Metall.*, **35**, 845, 1987.

7.18 R. Labusch, *Acta Metall.*, **20**, 917, 1972.

7.19 R. Labusch, *Czech. J. Phys. B*, **38**, 474, 1988.

7.20 R. Ebeling and M. F. Ashby, *Philos. Mag.*, **13**, 805, 1966.

7.21 E. Nembach and C.-K. Chow, *Mater. Sci. Eng.*, **36**, 271, 1978.

7.22 B. Reppich, W. Kühlein, G. Meyer, D. Puppel, M. Schulz, and G. Schumann, *Mater. Sci. Eng.*, **83**, 45, 1986.

7.23 D. J. Chellman, A. J. Luévano, and A. J. Ardell, *Proceedings of the 9th International Conference on the Strength of Metals and Alloys*, Vol. 1 (edited by D. G. Brandon, R. Chaim, and A. Rosen), p. 537, Freund Publishing House, London, 1991.

7.24 D. J. Chellman and A. J. Ardell, *Proceedings of the 4th International Conference on the Strength of Metals and Alloys*, Vol. 1 (edited by Laboratoire Physique du Solide), p. 219, Nancy, France, 1976.

7.25 E. Nembach, *Acta Metall. Mater.*, **40**, 3325, 1992.

7.26 W. Mangen and E. Nembach, *Acta Metall.*, **37**, 1451, 1989.

7.27 E. O. Hall, *Proc. Phys. Soc., Sect. B*, **64**, 747, 1951.

7.28 N. J. Petch, *J. Iron Steel Inst.*, **174**, 25, 1953.

7.29 G. I. Taylor, *J. Inst. Met.*, **62**, 307, 1938.

7.30 E. Nembach, *Scripta Met. Mater.*, **24**, 787, 1990.

CHAPTER 8

The Effects of Second-Phase Particles on the Coercive Field of Magnetic Materials

A ductile material is called *mechanically hard* if high stresses are required to deform it plastically. The term "hard" is also used to characterize the magnetization behavior of ferromagnetic and ferrimagnetic materials. They are called *hard* if strong magnetic fields are needed to reverse their magnetization. In Figure 8.1a a hysteresis loop is sketched. In Figure 8.1b it is simplified to a rectangular shape. Throughout this chapter, the hysteresis loop will be supposed to be of the latter type. Long, thin cylinders orientated along an easy magnetic direction approximately have such hysteresis loops [8.1–8.11]. First the specimen is saturated in a negative external magnetic field; the magnetization M equals $-M_s$ throughout the entire specimen. M_s is the saturation magnetization. If the field H is raised, M remains unchanged until H reaches $+H_c$. H_c is the coercive field, an alternative term is coercivity. At $H = +H_c$ a new domain with $M = +M_s$ starts to expand. A Bloch [8.12] wall separates this new domain from the remainder of the specimen. Often this wall is referred to as a 180° domain wall because the directions of the magnetization on either side of the wall differ by 180°. The further reversal of the magnetization proceeds by the motion of the 180° wall through the specimen [8.1–8.11, 8.13]. The nucleation of a reversed domain resembles that of a new phase in a supersaturated solid solution (Section 3.2.1.1). The 180° domain wall takes the place of the phase boundary.

The coercive field may be governed by two processes: (1) difficulties to create the 180° wall that surrounds the reversed domain and (2) pinning of the 180° wall by inhomogeneities present in the specimen. This is exactly the analogous situation described in Section 4.2 for the critical resolved shear stress (CRSS): The creation of dislocations and their mobility may govern the CRSS. The CRSS of ductile particle-strengthened materials is determined by the latter process: Particles interact with dislocations and thus reduce their mobility. The two most advanced hard magnetic materials are examples for either process [8.14]: H_c of iron–neodymium–boron sintermagnets is determined by the ease with which a reversed domain—that is, the 180° wall surrounding it—can be created. H_c of cobalt–samarium ($Co_{17}Sm_2$) magnets is governed by pinning of

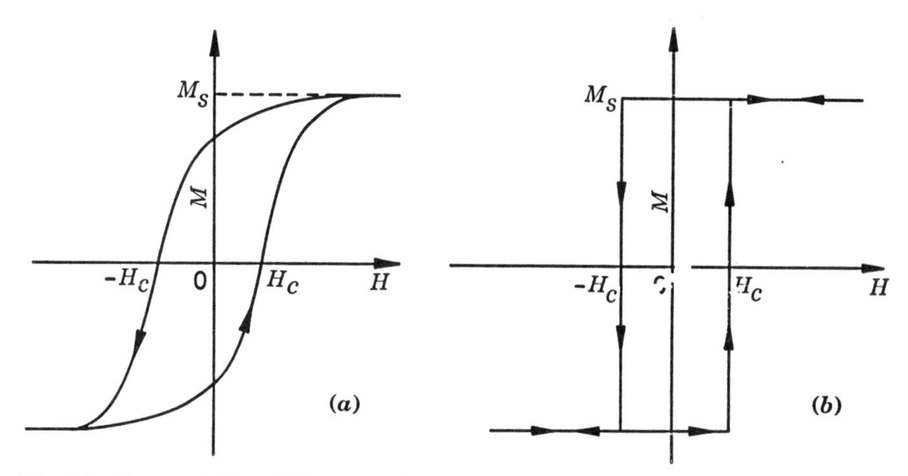

Fig. 8.1 Hysteresis loop. The arrows indicate the direction of the sweep. (*a*) Schematically, (*b*) idealized.

180° walls by inhomogeneities present in the material. Below only such materials will be discussed in which the latter mechanism is the decisive one. This corresponds to the treatments of particle hardening presented in Chapters 4 and 5.

8.1 MODELING OF THE COERCIVE FIELD

In Figure 8.2, a piece of a 180° domain wall is sketched. It is flexible, but only such that it stays parallel to the magnetization. More-dimensional bowing leads to the creation of free poles, whose energy is prohibitive. Thus the wall has the same type of flexibility as a sheet of paper. In Figure 8.2 the geometry and the coordinate system to be used below are shown. The wall, which is approximately parallel to the x–z plane, is driven by the external magnetic field H towards $+y$. By virtue of the limited flexibility of the wall, the length L_z of it can be projected into the x–y plane. Hence the two-dimensional wall is reduced to a one-dimensional, string-like entity. The obstacles to the motion of the 180° wall are also projected into the x–y plane. If there are n_v of them per unit volume, their density per unit area in the x–y plane is $n_v L_z$. Thus the geometry of the present problem is the same one as that of a dislocation gliding through a planar array of shearable obstacles (Chapter 4). The line energy $E_A L_z$ of the projected wall is identified with its line tension. E_A is the wall's energy per unit area. If a wall of area A moves over the distance dy, the field H does the work dE:

$$dE = 2\mu_0 M_s H A \, dy \tag{8.1}$$

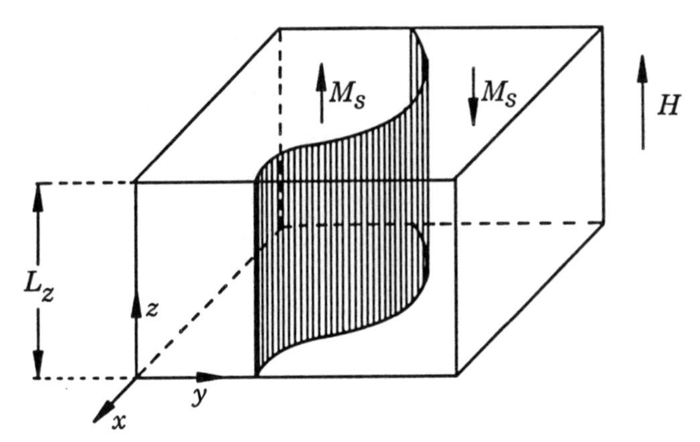

Fig. 8.2. Sketch of a 180° domain wall (hatched).

where μ_0 is the permeability of vacuum $= 4\pi \times 10^{-7}$ Vs/(Am) and M_s is the saturation magnetization of the material. The factor 2 in Eq. (8.1) is due to the change of the magnetization from $-M_s$ to $+M_s$—that is, by $2M_s$. Thus the force dF exerted by H on the length ds of the projected wall is given by

$$dF = H(2\mu_0 M_s L_z)\, ds \tag{8.2}$$

A in Eq. (8.1) has been replaced by $L_z ds$. Equation (8.2) is analogous to Eq. (2.9b) meant for dislocations. Evidently the following transcriptions can be made:

Dislocation	**180° Domain wall**	
b (length of the Burgers vector)	$\leftrightarrow 2\mu_0 M_s L_z$	(8.3a)
τ (resolved shear stress)	$\leftrightarrow H$ (magnetic field)	(8.3b)
τ_p (CRSS)	$\leftrightarrow H_c$ (coercive field)	(8.3c)
S (line tension)	$\leftrightarrow E_A L_z$	(8.3d)
$n_s = 1/L_{\min}^2$	$\leftrightarrow n_v L_z$	(8.3e)
F_0	$\leftrightarrow F_0$	(8.3f)
L_c	$\leftrightarrow L_c$	(8.3g)
w_y	$\leftrightarrow w_y$	(8.3h)

F_0 stands for the maximum of the absolute value of the interaction force between a particle (obstacle) and a dislocation or a projected wall. The analogue holds for the particle (obstacle) spacing L_c along the dislocation or the projected wall in the critical moment, when the dislocation/wall breaks free from a particle (obstacle). w_y is the range of the interaction force (Section 4.1.1).

Below the coercive field will be calculated for a ferromagnetic material that contains nonmagnetic spherical particles of radius r and volume fraction f. All particles are assumed to have the same radius r. Because energy is stored in 180° walls, they are attracted by the nonmagnetic particles. If the 180° wall coincides with the x–z plane, the wall's energy $e_m(y)$ per unit volume can be approximated by Eq. (8.4) [8.15, 8.16]; the subscript m of e_m stands for "magnetic," and the same holds for E_m and F_m introduced below:

$$e_m(y) = \frac{2E_A}{\pi q} \frac{1}{\{1 + (y/q)^2\}^2} \tag{8.4}$$

The parameter q is a measure of the thickness of the 180° domain wall. Normally q lies between 1 nm and 100 nm. The interaction energy E_m between the particle and the wall is obtained by integrating $e_m(y)$ over the volume of the particle:

$$E_m = -\int_p e_m \, dV \tag{8.5a}$$

The interaction force $F_m(y)$ is gained by differentiating Eq. (8.5a) with respect to y:

$$F_m(y) = -\partial E_m/\partial y \tag{8.5b}$$

There is a close analogy between Eqs. (8.4), (8.5a), and (8.5b) meant for magnetic hardening and Eqs. (5.8), (5.10), and (5.11), respectively, which apply to modulus mismatch strengthening. Both sets of equations are based on the straight-line approximation introduced in Section 4.1.1. The above-described interaction between a nonmagnetic particle and a 180° domain wall is sometimes called the "surface tension" or "foreign body" interaction [8.15–8.17]. Besides this mechanism, there is another one: a magnetostatic one. Schröder [8.16] has shown that it is negligible. The particles must not produce any strain in the matrix, because this may lead to a magnetostrictive particle–wall interaction.

The "surface tension" interaction is elastic: The 180° wall does not alter the nonmagnetic particle permanently. Because the normalized range η_0 [Eq. (4.58g)] is large, Schwarz and Labusch's [8.18] equation (4.65) is applicable. The term c_s/a^2 is replaced by $1/L_{\min}^2 = n_v L_z$, and the other transcriptions according to the listing (8.3) are used [8.15, 8.17, 8.19]:

$$H_c = C \frac{(n_v L_z)^{2/3} F_0^{4/3} w_y^{1/3}}{(2\mu_0 M_s L_z)(2E_A L_z)^{1/3}}$$

where C is a numerical constant. This relation can be slightly simplified:

$$H_c = C \frac{n_v^{2/3} F_0^{4/3} (w_y L_z)^{1/3}}{(2\mu_0 M_s L_z)(2E_A)^{1/3}} \tag{8.6}$$

A similar equation can be derived from Friedel's equations (4.26):

$$H_c = C' \frac{(n_v L_z)^{1/2} F_0^{3/2}}{(2\mu_0 M_s L_z)(2E_A L_z)^{1/2}}$$

$$H_c = C' \frac{n_v^{1/2} F_0^{3/2}}{(2\mu_0 M_s L_z)(2E_A)^{1/2}} \tag{8.7}$$

where C' is another numerical constant.

The above derivations suffer from the deficiency that due to the projection of the 180° domain wall and of the particles into the x–y plane, their density $n_v L_z$ per unit area is extremely high. Hence the wall is pinned not by individual particles but by clusters formed by statistical fluctuations [8.15]. They lead to effective obstacles [8.15, 8.20] with the effective parameters $F_{0\,\mathrm{eff}}$, $w_{y\,\mathrm{eff}}$, and $n_{v\,\mathrm{eff}}$:

$$F_{0\,\mathrm{eff}} = (L_z L_0 n_v I)^{1/2} \tag{8.8a}$$

$$w_{y\,\mathrm{eff}} = \pi (I/I_1)^{1/2} \tag{8.8b}$$

$$n_{v\,\mathrm{eff}} = 1/(w_{y\,\mathrm{eff}} L_z L_0) \tag{8.8c}$$

where the integrals I and I_1 are given by

$$I = \int_{-\infty}^{\infty} [F_m(y)]^2 \, dy \tag{8.8d}$$

$$I_1 = \int_{-\infty}^{\infty} [\partial F_m(y)/\partial y]^2 \, dy \tag{8.8e}$$

L_0 is a characteristic length that is governed by the flexibility of the 180° wall and by the force profile $F_m(y)$. Because the final Eqs. (8.9) and (8.11) do not involve L_0, there is no need to evaluate it. Replacing n_v, F_0, and w_y in Eq. (8.6) by $n_{v\,\mathrm{eff}}$, $F_{0\,\mathrm{eff}}$, and $w_{y\,\mathrm{eff}}$, respectively, yields

$$H_c = C \frac{n_v^{2/3} I^{1/2} I_1^{1/6} L_z^{1/3}}{\pi^{1/3} (2\mu_0 M_s L_z)(2E_A)^{1/3}} \tag{8.9}$$

Schröder [8.16] evaluated the integrals I and I_1 for the force profile $F_m(y)$ given by Eqs. (8.4) and (8.5). Instead of the radius r of the nonmagnetic spherical particles the reduced radius $\tilde{r}$ is used:

$$\tilde{r} = r/q \tag{8.10a}$$

The thickness q of the 180° domain wall has been defined in Eq. (8.4). The results for I and I_1 are

$$I(\tilde{r}) = 0.75\pi E_A^2 q^3[4\tilde{r}^3 \arctan(\tilde{r}) + 2\ln(1+\tilde{r}^2) - 5\tilde{r}^2 + 3\tilde{r}^2/(1+\tilde{r}^2)] \tag{8.10b}$$

$$I_1(\tilde{r}) = \frac{9\pi E_A^2 q \tilde{r}^6(9+5\tilde{r}^2)}{8(1+\tilde{r}^2)^3} \tag{8.10c}$$

In order to bring out the variation of H_c with r and the volume fraction f of the particles, n_v in Eq. (8.9) is replaced by $3f/(4\pi r^3)$ [Eq. (3.2)]:

$$H_c(r, f) = C^* f^{2/3} \frac{1}{r^2} I(r/q)^{1/2} I_1(r/q)^{1/6} \tag{8.11}$$

The parameter C^* is independent of r and f but involves M_s, L_z, E_A, and q.

8.2 COMPARISON WITH EXPERIMENTAL DATA

Nembach and coworkers [8.15, 8.17] measured the coercive force of the nickel-base superalloy NIMONIC PE16 at 77 K. The authors varied f and r of the long-range-ordered γ'-particles, which precipitate in this alloy. Their effects on the critical resolved shear stress of NIMONIC PE16 have been discussed in Sections 5.5.4, 6.2.2, and 7.2. Because the matrix is ferromagnetic below about 160 K and the γ'-particles are nonmagnetic [8.15], the equations derived in the preceding section are expected to be applicable. Some experimental results for the total coercive field $H_{ct\,\mathrm{exp}}(r, f = 0.06)$ are shown in Figure 8.3. $H_{ct\,\mathrm{exp}}(r, f)$ is actually a combination of the γ'-particles' contribution to the coercive field and of H_{c0}, which is the coercive field of the homogenized, γ'-free NIMONIC PE16 specimen. This superposition is assumed to follow a linear equation analogous to Eq. (7.3a):

$$H_{ct\,\mathrm{theo}}(r, f) = H_{c0} + H_{c\,\mathrm{theo}}(r, f) \tag{8.12}$$

$H_{c\,\mathrm{theo}}(r, f)$ is given by Eq. (8.11). The subscripts "theo" and "exp" refer to "theory" and "experiment," respectively. The curve in Figure 8.3 represents Eqs. (8.11) and (8.12) fitted to the experimental data $H_{ct\,\mathrm{exp}}(r, f = 0.06)$. The two adjustable parameters are q and a combination of C^* and E_A. Evidently Eq. (8.11) describes the experimentally observed variation of the coercivity with r well. The width q of the 180° domain wall turned out to be 11 nm. This is quite reasonable. Replacing the linear equation (8.12) by a nonlinear one

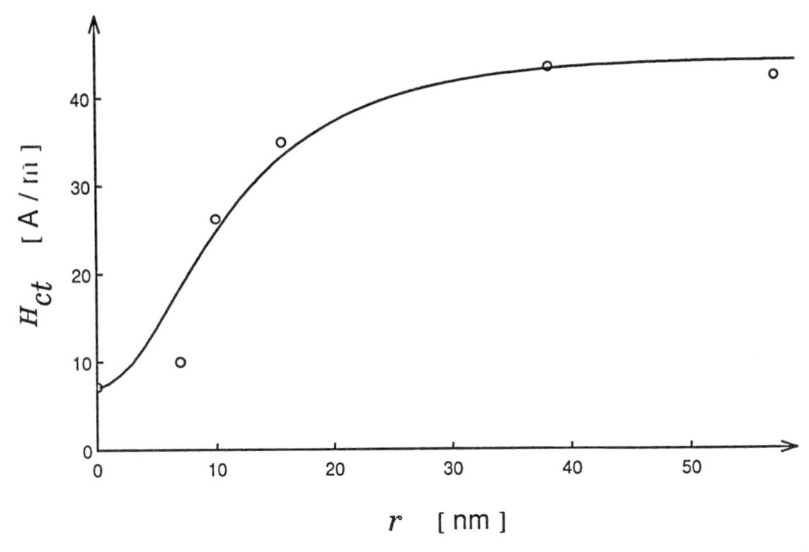

Fig. 8.3. Coercive field H_{ct} of a NIMONIC PE16 single crystal versus the average radius r of the γ'-precipitates. Their volume fraction is 0.06 [8.17, 8.21], and the data have been taken at 77 K. The curve represents Eqs. (8.11) and (8.12) fitted to the data. (After Ref. 8.17.)

similar to Eq. (7.5) with $k = 1.5$ yields nearly the same results. The dependence of $H_{ct\,\mathrm{exp}}$ on f is, however, much weaker than predicted by Eq. (8.11). The exponent of f was found to be around 0.2 instead of 2/3. The reason for this discrepancy is not known.

8.3 SUMMARY

In Section 8.1, models developed to describe the effects of particles on the CRSS have been adapted to describe their effects on the coercive field. In 1949, a similar transfer of ideas from the field of mechanical strength to that of magnetic hardness had already been reported: A model published by Smith [8.22] for the description of mechanical transient creep in metals was subsequently adapted (by Street and Woolley [8.23]) to the problem of magnetic creep in the hard magnetic material Alnico. Later on, Haasen [8.24, 8.25] based his model of the coercive field on Labusch's [8.26] model of solid solution strengthening. The derivations presented in Section 8.1 are similar to Haasen's.

Equations (8.6) and (8.11) have been applied to the nickel-base alloy NIMONIC PE16. Its matrix is ferromagnetic and the coercive field H_c is governed by the nonmagnetic γ'-precipitates of radius r and volume fraction f. Equations (8.6) and (8.11) describe the experimentally observed variation of H_c with r well, but they overestimate the variation with f.

REFERENCES

8.1 E. Kneller, *Ferromagnetismus*, Springer-Verlag, Berlin, 1962.

8.2 R. M. Bozorth, *Ferromagnetism*, 8th printing, Van Nostrand, Princeton, New Jersey, 1964.

8.3 A. H. Morrish, *The Physical Principles of Magnetism*, John Wiley & Sons, New York, 1965.

8.4 S. Chikazumi, *Physics of Magnetism*, 2nd printing, John Wiley & Sons, New York, 1966.

8.5 H. Träuble, in *Magnetism and Metallurgy*, Vol. 2 (edited by A. E. Berkowitz and E. Kneller), p. 621, Academic Press, New York, 1969.

8.6 D. J. Craik, *Structure and Properties of Magnetic Materials*, Pion, London, 1971.

8.7 B. D. Cullity, *Introduction to Magnetic Materials*, Addison-Wesley, Reading, Massachusetts, 1972.

8.8 J. Crangle, *The Magnetic Properties of Solids*, Edward Arnold, London, 1977.

8.9 C.-W. Chen, *Magnetism and Metallurgy of Soft Magnetic Materials*, North-Holland, Amsterdam, 1977.

8.10 J. P. Jakubovics, *Magnetism and Magnetic Materials*, The Institute of Metals, London, 1987.

8.11 J. Crangle, *Solid State Magnetism*, Edward Arnold, London, 1991.

8.12 F. Bloch, *Z. Phys.*, **74**, 295, 1932.

8.13 R. S. Tebble, *Magnetic Domains*, Methuen & Co., London, 1969.

8.14 H. Kronmüller, K.-D. Durst, and M. Sagawa, *J. Magn. Magn. Mater.*, **74**, 291, 1988.

8.15 H. G. Brion and E. Nembach, *phys. stat. sol.* (a), **26**, 599, 1974.

8.16 K. Schröder, *phys. stat. sol.*, **33**, 819, 1969.

8.17 D. Siemers and E. Nembach, *Acta Metall.*, **27**, 231, 1979.

8.18 R. B. Schwarz and R. Labusch, *J. Appl. Phys.*, **49**, 5174, 1978.

8.19 E. Nembach, *Proceedings of the 10th Risø International Symposium on Metallurgy and Materials Science* (edited by J. B. Bilde-Sørensen, N. Hansen, D. Juul Jensen, T. Leffers, H. Lilholt, and O. B. Pedersen), p. 473, Risø National Laboratory, Roskilde, Denmark, 1989.

8.20 K.-H. Pfeffer, *phys. stat. sol.*, **21**, 857, 1967.

8.21 E. Nembach and G. Neite, *Prog. Mater. Sci.*, **29**, 177, 1985.

8.22 C. L. Smith, *Proc. Phys. Soc.*, **61**, 201, 1948.

8.23 R. Street and J. C. Woolley, *Proc. Phys. Soc., Sect. A*, **62**, 562, 1949.

8.24 P. Haasen, in *Nachrichten der Akademie der Wissenschaften in Göttingen, II. Mathematisch-Physikalische Klasse*, No. 6, p. 1, 1970.

8.25 P. Haasen, *Mater. Sci. Eng.*, **9**, 191, 1972.

8.26 R. Labusch, *phys. stat. sol.*, **41**, 659, 1970.

AUTHOR INDEX

SUBJECT INDEX